Functional Plant Ecology

Functional Plant Ecology

Edited by **Clive Koelling**

SYRAWOOD
PUBLISHING HOUSE

New York

Published by Syrawood Publishing House,
750 Third Avenue, 9th Floor,
New York, NY 10017, USA
www.syrawoodpublishinghouse.com

Functional Plant Ecology
Edited by Clive Koelling

International Standard Book Number: 978-1-68286-108-0 (Hardback)

Printed in the United States of America.

Contents

Preface   IX

Chapter 1   **The slenderness of the softwood Riparian forest species *Salix alba* L. and *Salix fragilis* L. in the protected area of Nestos Delta, Greece**   1
George Efthimiou

Chapter 2   **Assessing ecosystem effects of small–scale cutting of Cameroon mangrove forests**   8
Longonje N. Simon and Dave Raffaelli

Chapter 3   **Population studies, habitat assessment and threat categorization of *Polygonatum verticillatum* (L.) Allioni in Kumaun Himalaya**   17
Nidhi Lohani, Lalit. M. Tewari, Ravi Kumar, G. C. Joshi, Jagdish Chandra, Kamal kshore, Sanjay Kumar and Brij Mohan Upreti

Chapter 4   **The invasive status of wild barley (*Hordeum spontaneum* Koch) in Iranian flora**   26
Reza Hamidi

Chapter 5   **Characterization and impact of wood logging on plant formations in Ngaoundéré District, Adamawa Region, Cameroon**   30
Tchobsala and Mbolo, M.

Chapter 6   **Predictive modelling of the distribution of two critically endangered Dipterocarp trees: Implications for conservation of riparian forests in Borneo**   43
Minerva Singh

Chapter 7   **The woodland tree *Brachystegia floribunda* facilitates the encroachment of forest tree species into miombo woodlands in northern Malawi**   49
Tomohiro Fujita

Chapter 8   **Habitat structure of flat-headed cusimanse (*Crossarchus platycephalus*) in Futa Wildlife Park, Ondo state, Nigeria**   56
Oguntuase B. G. and Agbelusi E. A.

Chapter 9   **Composition of understory vegetation in tree species of Cholistan desert, Pakistan**   62
Muhammad Farrukh Nisar, Farrukh Jaleel, Muhammad Waseem, Sajil Ismail and Muhammad Arfan

Chapter 10   **Reproductive aspects of common carp (*Cyprinus carpio* L, 1758) in a tropical reservoir (Amerti: Ethiopia)**   69
Mathewos Hailu

Chapter 11   **A new model: Herbaceous species diversity along the environmental gradient in the typical hilly areas of Henan Province**   74
Bing-Hua Liao, Pei-Song Liu, Zhen-Zhong Wen, Sheng-Yan Ding, Hai-Long Yu, Zhi-Chao Wang, Zhong-Kai Li, Huan-Xin Chu, Wen-Liang Li and Yi Shen

Chapter 12   **Phytosociology of some weeds of wheat communities around Kotli fields, Western Himalaya**   82
Zahid Hussain Malik, Muhammad Shoaib Amjad, Sidra Rafique and Nafeesa Zahid Malik

Chapter 13   **Study of flora of Miandasht Wildlife Refuge in Northern Khorassan Province, Iran (a)**   88
Rahimi A. and Atri M.

Chapter 14   **A preliminary simulation model of individual and synergistic impacts of elephants and fire on the structure of semi-arid miombo woodlands in northwestern Zimbabwe**   101
Isaac Mapaure

Chapter 15   **Assessment of biomass carbon stock in an *Ailanthus excelsa* Roxb. plantation Uttarakhand, India**   119
Nishita Giri and Laxmi Rawat

Chapter 16   **Influence of Zn stresses on growth and physiology in Khus-khus (*Vetiveria zizanoides* Nash.) and its essential sesquiterpene oil(s), in relation to roots diameter circumferential positions**   127
A. Misra, N. K. Srivastava and A. K. Srivastava

Chapter 17   **Comparison of the trees regeneration at different distances from Alang Dareh forest roads considering tourist pressure**   131
Adini Parsakhoo, Mohammad Hadi Moayeri and Majid Poursadeghi

Chapter 18   **Density and distribution of bongos (*Tragelaphus eurycerus*) in a high forest zone in Ghana**   137
Kwaku Brako Dakwa, Kweku Ansah money and Daniel Attuquayefio

Chapter 19   **Regulation of usages and dependency on indigenous fruits (IFs) for livelihoods sustenance of rural households : A case study of the lvindo National Park (INP), Gabon**   148
Mikolo Yobo Christian and Kasumi I. T. O

Chapter 20   **Ethinic-based diversity and distribution of enset (*Ensete ventricosum*) clones in southern Ethiopia**   164
Z. Yemataw, H. Mohamad, M. Diro, T. Addis and G. Blomme

Chapter 21   **Vegetation regeneration in formerly degraded hilly areas of Rwampara, South Western Uganda**   172
Juliet Kyayesimira and Julius B. Lejju

Chapter 22   **A study on ecological distribution and community diversity of spiders in Gulmarg Wildlife Sanctuary of Kashmir Himalaya**   180
Mansoor Ahmad Lone, Idrees Yousuf Dar and G. A. Bhat

Chapter 23   **Primary conifer succession on a 1915 mudflow in**
Lassen Volcanic National Park, California   **186**
Glenn Clinton Kroh, Rebecca Laura Upjohn and John Edgar Pinder III

Permissions

List of Contributors

Preface

This book has been a concerted effort by a group of academicians, researchers and scientists, who have contributed their research works for the realization of the book. This book has materialized in the wake of emerging advancements and innovations in this field. Therefore, the need of the hour was to compile all the required researches and disseminate the knowledge to a broad spectrum of people comprising of students, researchers and specialists of the field.

Plant ecology aims to study the distribution of various plant species across the globe, along with the several factors that affect plant biology and biodiversity. This book provides a comprehensive understanding of the field with the help of topics such as distribution of plants, conservation of endangered species, biological interactions, etc. The researches and case-studies included in this book are provided by eminent experts from around the world. It is an essential guide for students, academicians and those who wish to pursue this discipline further.

At the end of the preface, I would like to thank the authors for their brilliant chapters and the publisher for guiding us all-through the making of the book till its final stage. Also, I would like to thank my family for providing the support and encouragement throughout my academic career and research projects.

Editor

The slenderness of the softwood Riparian forest species *Salix alba* L. and *Salix fragilis* L. in the protected area of Nestos Delta, Greece

George Efthimiou

Department of Forestry and Natural Environment Management, Technological Educational Institute of Larissa, 34100, Karditsa, Greece. E-mail: efthimiou@teilar.gr.

The slenderness (height/diameter or h/d ratio) is an important factor, which describes the type of stem that each forest species develops. It depends on the species, the tree age and the site conditions. It is a basic factor that characterizes the structure and stability of the stand and a means for the assessment of the dynamics of height course. The objective of the present study is to compare the slenderness (h/d ratio) of the riparian forest species *Salix alba* L. and *Salix fragilis* L., which are found in mixed and pure stands in a protected area of international importance. 25 sampling plots (5 replicates in five different types of forest structure) were established in the riparian forest of Nestos. The structure and the dynamics of the stands were studied according to the IUFRO classification using slenderness as the overarching instrument. While the stands of both willow species have a high forest structure, slenderness values of *S. fragilis* stands were lower (degree of slenderness 40), which in turn indicates a more stable stand structure, in younger stages compared to *S. alba*, which attains similar values at nearly twice as large the stem diameters.

Key words: Slenderness, height/diameter ratio, *Salix alba*, *Salix fragilis*, riparian forest, Nestos Delta.

INTRODUCTION

The study of the h/d ratio (height/diameter ratio) or slenderness of riparian tree species is particularly important since riparian forests are among the most dynamic natural ecosystems (Dafis, 1992) as well as among the least studied ones in Greece (Efthimiou, 2000). Riparian forests are important elements of the European and global natural heritage because of their biological wealth. This wealth consists of their genetic, floristic and ecological diversity and high aesthetic, recreational, environmental, scientific and ecological value (Efthimiou, 2000; Efthimiou and Smiris, 2002). Depending on their species composition, structure, dynamics and ecological conditions riparian forests are distinguished in two types: softwood forest and hardwood forest (Yon, 1980; Mayer, 1984; Dister, 1988; Wenger et al., 1990; Kuhn, 1991). Riparian forests in Europe were excessively destroyed during the 20th Century. For example, up to 90% of riparian forests at the Upper Rhine was destroyed at the beginning of the second half of the 20th Century (Carbiener, 1974), while 60% of them was destroyed during the period 1955 to 1977 (Dister, 1988). In South Rhine, only the 6% of natural riparian stands are preserved (Hugin, 1981). In Bavaria, almost 80% of natural riparian forests have been destroyed (Wenger et al., 1990). During the same period in Greece around 60% of the larger natural riparian forest, the Kotza Orman (Papaioannou, 1953; Efthimiou, 2000; Emmanouloudis et al., 2006) in Nestos Delta, was destroyed.

The degree of slenderness is determined for each tree, via the relation between the height (h) and stem diameter (d) and characterises approximately the type of its stem. Slenderness is characteristic for each tree species and reflects the stem variability (Smiris, 1987). For this reason, it varies depending on tree species, age and conditions of the site (Röhle, 1982, 1984). The slenderness of the trees was often used for the characterization of the structure and the assessment of the dynamics of height development (Leibundgut, 1959, 1978; Dafis, 1966;

Smiris, 1987). The degree of slenderness according to Assmann (1959; 1961) is an important factor for the important factor for the characterization of the stability of the stand (Röhle, 1982; 1984). More concretely low degrees of slenderness show constant conditions, while high degrees are characteristic for stands of non-constant structure (Röhle, 1982).

The species of the genera *Populus* and *Salix* colonized Europe as companion species of *Pinus* and *Betula pendula* at the beginning of the intergracial period (Firbras, 1949, 1952; Straka, 1957; Strassburger, 1978; Ellenberg, 1996; Jenik, 1998). *Salix alba* is considered the most competitive species in an active riparian forest (Carbiener, 1974), which forms the most important category of softwood riparian forest (Dister, 1988; Kuhn, 1991), with hardness value 206 kp/cm^2 (Vorreiter, 1949; Tsoumis, 1986). The bark of *S. alba* has pharmaceutical properties (Arabatzis, 1998) because of salicin that it contains. *Salix fragilis* is a photophilus species and belongs also to softwood riparian species. It is present almost in the whole of Greece and in central European riparian forests (Athanasiadis, 1986; Dister, 1988; Arabatzis, 1998). Arabatzis (1998) reports that it may coincide with "Eliki" that Theofrastus mentions.

Knowledge of the forest structure and dynamics is a necessary precondition for the knowledge of a forest in general, its production potential as well as for developing proposals for its cultivation and development as well as making relevant decisions (Heller, 1963; Dister and Drescher, 1987). Forest structure used to be presented partly in the form of drawings and partly in the form of numerical indexes or combinations of both. The method of Leibundgut that finally prevailed is now widely used and is known as the IUFRO classification (Smiris and Dafis, 1983, 1984; Smiris, 1985; Dafis, 1989). The aim of the present study was to explore and describe the degree of slenderness of the tree forest species *S. alba* and *S. fragilis* in the riparian forest of Nestos Delta with regard to the dynamics of the stands where they occur. They are both softwood species that occur under similar or even identical conditions (Dister, 1988; Kuhn, 1991). Dister (1988) reports differences in the dynamics between softwood and hardwood riparian forests.

In the frame of this study, it was asked whether differences occur in the structure and the dynamics of the stands in which the two willow species were present using slenderness as a criterion. The working hypothesis was that these closely related species with similar ecological characteristics should exhibit a comparable behavior with regard to the conditions of the different stands.

MATERIALS AND METHODS

Study site

The study site was the riparian forest at the Delta of River Nestos,

which is located in the north-eastern utmost of Greece. It occupies the flat part of the river course and it is extended in the area between the exit of the Thracian Gorge (north), the estuaries of Nestos River by the sea (south), the region of lagoons by Abdira (East) and the lagoons by N. Karvali (west). The Delta area is divided by the river in two parts, the West (of Kavala) and the East (of Xanthi). The altitude of the region ranges from zero (sea level) to 40 m (gorge exit). The inclination of the river is relatively high (0.725%) compared to other rivers. The total distance between gorge exit and the sea is 29 km. The study site is delimited by the following parallels: 40° 50' 52" and 41° 5' 4" N and 24° 42' 22" and 24° 51' 38" E.

Administratively the region belongs to the Region of Eastern Macedonia and Thrace and to two prefectures, the western part in the prefecture of Kavala and the Eastern in the prefecture of Xanthi. As far as management is concerned, the Forest service of Kavala is responsible for both parts of the riparian forest, in order to achieve uniform protection, administration and management.

Sampling and analyses

By using as a criterion of discrimination on a phytosociological basis, the composition of dominant woody vegetation, five different forest conditions were distinguished as types of structure (TS) in which willow species were present and were numbered from TS 1 until TS 5. 25 sampling plots (5 in each TS) of 0.1 acre extent for mixed and 0.05 acre for the monospecific forest stands were established in the riparian forest of Nestos. Every effort was made in order to ensure the uniformity of the sampling plots. In each sampling plot the following measurements and recordings were carried out:

1) Measurement of stem diameter at breast height (d) of all trees with diameter larger than 4 cm, with a precision of 1 cm.
2) Measurement of height of trees with a Haga hypsometer.

The degree of slenderness is determined via the relationship of height (h) and diameter at breast height (d) and is calculated as the quotient h to d (h/d ratio). A number of different equations were considered for describing the relationship between h and d for each species in each type of forest structure. The equation that provided the best fit was based on the model:

$$h/d = d/(a_0 + a_1 * d + a_2 * d^2)$$

For the estimation of the parameters, a_0, a_1, a_2 the following transformation (Prodan, 1968) was used:

$$d^2 / h = a_0 + a_1 * d + a_2 * d^2$$

Tree age was established by extracting increment cores by means of a corer and counting tree rings under a dissecting microscope (Husch et al., 1982; Speer, 2010).

RESULTS

In the riparian forest of Nestos, different types of structures were distinguished based on the composition of the dominant woody vegetation. The willow species *S. alba* and *S. fragilis* were found in five types of forest structure (TS), namely:

Type of Structure (TS 1): *S. alba – Populus alba – Alnus*

glutinosa Type of Structure (TS 2): *P. alba – S.fragilis-A. glutinosa.*
Type of Structure (TS 3): *S. alba – A. glutinosa*
Type of Structure (TS 4): *S. fragilis – A. glutinosa - Salix amplexicaulis*
Type of Structure (TS 5): *S. alba*

The estimated equations describing slenderness for the species *S. alba* and *S. fragilis* for every type of structure (TS) are presented in Figures 1 and 2. The respective equation parameters are shown in Table 1.

S. alba stands

S. alba was present in three types of structure of the riparian forest of Nestos, namely TS 1, TS 3 and TS 5. TS 1 consisted of mixed stands of *S. alba*, *P. alba* and *A. glutinosa*. *S. alba* was present at ages ranging from 21 to 28 years old, with a frequency peak at the diameter group of 26 cm. More generally, this species reached its highest frequency at the sapling and mature stages. The curves of the degree of individual tree slenderness are presented in Figure 3. It is apparent that *S. alba* exhibited the lowest degree of slenderness for d<10 cm compared to the other tree species while for d>35 cm the degree of slenderness was lower than the one of *P. alba* but higher than the one of *A. glutinosa*.

TS 3 consisted of mixed stands of *S. alba* and *A. glutinosa*, where *S. alba* was present at an age range of 20 to 33 years, while the frequency of its diameter groups exhibited two maximums for the groups of d=22 cm and d=30 cm and a numerical dominance of the young stage. The degree of slenderness of *S. alba* (Figure 4) below the diameter of 20 cm was lower than that of *A. glutinosa*, while for d>20 cm the degree of slenderness for *S. alba*, was slightly higher than that of *A. glutinosa*. Both exhibited a declining trend.

TS 5 consisted of pure stands of *S. alba*, which were found directly on the river bank, at an age range of 30 to 40 years old. They were even aged stands with the higher numerical frequency observed for the diameter group of d=18 cm. The degree of slenderness of *S. alba* (Figure 5) exhibited an increasing trend up to a diameter of 15 cm, where the maximum values were observed. Afterwards a mild declining trend with increasing diameter was observed.

S. fragilis stands

S. fragilis was present in two types of structure (TS 2 and TS 4) in the riparian forest of Nestos. In TS 2, which consisted of mixed stands of *P. alba*, *S. fragilis* and *A. glutinosa*, *S. fragilis* was present on a large island in the River Nestos. Individuals were aged 25 to 30 years old, with a numerical peak at the diameter group of d=22 cm. *S. fragilis* exhibited the highest degree of slenderness

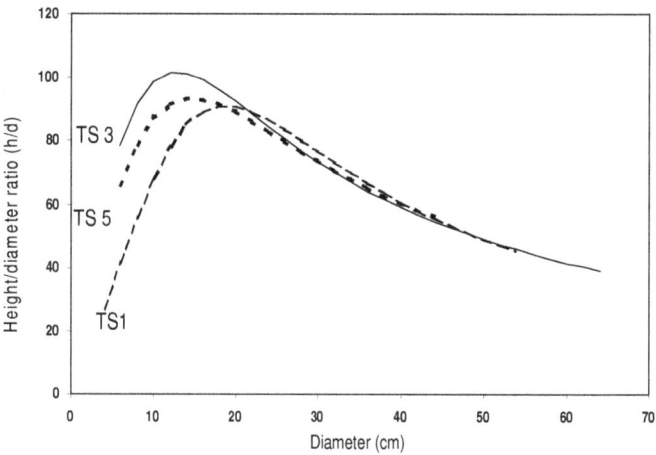

Figure 1. Slenderness curves of *S. alba* in different types of forest structure (TS).

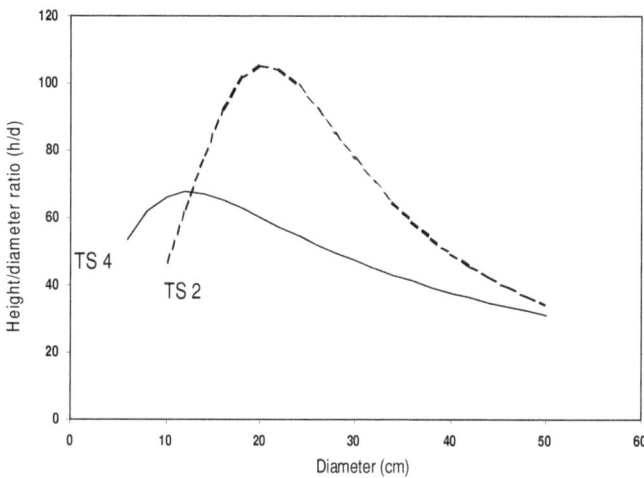

Figure 2. Slenderness curves of *S. fragilis* in different types of forest structure (TS).

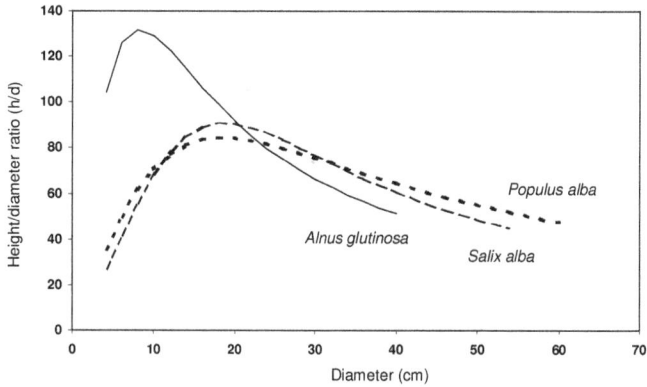

Figure 3. Slenderness curves of species *S. alba - Populus alba - Alnus glutinosa* in structure type TS1.

Table 1. Slenderness equation parameters of *S. alba* and *S. fragilis* in stands of the different structure types (TS) of the Nestos Riparian Forest.

Parameter	R^2 (Adj. R^2)	S.E.	Sign. F	Equation parameter	S.E.	Sign. T
S. alba TS1	0.944 (0.943)	5.6718	<0.001	a_0=17.0879 a_1= -0.7248 a_2= 0.4854	3.3487 0.2423 0.0039	0.0034 <0.001 <0.001
S. alba TS3	0.862 (0.861)	10.132	<0.001	a_0= 6.2815 a_2= 0.0385	0.9738 0.0009	<0.001 <0.001
S. alba TS5	0.774 (0.770)	7.9259	<0.001	a_0= 7.8311 a_2= 0.0366	1.7792 0.0026	<0.001 <0.001
S. fragilis TS2	0.901 (0.896)	8.609	<0.001	a_0= 46.3881 a_1=-3.5982 a_2= 0.1114	10.592 0.7559 0.0128	<0.001 <0.001 <0.001
S. fragilis TS 4	0.803 (0.802)	14.356	<0.001	a_0=9.03348 a_2=0.06062	1.6464 0.0023	<0.001 <0.001

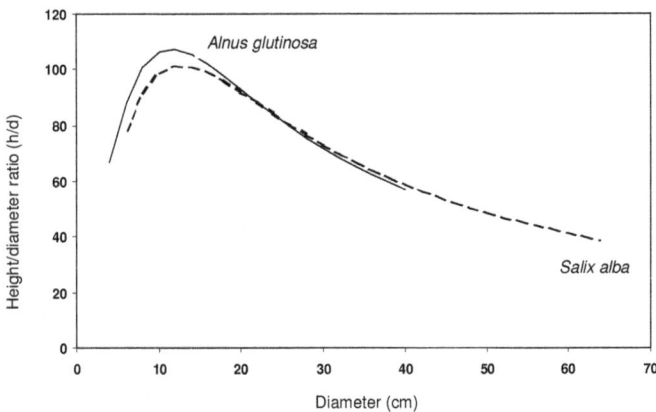

Figure 4. Slenderness curve of *S, alba – Alnus glutinosa* in structure type TS3.

(Figure 6) for a diameter of 20 cm, while within the range 20 cm<d<30 cm it had a higher degree of slenderness in comparison with the two other species. On the contrary for d>40 cm it had the lowest degree.

TS 4 consisted of mixed stands of *S. fragilis* and *A. glutinosa*. *S. fragilis* was present in all the layers of the mixed stands along with *A. glutinosa* and *S. amplexicaulis*. The largest proportion of *S. amplexicaulis* individuals was found in the understorey. The age of *S. fragilis* individuals laid between 15 and 25 years and the dominant diameter groups were those with d=18 cm and d=26 cm. *S. fragilis* slenderness degree (Figure 7) up to a diameter of d=20 cm was lower compared to the two other tree species. From d=20 cm onwards the degree followed a mild declining trend, but it still remained higher than the degree of *A. glutinosa* and *S. amplexicaulis*. At

the diameter of about 29 cm there were no differences in the slenderness degree of the three species.

DISCUSSION

The knowledge of forest structure and dynamics is a necessary precondition of the forest's knowledge in general and of forestry capacity as well as decision making regarding their management in particular (Heller, 1963, 1969; Dister and Drescher, 1987). Even today studies on the structure and dynamics of riparian forests remain limited. The degree of slenderness according to Assmann (1961) is an important factor for the characterization of the stability of the stand (Röhle, 1982; Petras and Mecko, 2010). Low values of slenderness show stable conditions, while high values indicate stand of nonconstant structure (Röhle, 1982). Values of slenderness equal to hundred mean that the height of tree in m is equal with its diameter in cm. The degree of slenderness in dense young stands may exhibit values higher than one (Röhle, 1982), something that is observed in many stands in the riparian forest of Nestos.

Numerical domination of *Salix* species (Heller, 1963) in the overstorey in mixed stands with *A. glutinosa* was observed in riparian forests in Switzerland and Austria as well as in mixed stands with *Alnus incana* and *Fraxinus excelsior*, in Switzerland and Yugoslavia. The degree of slenderness of *Salix* is lower than that of *Alnus* for the same diameter group (Heller, 1963). From Figures 1 to 7, it is evident that the slenderness curve is different even for the same species in the different types of forest structure where it occurs. Slenderness is related to the conditions of growth of the trees, the stand structure (density) and their age. In Figure 1 the slenderness

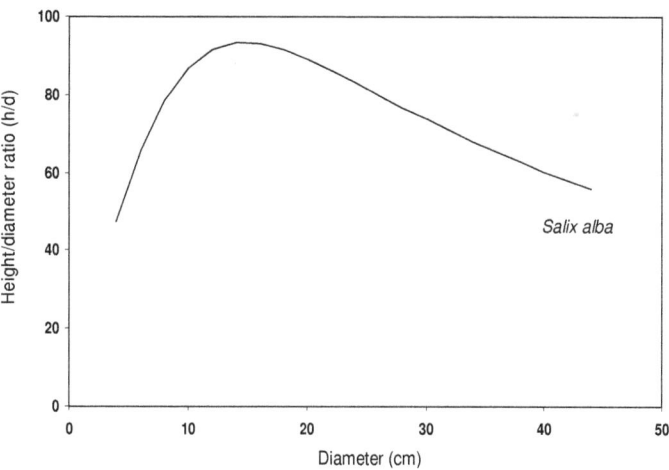

Figure 5. Slenderness curve of *Salix alba* in structure type TS 5.

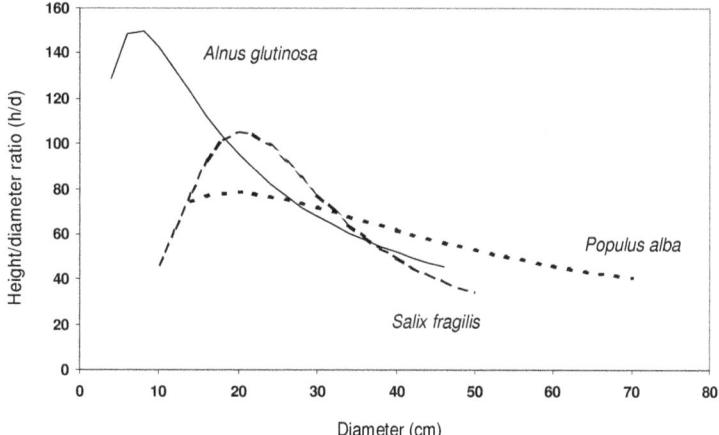

Figure 6. Slenderness curve of *Populus alba -Salix fragilis - Alnus glutinosa* in structure type TS2.

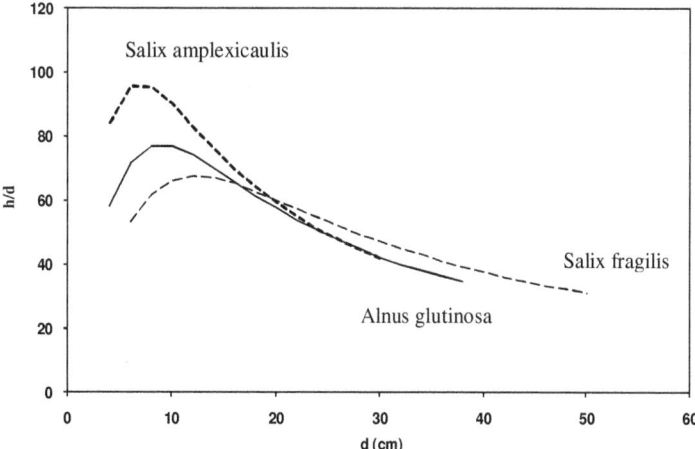

Figure 7. Slenderness curve of *Salix fragilis – Alnus glutinosa - Salix amplexicaulis* in structure type TS4.

curves for the three distinct structure types of *S. alba* are given. Up to the diameter of 20 cm the higher degree of slenderness is exhibited in mixed stands with *A. glutinosa* (TS 3), in which it dominates in the overstorey. The pure stands of *S. alba* (TS 5) follow with slightly lower values of slenderness. Individuals with diameter less than 20 cm exhibit the lowest slenderness values in mixed stands with *A. glutinosa* and *P. alba* (TS 1), where the higher percentage of *S. alba* individuals is found in the overstorey. The most constant conditions with regard to *S. alba* were observed in individuals with d>45 cm in which the lowest values of the slenderness degree were found.

With regard to *S. fragilis* considerably higher degree of slenderness (Figure 2), therefore more unstable conditions, and its highest dynamics are exhibited by the youngest individuals with diameter d<30 cm in mixed stands with *P. alba* and *A. glutinosa* (TS 2). More stable conditions, that is, a lower degree of slenderness of *S. fragilis* is found in mixed stands with *A. glutinosa* and *S. amplexicaulis* (TS 9). In these stands *S. fragilis* is numerically dominant in the overstorey and in the middlestorey and an almost stable degree of slenderness for diameters larger than 20 cm (Efthimiou, 2000) has been found. The most stable conditions for *S. fragilis* were observed for trees with d>50 cm. To sum up, the different behavior of the degree of slenderness of the riparian forest tree species *S. alba* and *S. fragilis* can be summarized as follows.

The degree of slenderness of *S. alba* young stands equals hundred. Young trees with d<25 cm appear to exhibit faster growth and dynamic evolution, while structure is stabilized around a tree diameter d>30 cm. In mixed stands (TS 3) co-domination with *A. glutinosa* in the overstorey dynamically tends to be replaced by the domination of *A. glutinosa* that numerically dominates in the middlestorey. In mixed stands (TS 1) *S. alba* numerically dominates in the overstorey along with *P. alba* and tends to be gradually displaced by *A. glutinosa* with which it co-dominates the middlestorey and competes at d<25 cm as it is shown by the high degree of slenderness of *S. alba*. *S. fragilis* exhibits more stable structure (low degrees of slenderness) in mixed stands with *A. glutinosa* and *S. amplexicaulis* (TS 4), at which it numerically dominates (frequency close to 69%) in the middlestorey and overstorey, a fact that indicates that it was established first on those surfaces and will be dynamically displaced by *A. glutinosa*, which numerically dominates the middlestorey and the understorey. In mixed stands with *A. glutinosa* and *P. alba* (TS 2), *S. fragilis* is numerically outnumbered by the other two species, while the largest percentage of the individuals is present in the overstorey and in the middlestorey. In those stands (TS 2), young trees (d<25 cm) of *S. fragilis* exhibit a higher dynamic and its structure is becoming stable for mature trees with d>40 cm. The dynamic of these stands leads to the replacement of the overstorey

domination of *S. fragilis* and *P. alba* by *A.glutinosa* domination, which numerically dominates already in the middlestorey.

S. fragilis exhibits a stable structure (degree of slenderness 40) for young trees of d=30 cm, while *S. alba* presents the same stability for mature trees of d>60 cm. This finding leads to a rejection of the working hypothesis of this study since the differences in the dynamics of the two willow species are profound. The remaining riparian forest areas in Greece are of small surface. The anthropogenous pressures (grazing, agriculture, fire, urban expansion) exerted continue to be multiple and severe, resulting in limited areas that are not suitable for management targeting economic returns. The sound management of Greek riparian ecosystems in the future should strictly target their preservation (Jerrentrup and Lösing, 1991) in a way that the remaining natural stands are allowed to resume their natural succession trajectories and dynamics. This is confirmed by the findings of this study, since it is evident that natural stands of *Salix* sp. attain a stable structure at high ages.

AKNOWLEDGEMENTS

V. Detsis is acknowledged for critically commenting on earlier drafts of this study.

REFERENCES

Arabatzis TH (1998). Bushes and trees in Greece. Publ. Ecological Movement of Drama and Technological Educational Institute of Kavala, Drama, Greece, 1: 292.

Athanasiadis H (1986). Forestal Botany II (Trees and shrubs of Greek Forests). Publ. Giachoudi-Giapouli, Thessaloniki, Greece, p. 309.

Assman E (1959). Height quality and real earnings power. Sci. Cent. For. Leaf., 78: 1-20.

Assmann E (1961). Forest Yield Science, Organic Production, structure, growth and yield of forest Passed. Publ. BLV, Munich, Bonn and Vienna, p. 490.

Carbiener R (1974). The left bank natural areas and forests and the protected areas of Rhinau Daubensand (Frankreich): a phytosociological and landscape ecological study. Nat. Lands B. Würt, 7: 438-535.

Dafis S (1966). Structure and growth analysis of natural pine forests. Beitr Geob. Land. .Switz, 1: 1-75.

Dafis S (1992). Stability, balance and self-regulation (omoiostasi) the forestal ecosystems. Scientific Annals of Dept. of Forestry and Natural Environment, of Aristotle Univ. of Thessaloniki, 5: 521-531.

Dister E, Drescher A (1987). The structure, dynamics and ecology of flooded hardwood riparian forest long at the lower March (Lower Austria). Ver. Ges. Ecol., 5: 295-302.

Dister E (1988). Ecology of riparian forests of central Europe. Gem Pen. For Pub. Heal. Hik, Nat. Her. Prot., 19: 6-26.

Efthimiou G (2000). Structure Analysis, Dynamic and Ecological meaning of the riparian forest of Nestos River. Ph.D. dissertation, Aristotle University of Thessaloniki, Thessaloniki, Greece.

Efthimiou G, Smiris P (2002). Management, Protection and Restoration measures of the riparian ecosystems of Nestos Delta. In Koungolos AG, Liakopoulos AB, Korfiatis GP, Koutsospyros AD, Katsifarakis KL, Demetracopoulos AD (eds) Protection and restoration of the Environment VI: Proceedings of an International Conference, held at Skiathos, Greece, Grafima, pp. 1871-1874.

Emmanouloudis D, Myronidis D, Panilas S, Efthimiou G (2006). The role of sediments in the dynamics and preservation of the aquatic forest in the Nestos delta (N. Greece). IAHS-AISH, 306: 214-222.

Ellenberg H (1996). Vegetation of Central Europe with the Alps. 5th ed. Eugen Ulmer, Stuttgart, pp. 1-1095.

Firbas F (1949). Spar and late-glacial forest history of Central Europe North of the Alps. Publ. G. Fischer, Vienna, 1: 480.

Firbas F (1952). Spar and late-glacial forest history of Central Europe North of the Alps. Publ. G. Fischer, Vienna, 2: 256.

Heller H (1963). Structure and dynamics of riparian forests. Geob. Beitr. Land. Switz., 42:1-75.

Heller H (1969). Living conditions and sequence of the floodplain vegetation in Switzerland. Mémoire de l'Institut suisse de recherches forestières, 45(1):1-124.

Hugin G (1981). The riparian forests of the southern Upper Rhine Valley - and its change by the Rhine endangerement Expansion. City Land, 13(2): 78-91.

Husch B, Miller CI, Beers TW (1982). Forest mensuration. 3rd edition. Publ. John Wiley and Sons, p. 402.

Jenik J (1998). Biodiversity of the Hercynian Mountains of Central Europe. JACA, 151/152: 83-99.

Jerrentrup H, Lösing J (1991). Situation of the flood plains in Greece. Akad Nat. Lands. (ANL), 4: 86-92.

Kuhn N (1991). The nature of the riparian forest as a habitat. Switz. J. For., 142(9): 731-749.

Leibundgut H (1959). About the purpose and methodology of the structure and growth analysis of native forests. Switz. J. For., 110: 111-124.

Leibundgut H (1978). About dynamics of European Natural forests. Gen. For. J. Mun., 33(24): 686-690.

Mayer H (1984). Forests in Europe. Publ. G. Fischer, Stuttgart, New York, USA, p. 691.

Papaioannou IK (1953). The forest Kotza – Orman. Publ. Eclogi. Athens, Greece, 88: 55-64.

Petraš R, Mecko J (2010). Stability of the development of basic stand parameters of beech yield tables constructed on the basis of short-term observations on research plots. J. For. Sci., 56 (7): 323-332.

Prodan M (1968). Forest Biometrics. Pergamon Press. First English Edition. Translated by S. H. Gardiner, p. 447.

Röhle H (1982). Structure and growth of mixed stands of oak-influenced groundwater locations in the riparian forest areas of southern Bavaria. For. Res. Rep., No. 52, Munich.

Röhle H (1984). Yield characteristics of oak-mixed stands on underground water influence locations in the riparian forest areas of southern Bavaria. FWCbl, 103(6): 330-349.

Smiris P (1987). The dynamic development of Structure in virgin forests of Paranesti-Dramas. Sci. Ann. Dept. For. Nat. Environ. Aristotle Univ. Thessaloniki, 13: 480-593.

Speer JH (2010). Fundamentals of Tree-Ring Research. University of Arizona Press, Tucson, Arizona, U.S.A.

Straka H (1957). Pollenanalysen.0.0 and vegetation history. The new Bremen bookstore, Wittenberg, Germany, p. 109.

Strassburger E (1978). Textbook of Botany. In: von Denffer D, Ehrendorfer F, Maegdefrau K, Ziegler H (eds), Publ. Fischer, Stuttgart, 31th ed., p. 895.

Tsoumis G (1986). Science and Technology of Timber. v. A' Structure and attributes. Publ. Aristotle Univ. of Thessaloniki, Thessaloniki. Greece, p. 351.

Vorreiter L (1949). Wood Technological Manual, Vol. I. Publ. G. PIOUS and Co, Vienna, p. 548.

Wenger EL, Zinke A, Gutzweiler KA (1990). Presend situation of the European floodplain forests. For. Ecol. Manag., 33/34: 5-12.

Yon D (1980). Strategic elements of conservation. Symp. Phyt., 9: 1-18.

Assessing ecosystem effects of small–scale cutting of Cameroon mangrove forests

Longonje N. Simon[1] and Dave Raffaelli[2]

[1]Department of Environment, University of Buea, P. O. Box 63 Buea, Cameroon.
[2]Environment Department, University of York, YO 105 DD United Kingdom.

One of the most universal forms of resource-use in the tropics is small-scale wood exploitation; but ecologists are only starting to study its effects. This paper examines the effects of small-scale wood harvesting on forest structure and composition of mangrove forests. A stratified sampling method was used to select the sample zone. The forest characteristics were assessed by employing the quadrat/census plot method (Cintron and Schaeffer, 1984). To assess canopy structure, plot perimeter was used as a basis for line intercept sampling (Lertzman et al., 1996). Two-thirds of all canopy gaps were caused by human activities and this might have dramatic effects on regeneration because there were significantly more seedlings in canopy gaps compared with closed canopy areas. *Rhizophora* was the dominant species and formed a virtually monospecific stand in the coastline zone with a gradual transition to a mixed forest of *Laguncularia, Avicennia* and *Rhizophora*. Ecological characteristics such as mean tree density, seedling density, mean diameter at breast height, basal areas and gap sizes differed among seaward, middle and landward zones. The findings from the present study highlight that the ecological effect of small scale wood exploitation is a potential threat to mangrove forest ecosystem health.

Key words: Cameroon, mangrove ecology, forest ecosystem health.

INTRODUCTION

Mangrove forests like most other ecosystems provide a full range of goods and services. They play an important role in maintaining a healthy coastal ecosystem by providing far reaching direct and indirect services (Dahdouh-Guebas, 2001; Walter et al., 2008). The social, ecological and economic importance of mangrove forests is enormous. They are among the world's most productive systems, have a high primary production, high rates of recycling and provide a high supply of nutrient source that supports many complex food chains (Clough, 1993; Lefebvre and Poulin, 2000). These features make mangrove systems perfect as breeding and nursery grounds for many marine species including commercially important fishes (Baran, 1999; Alongi, 2002). Mangroves

also contribute significantly to the global carbon cycle.

Total global mangrove biomass is approximately 8.7 gigaton dry weight, that is 4.0 gigaton of carbon (Clough, 1993; Twilley et al., 1992). Mangroves forests are reported to have historically provided a variety of renewable products including timber, food, charcoal, firewood and medicine to many local communities worldwide (Primavera, 1995; Dahdouh-Guebas, 2001; Walters, 2003). Mangrove forests are subject to a number of natural and anthropogenic threats. Though there has been considerable attention paid to natural disturbances of mangroves such as hurricanes and climate change (Roth, 1992; Gilman et al., 2008), human activities in these coastal areas such as physical alteration of the habitat, over-exploitation of the resources and pollution cause significant pressure on the environment. These pressures have increased steadily as the human population increases. For several decades, mangrove forests have been cleared and degraded on an alarming

*Corresponding author. E-mail: nlongonje@yahoo.com.

scale worldwide (Hamilton and Snedaker, 1984; Aksornkoae et al., 1992). Mangroves in many parts of the world are also affected by local-scale exploitation. The negative impact of local-scale exploitation on mangrove forest health varies from place to place and although the potential impacts are huge, formal assessments of the effects are uncommon (Walters, 2005a). The impacts are likely to be complex and may include social, economic and environmental dimensions. Frequent, but low inten-sity, small-scale mangrove exploitation has significant impact on forest structure, but limited information is available on how mangrove exploitation affects forest composition and regeneration (Eusebio et al., 1986; Smith and Berkes, 1993; Barnes, 2001). Cameroon has a growing coastal population as a result of which increasing use of the country's natural resources is endangering several ecosystems, especially estuarine systems. Mangroves are in decline in Cameroon mainly due to firewood extraction and the cutting of poles for construction (Longonje, 2002). The floristic composition of Cameroon mangrove is characteristic of the Atlantic mangroves of West Africa. It is dominated by *Rhizophora* and comprises mostly of three species *R. mangle, R. harrisonii* and *R. racemosa* (Spalding et al., 1997). Other mangrove species include *Avicennia germinans, Laguncularia reacemosa, Conocarpus erectus, Acrostichum aureum, Pandanus candelabrum* and *Nypa fructicans* (Spalding et al., 1997). The study area pro-vides several ecosystem services such as natural coastal barriers, recreation and fisheries. Local communities in and around the mangroves depend on the forest for their livelihood.

One major socio-economic activity in the mangroves is artisanal fishing; the fish catch is estimated between 76 and 106 tons per year (Gabche, 1997). Some of the densely populated and industrial towns are located at the fringe of mangrove forests, notably Douala (the economic capital), Limbe and Tiko. The Cameroon mangrove is biologically diverse. Apart from the different species of fishes and birds, many endangered species such as marine turtles (*Lepidoshelys olivacea*), dwarf crocodile (*Crocodylus cataphractus*) and West African manatee (*Trichechus senagalensis*) can be found.In this paper we examine disturbance associated with local community forest exploitation. The goal of the present study was to improve understanding of the effects of small-scale wood harvesting on forest structure and composition of mangrove forests for better resource management deci-sions. Our objectives were to quantify (1) the structure of the mangrove forest in Cameroon estuary, and (2) the changes in the mangrove forest in response to small-scale harvesting.

METHODS

Field work for this study was undertaken from 2008 to 2009. The forest characteristics were assessed by employing the quadrat/census plot method (Cintron and Schaeffer, 1984).

Study area

This study was carried out in the Cameroon Estuary mangrove (Figure 1) located in the South–Western part of Cameroon between latitude 3° 83' to 4° 10' N and longitude 9° 25' to 10° 00' E. It is a large forest of approximately 1,750 km^2 and is representative of the bigger mangrove areas in Cameroon. The coastal and marine environment of Cameroon forms part of the southern section of the Gulf of Guinea Large Marine Ecosystem (Price et al., 2000). The coastline stretches from the Equatorial Guinea border at latitude 2° 30' to 4° 67' N at the Nigeria border and it is estimated at about 400 km in length (Price et al., 2000).

Plot description

Two different quadrat sizes (50 × 50 m and 10 × 10 m) were used, with the corners and boundaries of plots marked using calibrated measuring ropes and tapes. The larger plot size was used in some of the stands surveyed because trees were typically large and sparsely located. The smaller plot size was used where stands had typically small and densely crowded trees. A stratified sampling method was used to select the sample zones in the Cameroon Estuary. The mangrove forest was divided into 3 zones: seaward (coastline forest), middle (interior forest) and landward (fringe forest). Nine sites were selected randomly to capture representative forest structure and characteristics and these sites were distributed equally between seaward (coastline forest), middle (interior forest) and landward (fringe forest) zones (Figure 1). Thirty-one plots were sampled randomly for floristic composition, stand and canopy structure with almost equal effort in coastline, interior and fringe forest. To assess the floristic composition and stand structure, data were collected on tree species composition, diameter at breast height (dbh), tree height, canopy cover, numbers of seedlings, gaps, gap size, stumps and snag (dead stems). In each plot, every tree was numbered, marked and measured (> 1.0 m tall) and seedlings (< 1.0 m) recorded (Walter, 2005b).

The diameter at breast height (dbh) of each tree stem was measured at 1.3 m or above the highest prop root following Cintron and Schaeffer (1984). Tree height was measured using marked bamboo poles and clinometers.

Canopy assessment

To assess canopy structure, I used the plot perimeter as a basis for line intercept sampling (Lertzman et al., 1996), providing a transect length of 200 m for the 50 × 50 m plots and 40 m for the 10 × 10 m plots. At 2.5 m intervals along each transect, the canopy was viewed vertically upward, and scored as CC (closed canopy) or CG (canopy gap) defined as an area where the canopy is noticeably reduced compared to adjacent areas (Runkle, 1992). All gaps were classified as either natural (that is not caused by human activity; for example gaps caused by fallen tree due to strong wind) or induced (that is caused by humans; for example gaps caused by human clearing or wood exploitation). Gap age was estimated from the stage of decomposition of the gap-making tree, as fresh (recent gap), old (dry trees, but no sign of decomposition) and very old (decomposing trees). To estimate gap size, the distance and angle from the centre of the gap to the gap edge was measured in each of 8 (45°) sectors and the area of the triangles summed (Lertzman et al., 1996).

Statistical analysis

Descriptive statistics (histograms) were used to analyse species composition, distribution and utilization. To test whether or not,

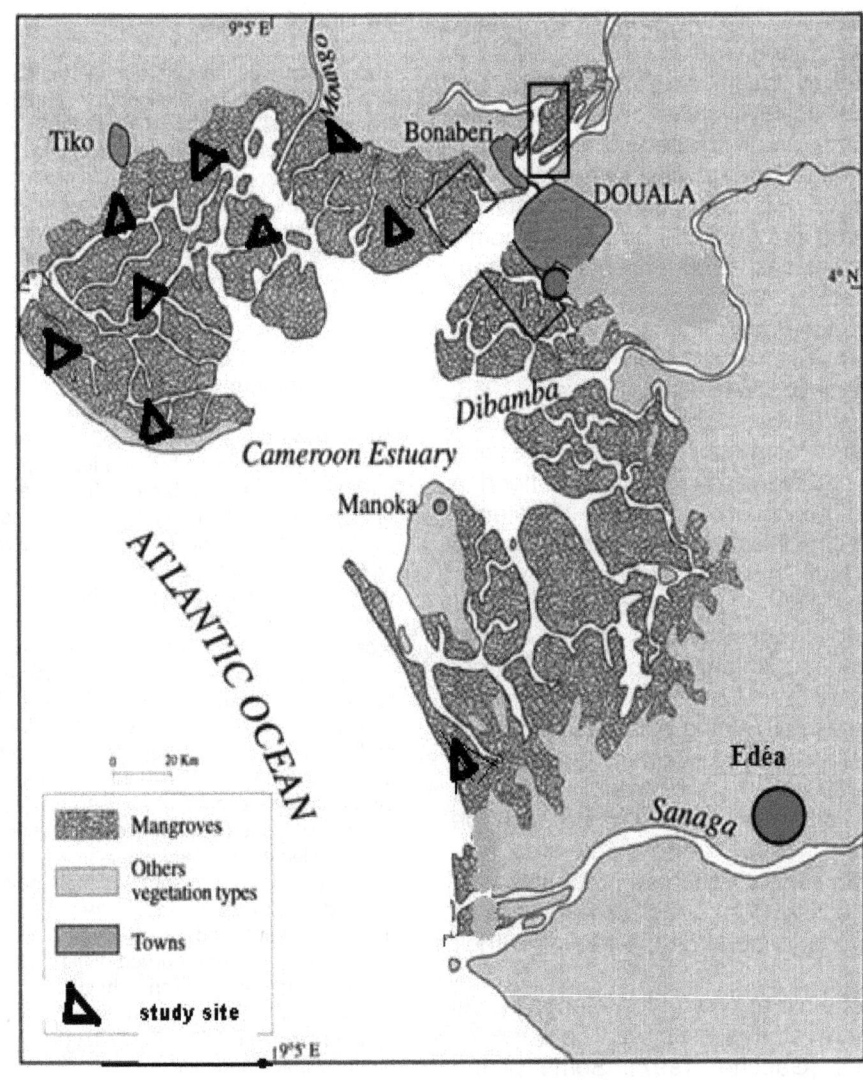

Figure 1. Study site and locations of plots used to evaluate forest structure.

there was a significant difference between selected forest characteristics in coastline, interior and fringe zones, the non parametric Kruskal-Wallis ANOVA was performed. This was because the data were heteroscedatic and transformation (square root and log) did not normalize the data. Nonparametric t- tests were employed to test for differences in seedling abundances in canopy gaps and closed canopy. All of the aforementioned analyses were performed using SPSS and PRISM 5.

RESULTS

Thirty one plots were sampled for forest structure and composition, 12 within the coastline, 8 within the interior forest and 11 within the forest fringe. A total of 3167 individual trees, 423 stumps and 103 snags were recorded. *Rhizophora* (red mangrove) was the dominant species (83.6%) followed by *Avicennia* (black mangrove) at 9.1% and *Laguncularia* (white mangrove) at (7.1%). Virtually, monospecific stand of *Rhizophora* occupied in

the coastline zone, with a gradual transition to a mixed forest of *Laguncularia, Avicennia* and *Rhizophora* occupying the interior and fringe zones. Although, *Avicennia* and *Laguncularia* were abundant in interior and fringe zones, they were not the dominant species (Figure 2).

Forest structure

Coastline forest had lower mean tree density and seedling density compared to interior and fringe forest, but the difference was not significant. The mean diameter at breast height (dbh) and basal areas were higher in coastline forest compared to interior and fringe forest, but the difference was not significant (Tables 1 and 2). When comparing the seedling density between the forest zones, the interior had the highest density, but the difference was not significant (Tables 1 and 2).

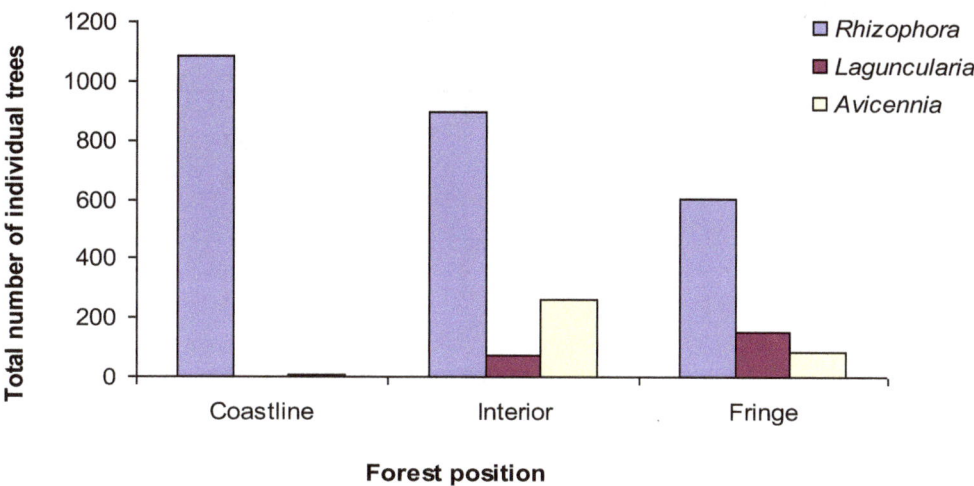

Figure 2. Mangrove tree species distribution within the 3 major zones.

Table 1. Summary of selected ecological characteristics of the mangrove stands studied (means and standard deviation: for numbers of replicate plots).

Characteristics	Coastline plots	Interior plots	Fringe plots	Average
Tree density (n/100 m^2)	11.9 (19.2)	15.4 (21.1)	20.7 (24.8)	16.0 (20.2)
Diameter at breast height (dbh) of stem (cm)	27.6 (8.7)	24.7 (7.7)	19.2 (7.8)	23.8(19.7)
Stem basal area (m^2/ha)	74.2 (38.8)	59.5 (38.1)	46.6 (5.2)	60.1 (29.8)
Seedling density (n/100 m^2)	24.8 (39.7)	31.4 (45.4)	14.3 (14.6)	23.5(40.1)

Canopy gaps

257 gaps were recorded during the survey, two-thirds (66%) of which were caused by human influence, whilst one-third (34%) was due to natural factors (Table 3). Human influence was responsible for most of the gaps created in fringe, coastline and interior forest (73, 72 and 54% respectively) (Table 3). Average gap size of 3.1m^2 was recorded and gap size differed significantly between zones (ANOVA, P = 0.001) (Table 2). The coastline and fringe gap size were both significantly different from interior (Figure 4). The average gap density of 27.4 was recorded overall (Table 3), but there were significant differences between zones (ANOVA, P = 0.02) (Table 2), with the fringe, interior and coastline canopy gap density differing significantly from one another (Figure 3). Seedling density was not significantly different between zones (Table 2). The relationship between seedlings and canopy was examined as an alternative way to estimate the effect of exploiting forest on mangrove regeneration. Significantly, more seedlings were observed in canopy gaps compared to closed canopy areas (t = 3.5, P = 0.01, Table 4). *Rhizophora* seedlings were more abundant in canopy gap than in closed canopy areas (t = 2.4, P = 0.04), whilst *Avicennia* and *Laguncularia* were not profuse in canopy gap.

Forest species composition

The size-frequency distributions of all mangrove species are represented in Figure 5. All three species showed a higher abundances of stems in small size classes (<25 cm). In contrast to *Rhizophora*, *Avicennia* is completely absent from size classes greater than 95 cm and *Laguncularia* from size classes more than 25 cm. For *Rhizophora*, coastline forest plots had many large stems greater than 25 cm, whilst the interior forest plots had few medium sizes stems, and lack very large stems and fringe forest plots only supported small stems (≤25 cm: Figure 6).

DISCUSSION

According to Walter (2004), some mangrove forests have been dramatically altered through small-scale cutting and deliberate planting of trees by local communities. Few studies have examined the ecological impacts of small–scale exploitation of mangrove with the aim of assessing forest health. According to Smith and Berkes (1993), small-scale cutting of mangrove in the Caribbean reduces the abundance of large trees, but greatly increase the density of smaller trees. Esusebo et al. (1986) found that

Table 2. ANOVA results of selected ecological characteristics in coastline, interior and fringe zones.

Characteristics	Source of variation	SS	DF	MS	F	P-value
Tree density (100 m²)	Between groups	582	2	291.05	0.69	0.51
	Within groups	11236	27	416.16		Not significant
	Total	11818	29			
Canopy gap density (100 m²)	Between groups	188	2	94.19	4.72	0.017
	Within groups	558	28	19.96		Significant
	Total	747	30			
Diameter at breast height (dbh) of stem (cm)	Between groups	92563	2	46281.74	78.66	0.58
	Within groups	1863447	3167	588.39		Not significant
	Total	1956010	3169			
Canopy density (100 m²)	Between groups	144	2	72.27	0.74	0.48
	Within groups	2654	27	98.31		Not significant
	Total	2798	29			
Stem basal area (m²/ha)	Between groups	82	2	41.24	75.59	0.32
	Within groups	1744	3167	0.55		Not significant
	Total	1826	3169			
Gap size (m²)	Between groups	1	2	0.48	10.43	0.001
	Within groups	1	27	0.046		Significant
	Total	2	29			
Seedling density (100 m²)	Between groups	51	2	25.97	0.014	0.9
	Within groups	46779	27	1732.56		Not significant
	Total	46831	29			

Table 3. Canopy gaps and their causes in the 3 forest zones.

	Coastline	Interior	Fringe	Average
Canopy gap density (m²/100 m²)	3.7	23.8	54.7	27.4
Gap size (m²)	3.5	1.1	4.7	3.1
Human cause (%)	72	54	73	66.3
Non-human cause (%)	28	46	27	33.6

cutting of mangroves in the Philippines resulted in stunted and shrubby tree growth, but other studies have shown otherwise. For instance, Nurkin (1994) suggests that small-scale mangrove exploitation has an insignificant effect on mangrove forest structure. In the present study, the canopy gaps created were relatively small, the largest gap size measured was 72.2 m², but the mean gap size was much smaller at 3.1 m², relatively small when compared to findings from other mangrove studies. For example, Ewel et al. (1998) recorded a mean gap size of 158 m² for mangrove in Kosrae Micronesia, though the author deliberately ignored gap sizes less

than 10 m². Smith (1992) observed gap sizes of mature mangrove forest in Australia of 40 to 120 m², but it is possible that he overlooked gaps of less than 10 m². By contrast, Walter (2005a) found a smaller mean gap size of 2.6 m² for Philippines mangroves and studies of other forest types have shown that such small canopy gaps have an important effect on the forest structure (Feller and Mckee, 1999; Kennedy and Swaine, 1992). Exploitation of mangrove wood product was not completely species-selective in this study, but *Rhizophora* was the preferred species. There is evidence that wood exploitation might have changed *Rhizophora* stem-size

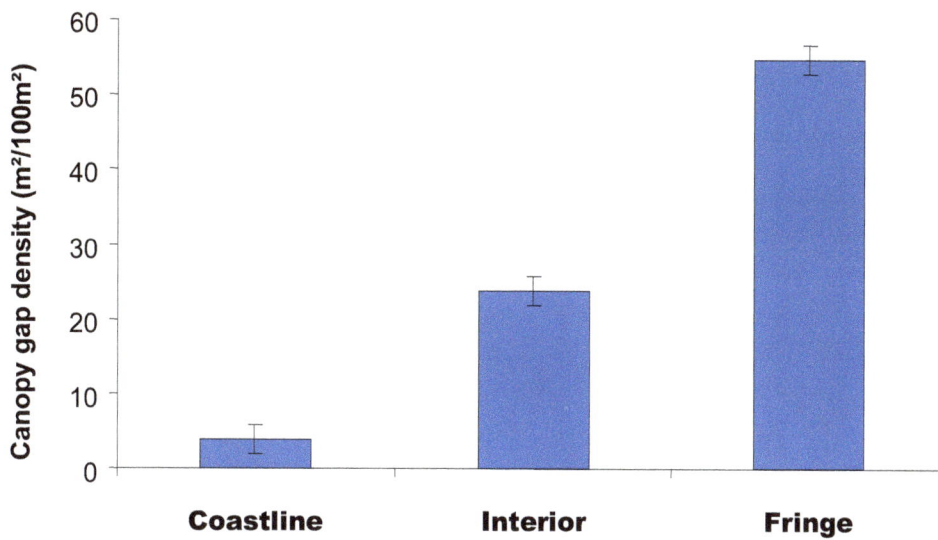

Figure 3. Comparison of canopy gap density within zone.

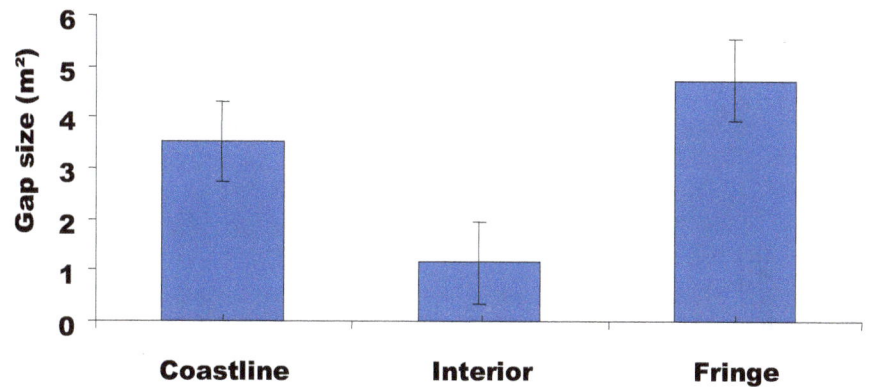

Figure 4. Comparison of gap size within zone.

Table 4. Seedling abundance (total count) of different mangrove species in open and closed canopies.

Species	Canopy gap	Closed canopy
Rhizophora	863	375
Laguncularia	220	59
Avicennia	161	93
Total	1244	527

distribution. Coastline forest (least accessible) is characterised by *Rhizophora* with large stem size, whilst the interior forest has medium range stem sizes and the fringe forest have relatively small stem sizes (Figure 6). This study suggests that mangrove forests differed structurally from the fringe to the coastline, due to a combination of anthropogenic and natural factors (Table 1).

Also, the further one moves away from the residential areas, the bigger the tree sizes, although other factors such as soil salinity and nutrient concentrations are known to influence tree size (Calumpong and Menez, 1997). However, *Avicennia* and *Laguncularia* species did not show clear patterns of size class distribution (Figure5). Mangroves are thought to recover quickly after disturbance (Smith and Berkes, 1993), but the evidence

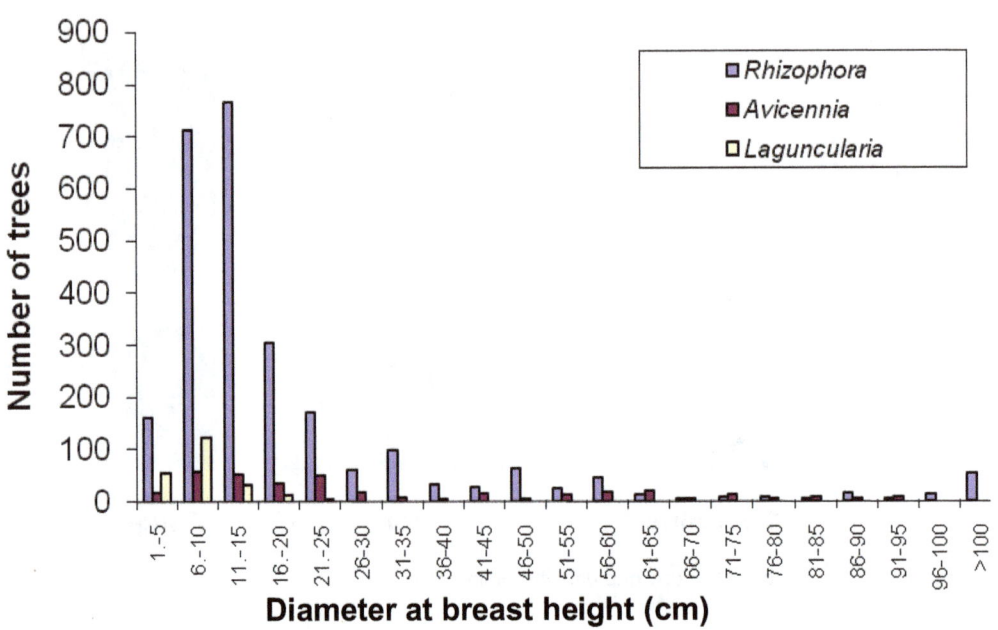

Figure 5. Size-frequency distribution of (dbh) of *Rhizophora*, *Avicennia* and *Laguncularia* species (all 3 zones combined).

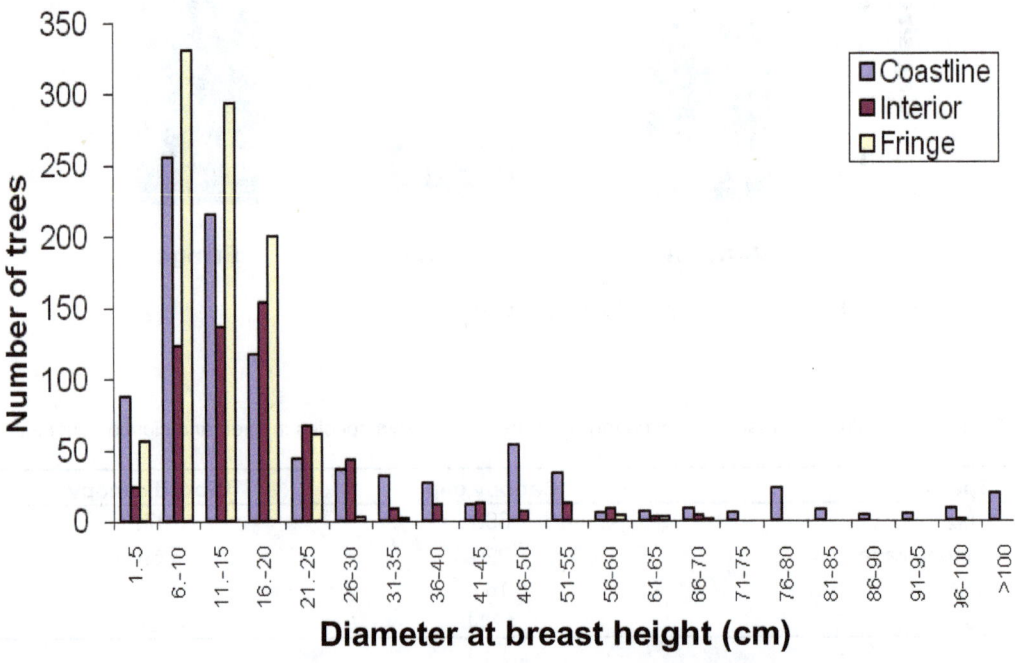

Figure 6. Size-frequency distribution of dbh of *Rhizophora* in coastline, interior and fringe forest plots.

is mixed. Thus, Ewel et al. (1998) found no differences in gap regeneration as a result of selective logging in Kosrae. Clarke and Kerrigan (2000) found that canopy gap had a strong influence on the abundance of mangrove seedlings. In the present study, the most sensitive species in producing the most seedlings was *Rhizophora* which shows a vital difference in gap regeneration (Table 3). Smith (1987) observed significant recruitment of *Rhizophora* species in gaps. According to Feller and Mckee (1999), gap size do not influence *Rhizophora* regeneration. According to Smith (1987), mangrove seedlings regenerate quickly in large numbers in canopy openings. In the present study, the seedling density is relatively low; this might suggest that the

Cameroon mangrove canopy is relatively closed and the forest structure is relatively healthy. Local communities in many tropical coastal regions have exploited mangroves for many years, but studies which examine local-wood utilisation and its ecological effects are uncommon. Policy makers and researchers alike have overlooked local-scale wood harvesting in mangrove forests. Management strategies are thus often developed without regards to either the ecological or economic significance of such attributes. Where such harvesting is significant, conservation efforts may encounter much opposition from the local communities. Forest biodiversity may also be eroded over the long term by continued selective removal of some species more than the others, and by the varied responses of species to cutting distribution (Walters, 2005a). At the same time, understanding patterns of wood use can inform management planning so that it is compatible with existing resource use practices (ITTO, 2002). For example a well- managed mangrove forest plantations provide abundant construction wood that can reduce harvesting pressures on natural forests, so long as the plantations are not permitted to encroach too much into natural forest (Walters, 2004).

A great deal of ecological research has been done on mangroves. Given this, and the fact that we know mangroves are harvested by local people in many tropical, coastal regions (Diop, 1993), it is remarkable that barely a few of published studies have examined local-scale wood use and its ecological effects on these forests. Study findings presented elsewhere (Walters, 2004) show how some mangrove forests have been dramatically altered through deliberate planting of trees by local people. Likewise, findings presented here demonstrate that small-scale, local wood cutting can be a significant form of ecological disturbance in mangroves. Forest structure was dramatically altered by cutting, but impacts on composition and regeneration were also detectable. Most notably in this respect is the finding that most mangrove species appeared to respond to small-scale cutting by significant recruitment of new species. In fact, it is plausible that mangrove forests in many places have already experienced significant changes to species composition as a result of past cutting and other anthropogenic influences (Walters, 2003). Efforts to understand these unique forests and ensure their long term conservation will depend in many cases on under-standing and effectively managing such small scale forest cutting.

ACKNOWLEDGEMENTS

I would like to express my thanks to University of York, Department of Environment for contributing to the project. The manuscript was greatly improved by the critical comments of anonymous reviewers, with statistical evaluation by Ajonina Samuel.

REFERENCES

Alongi DM (2002). Present state and future of the world's mangrove forest. Environ. Conserv., 29: 331-349.

Aksornkoae S, Khemnaerk C, Mellink WHH (1992). Mangrove foe charcoal. Regional wood Energy Development Program in Asia. Field document No. 30. Food and Agricultural Organisation of the United Nations, Bangkok. p. 40.

Baran E (1999). A review of quantified relationships between mangroves and coastal resources. Phuket Mar. Biol. Center Res. Bull. 62: 57-64.

Barnes DKA (2001). Hermit crabs, humans and Mozambique mangroves. Afr. J Ecol., 39: 241–248.

Calumpong HP, Menez EG (1997). Field Guide to the Common Mangroves, Sea grasses and Algae of the Philippines. Makati City, Philippines: Bookmark.

Cintron G, Schaeffer NY (1984). Method for studying mangrove structure. In: Snedaker,S. C., Snedaker, J. G. (Eds.), The mangrove Ecosystem: Research Methods. UNESCO Paris. pp. 91-113.

Clough BF (1993). The status and value of mangrove in Indonesia, Malaysia and Thailand: Summary . The economic and environmental value of mangrove and their present state of conservation in the south East Asia/Pacific Region. pp. 1-10. American Geophysical Union, Washington D C, USA.

Dahdouh-Guebas F (2001). Mangrove Vegetation Structure Dynamic and Regeneration. (PhD Thesis) Free University of Brussels, Belgium.

Clarke PJ, Kerrigan RA (2000). Do forest gaps influence the population structure and species composition of mangrove stands in Northern Australia. Biotropica, 32: 642-652.

Diop ES (Ed.) (1993). Conservation and Sustainable Utilization of Mangrove Forests in Latin America and African Regions (Part 2: Africa). Mangrove Ecosystem Technical Report 3. International Society for Mangrove Ecosystems and International Tropical Timber Organization, Tokyo.

Eusebio MA, Tesoro FO, Cabahug DM (1986). Environmental impact of timber harvesting on mangrove ecosystem in the Philippines. In: National Mangrove Committee (Ed.), Mangrove of Asia and the Pacific: Status and Management. pp. 337-354.

Ewel KC, Zheng S, Pinzon Z, Bourgeois JA (1998). Environmental effects of canopy formation in high-rainfall mangrove forests. Biotropica, 30, 510-518.

Feller IC, Mckee KL (1999). Small gap creation in Belizean mangrove forests by a wood-boring insect. Biotropica 31: 607-617.

Gabche CE (1997). An appraisal of fisheries activities and evaluation of economic potential of the fish trade in the Douala – Edea reserve – Cameroon. Cameroon Wildlife and conservation society consultancy report, June 1997, p.39.

Gilman E, Ellison JC, Duke N, Field CD, Fortuna S (2008). Threats to mangroves from climate change effects and natural hazards and mitigation opportunities. Aquatic Bot., 89 (2): 237-250.

Hamilton LS, Snedaker SC (1984). Handbook for mangrove area management. IUCN/UNESSCO/UNEP East –West Centre, Honolulu, Hawaii.

Kennedy DN, Swaine MD (1992). Germination and growth of colonizing species in artificial gaps of different sizes in dipterocarp rain forest. Philosophical Transactions: Biol. Sci., 335:357-368.

Lertzman KP, Sutherland GD, Inselberg A, Saunders SC (1996). Canopy gaps and landscape mosaic in a coastal temperate rain forest. Ecol., 77: 1254 -1270.

Lefebvre G, Poulin B (2000). Determinant of avian diversity in neotropical mangrove forest. In: Biodiversity in wetland: assessment, function and conservation, volume 1. B. Gopal, W. J. Junk &J.A Davies, eds., Bachuys publishers, Leiden, The Netherlands.

Longonje SN (2002). Utilisation of mangrove wood product amongst subsistence and commercial users. M.Sc thesis unpublished . Free University of Brussels (VUB).

Nurkin B (1994). Degradation of mangrove forests in South Sulawesi, Indonesia. Hydrobiologia, 285: 271-276.

Price ARG, Klaus R, Sheppard CRC, Abbiss MA, Kofani M, Webster G (2000). Environmental and economic characterisation of coastal and marine system of Cameroon, including risk application of the Chad-

Cameroon pipeline project. Aquatic Ecosyst. Health Manage., 3: 137-161.

Primavera JH (1995). Mangroves and brackish water pond culture in the Philippines. Hydrobiologia, 295: 303–309.

Roth LL (1992). Hurricanes and mangrove regeneration-effects of hurricane Joan, October 1988, on the vegetation of Isla del Venado, Bluefields, Nicaragua. Biotropica, 24:375 –384.

Runkle JR (1992). Pattern of disturbance in some old –growth mesic forest of eastern North America. Ecol., 63: 1533-1546.

Smith TJ (1987). Effects of light and intertidal position on seedling survival and growth in tropical tidal forests. J Exp. Mar. Bio. Ecol., 110: 133-146.

Smith TJ (1992). Forest structure. In Robertson, A. I., Alongi, D. M. (Eds.), Tropical Mangrove Ecosystems. American Geophysical Union, Washington, DC. pp. 101-136.

Smith AH, Berkes F (1993). Community-based use of mangrove resources in St. Lucia. Int. J Environ. Stud., 43:123-131.

Twilley RR, Chen R, Hargis T (1992). Carbon sinks in mangroves and their implication to carbon budget of tropical ecosystem. Water, Air Soil Pollut., 64: 265-288.

Walter BB (2003). People and mangroves in the Philippines: fifty years of coastal environmental change. Environ. Conserv., 30: 293–303.

Walter BB (2004). Local management of mangrove forests: effective conservation or efficient resource exploitation? Human Ecol., 32: 177-195.

Walter BB (2005a). Ecological effect of small scale-cutting of Philippine mangrove forests. For. Ecol. Manage., 206: 331-348.

Walter BB (2005b) Pattern of local wood use in cutting of Philippine mangrove forest. Econ. Bot., 59: 66–76.

Walter BB, Ronnback P, Kovacs JM, Crona B, Hussain SA, Badola R, Primavera JH, Barbier E, Dahdouh-Guebas F (2008). Mangrove ethnobotany, socio-economics and management: A review. Aquatic Bot., 89:220-236.

3

Population studies, habitat assessment and threat categorization of *Polygonatum verticillatum* (L.) Allioni in Kumaun Himalaya

Nidhi Lohani[1], Lalit. M. Tewari[1], Ravi Kumar[2], G. C. Joshi[2], Jagdish Chandra[2], Kamal kshore[1], Sanjay Kumar[1] and Brij Mohan Upreti[1]

[1]Department of Botany, D.S.B. Campus, Kumaun University, Nainital Uttarakhand, India.
[2]RRIHF CCRAS, Ranikhet, Uttarakhand, India.

Natural populations of *Polygonatum verticillatum* in Kumaun Himalaya were surveyed for population studies, habitat assessment and threat status. This research reveals density of individuals and area occupied were low as compared to other species of the region, indicating habitat loss and heavy exploitation. Status was determined on a site-to-site basis for the entire Kumaun region. Based on species occurrence in selected areas, the species were identified as critically endangered to endangered in different areas. Frequency of *P. verticillatum* ranged between 50 and 80% at different population sites. Distribution of the species was between 50 and 80% indicating contiguous distributional range at most of the sites and random distributional range at Mukteshwer and Gagar. Density of *P. verticillatum* was highest (4.40 plant m^{-2}) in way to Kafani and lowest (2.60 plant m^{-2}) in Bhaman gupha. Total basal cover (TBC) was also found highest (0.91 cm^2 m^{-2}) in way to Kafani and lowest (0.35 cm^2 m^{-2}) in Bhaman gupha. Important value index (IVI) was found highest (51.68) in Munsyari and lowest (28.84) in Khati. Concentration of dominance (Cd) for the region showed a slight variation, with a range between 0.30 and 0.10. This may be attributed to the narrow range of distribution, habitat restriction and dominance of some species.

Key words: Population, habitat, threat, *Polygonatum verticillatum*.

INTRODUCTION

The Indian Himalayan region (IHR) is one of the most astonishing physical features on the surface of the earth. Among global mountain systems, the IHR is well known for its diverse landscapes and aesthetic, cultural and biological values (Samant and Dhar, 1997). This richness accompanied with uniqueness (endemism), sensitivity (rarity) and economic value make the biological resources of the region important from different perspectives (Dhar, 2002). The resources are used by inhabitants for various purposes such as fodder, fuel, timber, medicinal, wild edible, etc. (Samant and Dhar, 1997).

Out of 7000 endemic species of plants found in India, over 3000 grow in the Himalayan region (Chatterjee, 1980). In the last few decades, Himalayan ecosystems faced loss of forest lands due to increasing biotic pressure and exploitation for many valuable medicinal plants, which have been used for time immemorial. These plants have been mentioned in literature (Samant et al., 1997) and folk-lore, yet are no longer found in accessible habitats in large quantities. Many species have become rare in several tracts and are found only in inaccessible hilly areas, while a few others have been listed as endangered species.

Himalayan regions are considered as a primary source for collection of important medicinal plants. Several species including *Podophyllum hexandrum*, *Nardostachys jatamansi*, *Picrorhiza kurrooa*, *Aconitum* spp., *Saussurea ovallata*, *Saussurea lappa*, *Rheum* spp., *Polygonatum cirrhifolium*, *Polygonatum verticillatum* and *Angelica* spp. from the alpine areas are well known for their medicinal values. In India, 814 plant species have been identified as threatened; of these, over 113 taxa occur in the Indian Himalayas. Many of these medicinally important plant species are restricted to small pockets of habitat and their population size is decreasing at an alarming rate (Nayar and Shastry, 1987, 1988, 1990). Environmental degradation and loss of biodiversity as a result of excessive anthropogenic pressure, particularly in the fragile Himalaya have caused much concern among conservationists in the recent years (Kant, 1989; Saraswat and Thakur, 1998).

Kumaun region of Uttarakhand State lies between the latitudes of 28°-44' and 30°-49'N and longitudes of 78°-45' and 81°-5'E (Joshi et al., 1983) and is a hilly region containing a diverse physiographic, orographic and demographic mosaic. Its uniqueness supports rich biodiversity. A number of medicinal plant species have been reported from the Kumaun Himalaya which are facing a risk of extinction. *P. verticillatum* is one of these species. This species is widely used in traditional systems of medicine.

P. cirrhifolium (Wall) Royle belongs to the family Liliaceae, and commonly known as Mahameda (English name: Solomon's seal), and is a tall, erect, weak herb with stout, creeping rhizomes and a stem 60 to 120 cm high that is terete and grooved. Leaves occur in whorls of 3 to 6 and are linear to narrowly lanceolate, 6 to 15cm long, with margins enrolled, apex coiled and tendrils like. Flowers are white, tinged purple or green, in short stalked clusters of 2 to 4, and arise from the leaf axils. The perianth is 6-parted and somewhat reflexed (Gaur, 1999). It is found in rare, moist-shady localities of montane forests distributed in Himalaya, H.P. to Khasia hills (Gaur, 1999), specifically in temperate Himalayas at an altitude ranging between 2000 and 3000 m (Anonymous, 2006; Garg, 1996). It is an important ingredient of Astavarga (Singh, 2006); a medicine used for diseases of children, burning sensations, fever, jaundice, bleeding disorders, blood disorders and debility due to chest injuries (Anonymous, 1999, 2006). Leaves are eaten as vegetables; roots infused with milk are used as an aphrodisiac and blood purifier; and a paste is used in cuts and wounds (Gaur, 1999). It is also used in Jvara, Raktavikara, Ksaya, Daha, Raktapitta, Balaroga, Kamala, Ksata and Ksina (Anonymous, 2006). It is considered as a main constituent of 'Astavarga', a group of eight drugs, which forms an important base for a number of Ayurvedic preparations.

Keeping in mind the importance of *P. verticillatum,* this study attempted to determine population status, habitat assessment and threat categorization of this herb in the Kumaun region.

MATERIALS AND METHODS

Study area

The study was conducted in temperate regions of Kumaun. Five districts of Kumaun including Almora, Bageshwar, Champawat, Nainital and Pithoragarh were selected for population studies of the four herbs: *P. verticillatum*, *P. hexandrum*, *N. jatamansi* and *P. kurrooa* as these five districts cover temperate region of Kumaun. The study areas were surveyed extensively and a total 17 sites for *P. verticillatum* were identified on the basis of (a) habitat attributes (altitude/slope/aspect), (b) population size and (c) accessibility for data collection. A district-wide geographical description of each study area is given in Table 1.

Population studies, habitat assessment and threat categorization

In nature, *P. verticillatum* sprouts between April and May and reaches senescence by the end of October. Considering this, a phytosociological study was carried out in August-September after all species had attained maximum growth. For population studies, temperate regions of Kumaun were visited at regular intervals during three consecutive years (2007to 2010). Plots of 100 x 100 m were identified and marked on each region of *P. verticillatum*. Vegetation sampling was conducted through vertical belt transects (Michael, 1990). Since the distribution range is narrow and topography is very diverse, approximately 60 m long and 30 m wide transects were laid across each plot. Transects were divided into three stands of 20 x 10 m size as replicates and ten quadrats of 1 x 1 m size were laid randomly in each stand. Individuals of all species were counted in each quadrat. To determine status of the species, mean values of each quantitative parameter of three stands of transect were considered for further interpretation.

Data were analyzed for the population study as frequency (%F), density (D, plant m^{-2}), A/F ratio, relative frequency, relative density, relative dominance and total basal cover (TBC, $cm^2 m^{-2}$) and calculated following Misra (1968):

$$\text{Frequency} = \frac{\text{Total number of quadrats in which species occurred}}{\text{Total number of quadrats studied}} \times 100$$

$$\text{Density} = \frac{\text{Total number of individuals of a species in all quadrats}}{\text{Total number of quadrats studied}}$$

$$\text{Abundance} = \frac{\text{Total number of individuals of a species in all quadrats}}{\text{Total number of quadrats in which the species occurred}}$$

$$\text{A/F ratio} = \frac{\text{Abundance}}{\text{Frequency}}$$

Distribution pattern of the species was analyzed on the basis of abundance to frequency (A/F) ratio. Value of A/F < 0.025 was categorized as regular, between 0.026 to 0.050 as random and >

Table 1. Districts wise studied sites

S/N	District	Study site
1.	**Almora** Located between 29° 36' North Latitude and 79° 30' East Longitude at an altitude of 1638 m sea level (msl).	Bhatkot, Vinayak, Balloni.
2.	**Bageshewar** Located between 29°42'40" to 30°18'56" North Latitude and 79°23' to 80.9° East Longitude. The district lies at an altitude of 1646 msl.	Way to Sunderdhunga, Way to Kafni, Khati, Phurkia.
3.	**Champawat** Located between 29°5'and 29°30' in Northern Latitude and 79°59' and 80°3'at the center of Eastern Longitude with an altitude of 1615 msl.	Vanasur, Debidhura, Khetikhan.
4.	**Nainital** Located between29°23' North Latitude and 79°30' East Longitude at a height of 1939 msl.	Ramgarh, Mukteshwer, Gagar.
5.	**Pithoragarh** Located between 29.4° to 30.3° North Latitude and 80° to 81° East Longitude at a height of 1645 msl.	Lilam, Thal, Munsyari, Bhaman gupha.

0.050 as contiguous types of distribution (Kershaw, 1973). Similarly, relative values of frequency, density and dominance were calculated following the methods of Misra (1968) and Kershaw (1973) as:

$$\text{Relative frequency} = \frac{\text{Percent frequency of species}}{\text{Total percent frequency of the community}} \times 100$$

$$\text{Relative density} = \frac{\text{Density of species}}{\text{Total density of the community}} \times 100$$

$$\text{Relative dominance} = \frac{\text{Total basal cover of species}}{\text{Total basal cover of the community}} \times 100$$

$$\text{Basal cover} = \frac{(Cbh)^2}{4\pi}$$

Total Basal Cover (TBC) = Mean basal cover × Density

Importance Value Index (IVI) = relative frequency + relative density + relative dominance.

The concentration of dominance (CD) was computed by Simpson's Index (Simpson, 1949). For threat assessment, two criteria, that is, population estimation (density and number of mature individuals) and extent of occurrence (number of populations/plots) were used as per IUCN Red List Categories (IUCN, 1993). During the study, only flowering plants were considered as mature individuals and taken further for population estimation. Species having mature individuals <250 was considered as critically endangered, <2,500 as endangered and <10,000 as vulnerable. Similarly, species

having a single population was categorized as critically endangered, <5 populations as endangered and <10 populations as vulnerable. Furthermore, status was assigned separately for each natural site as well as for the entire Kumaun region.

RESULTS

The frequency of *P. verticillatum* was found highest (80%) in Mukteshwer and Gagar and lowest (50%) in Khati and Bhaman gupha. Density was highest (4.40 plant m^{-2}) in way to Kafani and lowest (2.60 plant m^{-2}) in Bhaman gupha. Abundance was found highest (7.33) in way to Kafani and lowest (3.50) in Mukteshwer. Total basal cover (TBC) was highest (0.91 cm^2 m^{-2}) in way to Kafani and lowest (0.35 cm^2 m^{-2}) in Bhaman gupha. Important value index (IVI) was found highest (51.68) in Munsyari and lowest (28.84) in Khati. Concentration of dominance (Cd) was highest (0.30) in way to Sunderdhunga and lowest (0.10) in Mukteshwer and Gagar. Distribution pattern (R/F ratio) of the species was found contiguous in all sites except Mukteshwer and Gagar, where distribution was found to be random. As per IUCN Red List Categories, data on extent of occurrence (number of populations/plots) indicated critically endangered status of the species in most of the sites studied except Phurkia, Ramgarh and Mukteshwer, where its status was endangered. Population estimation (density and number of mature individuals) indicated critically endangered status for the species in all the sites studied, and overall status for Kumaun region was found to be vulnerable and endangered (Table 2).

Moist grassy slopes, under canopies of *Cedrus deodara*

Table 2. Population Status and Assignment of Threat Categories of *Polygonatum verticillatum* in Kumaun Himalaya

District	Sites	F	Rfr	D	Rden	A	A/F	TBC	Rdom	IVI	D	Distribution	No. of population	No. of mature individuals	Satatus
Almora	Bhatkot	70	10	2.8	10.98	4	0.06	0.41	10.59	31.57	0.12	Contiguous	1	28	CR*, CR**
	Vinayak	70	9.59	2.9	11.98	4.14	0.06	0.53	12.91	34.48	0.11	Contiguous	1	29	CR*, CR**
1638m asl N 29° 36' E79°30	Balloni	60	9.38	3.7	11.71	6.17	0.1	0.73	14.75	35.84	0.18	Contiguous	1	37	CR*, CR**
Bageshwer	Way to sunderdhunga	70	11.29	4	10.72	5.71	0.08	0.63	20.52	42.53	0.3	Contiguous	1	40	CR*, CR**
	Way to Kafni	60	9.52	4.4	12.61	7.33	0.12	0.91	21.4	43.54	0.2	Contiguous	1	44	CR*, CR**
1646 asl	Phurkia	60	10.34	4.1	14.09	6.83	0.11	0.58	17.44	41.88	0.21	Contiguous	2	41	EN*, CR**
N 29°42'40" E 79°23'	Khati	50	7.81	3.2	9.09	6.4	0.13	0.46	11.94	28.84	0.2	Contiguous	2	32	EN*, CR**
Champawat	Vanasur	60	9.52	3.5	11.9	5.83	0.1	0.71	17.62	39.05	0.18	Contiguous	1	35	CR*, CR**
	Debidhura	70	10.77	3.9	13.36	5.57	0.08	0.5	20.36	44.48	0.22	Contiguous	1	39	CR*, CR**
1615 asl N 29°5' E 79°59'	Khetikhan	60	9.38	3	14.49	5	0.08	0.61	19.1	42.97	0.11	Contiguous	1	30	CR*, CR**
Nainital	Ramgarh	60	10	3	11.24	5	0.08	0.56	20.1	41.34	0.22	Contiguous	2	30	EN*, CR**
	Mukteshwer	80	8.7	2.8	8.89	3.5	0.04	0.58	14.17	31.75	0.1	Random	2	28	EN*, CR**
1939 asl N 29°23' E 79°30'	Gagar	80	11.27	2.9	12.24	3.63	0.05	0.55	10.17	33.68	0.1	Random	1	29	CR*, CR**
Pithoragarh	Lilam	70	10.14	3.7	17.79	5.29	0.08	0.62	22.82	50.75	0.12	Contiguous	1	37	CR*, CR**
	Thal	60	9.52	3.3	11.5	5.5	0.09	0.44	23.53	44.55	0.27	Contiguous	1	33	CR*, CR**
1645 asl	Munsyari	70	10.61	3.4	12.59	4.86	0.07	0.66	28.49	51.68	0.21	Contiguous	1	34	CR*, CR**
N 29.4° E 80°	Bhaman gupha	50	7.94	2.6	15.76	5.2	0.1	0.35	20.36	44.06	0.11	Contiguous	1	26	CR*, CR**
Overall status for Kumaun region															Vu*, EN**

*Based on extent of occurrence; **based on population estimation;

and *Quercus leucotrichophora* trees are the major habitats of *P. verticillatum*. In some places, it is also found under the canopies of *Myrica esculenta* and *Rhododendron arboreumn* with *Quercus leucotrichophora* and *Cedrus deodara*. Dominant associates of *P. verticillatum* at most sites were *Roscoea procera*, *Viola canesense*, *Thalictrum foliolosum*, *Rumex nepalensis* and *Oxalis corniculata*. Other dominant associates of the species were *Achyranthes bidentata*, *Paspalum scrobiculatum*, *Valariana wallichii* and *Polygonum* sp. The main threats to the species in most of the sites were habitat degradation, medicinal harvest, human inter-ference and over-exploitation. Other potential threats to the species include trade and grazing (Table 3). District wise phytographs of IVI, relative frequency, relative density and relative dominance of all four species are given in Figure 1.

DISCUSSION

Population studies

Over exploitation and habitat degradation have been causing decreases in the population of *P. verticillatum*. Research reveals that some of the medicinal plants of high therapeutic value are endangered and require protection. In general, endangered and rare species generally show low levels of morphological variations, so it is important to determine population polymorphism at the biochemical and genetic levels to study these varia-tions and develop appropriate conservation strategies.

It is indeed fascinating that two third of the world's population depend upon plant resources for their primary health care needs and a fairly large number of modern drugs have been derived from plant natural products, with many following leads provided by indigenous knowledge system. This has added to the popularity of herbal products as part of new health programs in developed countries, and combined with the traditional demand of third world nations has led to a steady increase in the market for medicinal plants worldwide.

Biodiversity conservation is a global issue, and special attention is being given to the conservation of endan-gered and threatened species. Most conservation programmes in India started on animal systems, however recently endangered plant species have received consi-derable attention. The importance of Himalayan medici-nal species of endangered or threatened status and an urgent need for their conservation has recently been emphasized by many workers (Khoshoo, 1993; Bhadula et al., 1996).

Frequency of *P. verticillatum* ranged between 50 and 80% at different population sites. Distribution of the species was between 50 and 80% indicating contiguous distributional range at most of the sites and random distributional range at Mukteshwer and Gagar. Density of *P. verticillatum* was maximum (4.40 plant m^{-2}) in way to

Kafani and minimum (2.60 plant m^{-2}) in Bhaman gupha. Total basal cover (TBC) was also found maximum (0.91 cm^2 m^{-2}) in way to Kafani and minimum (0.35 cm^2 m^{-2}) in Bhaman gupha. Important value index (IVI) was found maximum (51.68) in Munsyari and minimum (28.84) at Khati population. Concentration of dominance (Cd) of the region showed a slight variation. It ranged between 0.30 and 0.10. This may be attributed to narrow range of distribution, habitat restriction and dominance of some species.

Low density and relatively low dominance of the species in the present study may be due to specific microhabitat requirements of the species and over exploitation for illegal trade, etc. These factors are responsible for restricted distribution fragmented habitat and low populations of all species.

Low population density across the surveyed populations indicates poor availability of the species in the study area. However, random distribution and higher frequency of occurrence is indicative that the species have potential for better performance in these sites (habitats).

It is observed that the whole plant is used for medicinal properties (Murkherjee, 1953; Kirtikar and Basu, 1984) therefore individuals are uprooted indiscriminately. It is reported that harvesting of the whole plant is more destructive than the harvesting of fruits, seeds or leaves in isolation (Sheldon et al., 1997). Furthermore, the removal of the entire plant before seed maturation ceases the possibilities of development of future regeneration (Sheldon et al., 1997).

Threat categorization

The status of *P. verticillatum* was found critically endangered and endangered in all the sites studied (Figure 2). The principal reason for species endan-germent in Himalayan medicinal plants is human inter-ference in natural ecosystems, resulting in habitat destruction and a loss of other natural and biological factors. The use of wild plant resources and subsequent ecosystem alteration often leads to habitat fragmentation. Species susceptible to slack habitats are more fragile and have more difficulty in sustaining populations (especially small and narrowly distributed ones) and consequently this often leads to species endangerment. Habitat loss and degradation have been identified as the major fac-tors, threatening 91% of plant species globally (IUCN, 2000). In the 2000 IUCN Red List, India is ranked sixth for having the highest number of threatened plant species.

An area-specific threat categorization of species is very important for short- or long-term management planning. Various studies have been carried out to explore and identify the threatened plants of IHR (Pangtey and Samant, 1988; Samant et al., 1993, 1996a, b, 1998a, b, 2000a; Pandey and Well, 1997; Kala et al., 1998).

Table 3. Site characteristics of selected sites of *P. verticillatum*.

District		Altitude (m)	Habitat	Dominant species	Threat
Almora	Bhatkot	2950	Grassy slope with *Quercus leucotrichophora* A. Camus *and Myrica esculenta* Buch-Ham.	*Achyranthes bidentata* Blume, *Roscoea procera* Wall., *Oxalis corniculata* Linn., *Thalictrum foliolosum* D.C.	Habitat degradation, harvested for medicine
	Vinayak	2285	Moist gentle slope with *Quercus leucotrichophora* A. Camus and *Myrica esculenta* Buch- Ham.	*Arisaema speciosum* (Wall.)Mart., *Achyranthes bidentata* Blume, *Thalictrum foliolosum* D.C., *Rumex nepalensis* Spreng.	Habitat degradation, over exploitation
	Balloni	4000	Moist shady place with *Acer, Quercus leucotrichophora* A. Camus and *Cedrus deodara* Roxb. ex D.Don.	*Roscoea procera* Wall., *Oxalis corniculata* Linn., *Polygonum nepalense* Meisn., *Paspalum scrobiculatum* Linn.	Habitat degradation
Bageshwar	Way to Sunderdhunga	3350	Shady moist grassy slope	*Oxalis corniculata* Linn., *Thalictrum foliolosum* D.C., *Paspalum scrobiculatum* Linn., *Roscoea procera* Wall.,	Habitat degradation, over exploitation
	Way to Kafni	3900	Shady moist grassy slope	*Achyranthes bidentata* Blume, *Polygonum nepalense* Meisn., *Roscoea procera* Wall., *Paspalum scrobiculatum* Linn.	Over exploitation
	Phurkia	3300	Moist shady slope with *Quercus leucotrichophora* A. Camus and *Rhododendron arboretum*	*Rumex nepalensis* Spreng., *Oxalis corniculata* Linn., *Arisaema speciosum* (Wall.) Mart., *Achyranthes bidentata* Blume	Human interference
	Khati	2300	Grassy slopes with *Quercus leucotrichophora* A. Camus	*Thalictrum foliolosum* DC., *Achyranthes bidentata* Blume, *Polygonum nepalense* Meisn., *Oxalis corniculata* Linn.	Habitat degradation
Champawat	Vanasur	1920	Grassy slope with *Cedrus deodara* Roxb. ex D.Don.	*Roscoea procera* Wall., *Arisaema speciosum* (Wall.) Mart., *Oxalis corniculata* Linn., *Achyranthes bidentata* Blume	Habitat degradation
	Debidhura	1800	Grassy slope with *Quercus leucotrichophora* A. Camus and *Cedrus deodara* Roxb. ex D.Don.	*Paspalum scrobiculatum* Linn., *Roscoea procera* Wall., *Arisaema speciosum* (Wall.) Mart., *Oxalis corniculata* Linn.,	Habitat degradation
	Khetikhan	1850	Gentle grassy slope with *Quercus leucotrichophora* A. Camus, *Myrica esculenta* Buch- Ham. and *Rhododendron arboretum*	*Viola canescens* Wall., *Galingsoga parviflora* Cav., *Asparagus curillus, Thalictrum foliolosum* D.C.	Human interference, Habitat degradation
Nainital	Ramgarh	2040	Grassy, shady place with *Cedrus deodara* Roxb. ex D.Don.	*Roscoea procera* Wall., *Achyranthes bidentata* Blume, *Polygonum nepalense, Oxalis corniculata* Linn.	Over exploitation, harvested for medicine
	Mukteshwer	2180	Grassy, shady place with *Cedrus deodara* Roxb. ex D.Don.	*Oxalis corniculata* Linn., *Rumex nepalensis* Spreng., *Paspalum scrobiculatum* Linn., *Cynodon dactylon* (Linn.)Pers.	Human interference

Table 3. Contd

	Gagar	2100	Moist grassy slope with *Quercus leucotrichophora* A. Camus and *Cedrus deodara* Roxb. ex D.Don.	*Thalictrum foliolosum* D.C., *Polygonum nepalense* Meisn., *Oxalis corniculata* Linn., *Rumex nepalensis* Spreng.,	Human interference, Habitat degradation
	Lilam	1850	Moist, shady place with *Quercus leucotrichophora* A. Camus and *Cedrus deodara Roxb. ex D.Don.*	*Viola canesense* Wall., *Achyranthes bidentata* Blume, *Thalictrum foliolosum* D.C., *Polygonum nepalense* Meisn.	Over exploitation
	Thal	3150	Shady moist slope with *Quercus leucotrichophora* A. Camus	*Achyranthes bidentata Blume, Thalictrum foliolosum* D.C., *Paspalum scrobiculatum* Linn., *Oxalis corniculata* Linn.	Habitat degradation
Pithoragarh	Munsyari	2150	Moist grassy slope with *Quercus leucotrichophora* A. Camus and *Cedrus deodara* Roxb. ex D.Don.	*Viola canesense* Wall., *Thalictrum foliolosum* D.C., *Rubia cordifolia* Linn., *Oxalis corniculata* Linn.	Harvested for medicine, trade
	Bhaman gupha	3000	Moist grassy slope with *Quercus leucotrichophora* A. Camus and *Cedrus deodara* Roxb. ex D.Don.	*Valeriana wallichii* D.C., *Geum alatum* Wall., *Viola canesense* Wall., *Achyranthes bidentata* Blume	Habitat degradation

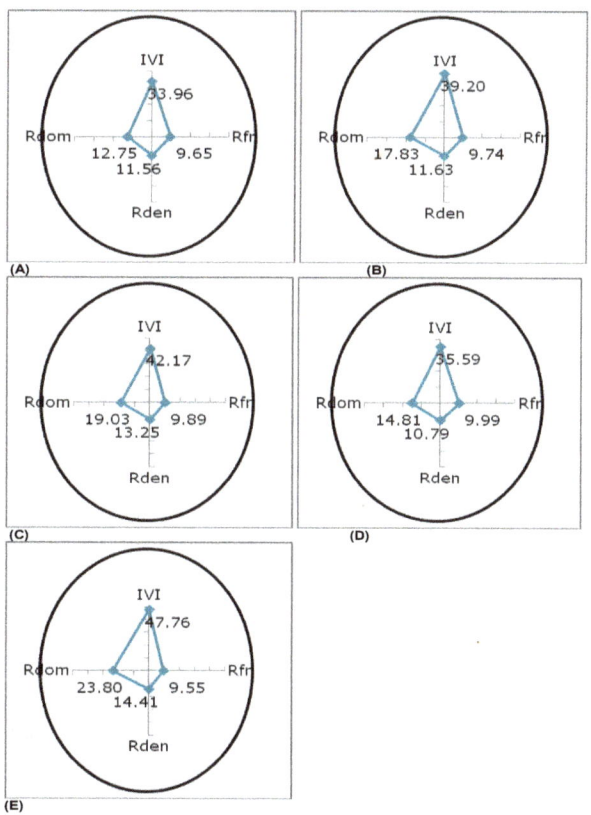

Figure 1. Phytograph of *P. verticillatum* in five districts of Kumaun (A) Almora (B) Bageshwar (C) Champawat (D) Nainital and (E) Pithoragar.

Figure 2. *P. verticillatum* in natural habitat.

Conclusions

If over-exploitation and habitat degradation of *P. verticillatum* continues, it may disappear from these areas within a few years. The patchy occurrence of critically endangered, endangered and vulnerable medicinal plants indicates that high anthropogenic pressure, over-exploitation, habitat degradation, habitat fragmentation and lack of awareness among inhabitants are the main causes of declining population of the species.

Population assessment of the species using standard ecological methods and recognition of key areas as medicinal plants conservation areas (MPCAs) for *in situ* conservation, including involvement of the Forest Department and tribal communities are suggested. Mass reproduction using conventional (vegetative and seeds) methods, establishment and maintenance of herbal gardens and medicinal plant nurseries for *ex situ* conservation, ensuring the availability of quality planting material for cultivation, and education and awareness programs for large-scale cultivation are also suggested.

The data from this study concerning population status, habitat preferences, and threat categorization for *P. verticillatum* may assist in understanding the ecology of the species and can be used in the development of a conservation plan. The study also recommends the collection of plant material in the senescence phase, which ultimately leads to sustainable utilization of the species

REFERENCES

Anonymous (1999). A Clear Mirror of Tibetan Medicinal Plants. Vol.1, Tibet Domani, Men-Tsee-Khang, Dharamsala, India.

Anonymous (2006). The Ayurvedic Phamacopoeia of India Part-I, Vol. V. The Controller of Publication, Civil Lines, Delhi.

Bhadula SK, Singh A, Lata H, Kuniyal CP, Purohit AN (1996). Genetic resources of *Podophyllum hexandrum* Royle, an endangered medicinal species from Garhwal himalaya, India. Plant Genet. Resour. Newsl. 106:26-29.

Chatterjee D (1980). Studies on the endemic flora of India and Burma. J. Royal Asiatic Soc. Bengal (n.s.) 5:19- 67.

Dhar U (2002). Conservation implication of plant endemism in high altitude Himalaya. Curr. Sci. 82 (2): 152-181.

Garg S (1996). Substitute and Adulterant Plants. Publication and Information Directorate, CSIR, New Delhi.

Gaur RD (1999). Flora of District Garhwal North West Himalaya. Transmedia Srinagar (Garhwal). pp.704-705.

IUCN (1993). Draft IUCN Red List Categories, Gland Switzerland.

IUCN (2000). Draft IUCN Red List Categories, Gland Switzerland.

Joshi SC, Joshi DR, Dani DD (1983). Kumaun Himalaya- A Geographical Perspective on Resource Development. Gyanodaya Prakashan, Nainital.

Kala CP, Rawat GS, Uniyal VK (1998). Ecology and Conservation of the Valley of Flowers, National Park, Garhwal Himalaya, Report. Wildlife Institute, Dehradun.

Population studies, habitat assessment and threat categorization of Polygonatum verticillatum (L.) Allioni in Kumaun Himalaya

25

Kant S (1989). Phytogeography of North Western Himalayas. The academy of Environmental Biology, India, Muzaffaenagar, India.

Kershaw KA (1973). Quantitative and Dynamic Plant Ecology. Edward Arnold Ltd., London. pp. 308.

Khoshoo TN (1993). Himalayan biodiversity conservation: an overview. In: *Himalayan Biodiversity: conservation Strategies* (ed.U.Dhar). G.B. Pant institute of Himalayan Environment and Development, Almora. pp. 5-38.

Kirtikar KR, Basu BD (1984). *Indian Medicinal Plants.* Volume III. Bishen Singh Mahendra Pal Singh, Dehradun. Misra R (1968). Ecological work book. Oxford and IBH.

Michael P (1990). *Ecological Methods for Field and Laboratory Investigations.*Tata Mcgraw Hill, New Delhi,India.

Misra R (1968). Ecology work book. Oxford and IBH Publishing Co., New Delhi. pp. 244.

Murkherjee B (1953). The Indian Pharmaceutical Codex. C.S.I.R., New Delhi.

Nayar MP, Shastry ARK (1987, 1988 and 1990). Red data book of Indian Plants. Vol. I, II and III. BSI, Calcutta.

Pandey S, Well MP (1997). Eco-development planning at India's Great Himalayan National Park for biodiversity conservation and participatory rural development. Biodivers. Conserv. 6:1277-1292.

Pangtey YPS, Samant SS (1988). Observation on the threatened, rare and endangered flowering plants and ferns in the flora of Kumaun Himalaya. Advances in Forestry Research in India. 3:65-74.

Samant SS, Dhar U (1997). Diversity, endemism and economic potential of wild edible plants of Indian Himalaya. Int. J. Sustain. Dev. World Ecol. 4:179-191.

Samant SS, Dhar U, Rawal RS (1993). Botanical hot spots of Kumaun Himalaya: Conservation perspectives. In: Dhar, U. (ed.), Himalayan Biodiversity Conservation Strategies. Gyanodaya Prakashan, Nainital. pp. 377-400.

Samant SS, Dhar U, Rawal RS (1996a). Natural resources use by some natives of Nanda Devi Biosphere Reserve in West Himalaya. Ethnobotany 8:40-50.

Samant SS, Dhar U, Rawal RS (1996b). Conservation of rare endangered plants: the context of Nanda Devi Biosphere Reserve. In: Ramakrishanan, P. S., Purohit, A. N., Saxena, K. G., Rao, K. S. and Maikhuri, R. K. (eds.). Conservation and Management of Biological Resources in Himalaya. Oxford and IBH Publication Co. Pvt. Ltd. New Delhi. pp. 521-545.

Samant SS, Dhar U, Rawal RS (2000a). Assessment of fuel resource diversity and utilization patterns in Askot Wildlife Sanctuary in Kumaun Himalaya, India, for Conservation and Management. Environment Conservation 27(1):5-13.

Samant SS, Dhar U, Palni LMS (1998b). Medicinal plants of Indian Himalaya: Diversity Distribution Potential Values. Gyanodaya Prakashan, Nainital.

Samant SS, Dhar U, Rawal RS (1998a). Biodiversity status of a protected area of West Himalaya: Askot Wildlife Sanctuary. Int. J. Sustain. Dev. World Ecol. 5: 192-203.

Saraswat CV, Thakur V (1998). Need to Conserve Biodiversity in Western Himalaya. In: Bawa, R. and Khosla, P. K. (eds.) Biodiversity of Forest species. Bishen Singh Mahendra Pal Singh, Dehradun. pp. 117-129.

Sheldon JW, Balick MJ, Laris SA (1997). Medicinal plants: can utilization and conservation coexist? Adv. Econ. Bot. 12:1-104.

Simpson HE (1949). Measurement of diversity. Nature 163:688.

Singh AP (2006). *Ashtavarga-* Rare Medicinal Plants. Ethnobot. Leaflets 10:104-108.

4

The invasive status of wild barley (*Hordeum spontaneum* Koch) in Iranian flora

Reza Hamidi

Crop Production and Plant Breeding Department, College of Agriculture, Shiraz University, Shiraz, Iran. E-mail: hamidi@shirazu.ac.ir.

The success of a weed's biological pathway is dependent on several factors including the mechanism(s) of dispersal, longevity of seeds in the soil seed bank, adaptation to varying environmental conditions and, competitive and reproductive abilities. In Near East barley, some races can be identified as wild but in other races wild and weed forms are confounded. Wild barley has wider distribution than the wild wheat and is spread over a wide area in the East Mediterranean basin and West Asiatic countries. In the Fertile Crescent countries, *Hordeum spontaneum* occupies a whole array of secondary man-made habitats. Invasive weeds possess a variety of characteristics that enable them to disperse rapidly into new areas. A variety of both aboveground and belowground plant traits can be used for identifying invasive species. Wild barley has some considerable traits including high genetic variation, high allelopathic potential, high relative growth rate, root characteristics, high competitive abilities and, became an invasive species in many habitats of Iran. Further studies are essential to evaluate the other mechanism(s) of invasiveness of wild barley in alien environment.

Key words: Wild barley, *Hordeum spontaneum*, habitat, distribution, invasive weed.

INTRODUCTION

As a result of the importance of weeds to agriculture and their probable roles in plant domestication, definition of weed is important. Some of the current definitions used in agronomic instruction, such as "a weed is a plant that does more harm than good", are clearly inadequate. A weed is much more than that, but, the implications of the term have changed over the years (Harlan, 1975). Harlan and de Wet (1965) defined weed as "a generally unwanted organism that thrive in habitats disturbed by man". Man has probably always caused some disturbance of habitats. Before he knew how to manipulate fire, man's disturbances were probably very minor and more or less limited to the vicinity of cave or camp. After he began to use the fire deliberately after the vegetation, his disturbances were more widespread and more intense. Still, his set fires were relatively causal compared to the habitats he created after developing an effective agriculture in which whole landscapes were churned up and entire floras destroyed and replaced by new vegetation. The species adapted to the new, artificial habitats are mostly crops or weeds. Because both are adapted to the same habitats, however, practices that tend to favor crops also tend to favor weeds.

Some crops undoubtedly originated from weed progenitors and some crops have degenerated into weed races. The evolution of weeds often parallels the evolution of crops and the same principles apply to both. Both weeds and crops often begin with a common progenitor, as in those complexes where each crop has a companion weed. There are weed and cultivated races of wheat, barley, sorghum, rice, oats, pearl millet, potato, tomato, pepper, sunflower, carrot, radish, lettuce, and many others (Klingman and Ashton, 1982). In Near Eastern barley, some races can be identified as wild, but in other races wild and weed forms are confounded. There is a small wadi race that appear to be truly wild. In more mesophytic races of barley, however, it is difficult to distinguish the weed from the wild (Harlan,1975).

Effects of seed dormancy on longevity of weed seeds

Seed dormancy, a major adaptive trait in plants facilitates the survival of weeds and provides resistance to preharvest sprouting in members of Poaceae family. Most weeds are able to germinate throughout the growing season and that is an important reason why they are successful, though each individual will experience different success (Gutterman and Nevo, 1993). Seeds exhibiting dormancy usually have to experience periods of favorable environmental conditions during a period called "after-ripening". After-ripening is a process whereby seeds are gradually able to germinate over a broad range of conditions (Baskin and Baskin, 1987). Hamidi et al. (2011a) showed that the major factor of seed dormancy in *Hordeum spontaneum* is glumellae. In fact, glumellae have either physical or chemical effects on the germination of caryopses and finally, on the longevity of this species in the soil seed bank.

H. spontaneum as an ancestor of *H. vulgare*

It is clear now that only a single genuinely wild species of barley is closely related to the various cultivated barley forms, and should be regarded as their sole ancestor. This is two-row brittle *H. spontaneum*. This plant has the same chromosome number as cultivated barley, *H. vulgare*. Both are diploid (2n = 14), hybrids between them are fully fertile and show regular pairing in meiosis. From a genetic point of view, *H. spontaneum* and various forms of cultivated barley did not diverge to the extent of representing fully independent, separated species (Harlan, 1966).

H. spontaneum distribution

Wild barley shows wider distribution than the wild wheats. It is spread over a wide area in the East Mediterranean basin and West Asiatic countries, penetrating east as far as Turkmenia and Afghanistan. Wild barley occupies, at present, both primary and segetal, man-made habitats. Its distribution center lies in the Fertile Crescent Belt, that is, in a wide arc, stating from Palestine and Transjordan in the south-west, stretching north towards South Turkey, and bending south-east towards Iraqi Kurdestan and South-west Iran. In this general area, and only here, *H. spontaneum* is massively and continuously spread over primary habitats. It constitutes an important annual component of open formations, and is particularly common in the summer-dry belt of the deciduous oak park forest, east, north and west of the Syrian desert and the Euphrates basin, and the slope facing the Jordan rift valley.

From here, wild barley spills over the drier and warmer deserts. In the Fertile Crescent countries, *H. spontaneum* also occupies a whole array of secondary man-made habitats, that is, opened-up Mediterranean maquis, abandoned cultivation, edges of fields and roadsides. Further west (Aegean region and Cyrenaica) and further east (North-east Iran, Central Asia and Afghanistan), *H. spontaneum* is rare and much more sporadic in its distribution, it rarely builds even local masses and seems to be completely restricted to segetal habitats or to sites which have been drastically churned by man's activity. Thus, in these peripheral areas, wild barley does not seem to be genuinely wild. As in the case of wild einkorn, it apparently spread to these locations as a weed, as a consequence of agricultural activity (Harlan, 1966). In general, wild barley does not tolerate extreme cold and it is only occasionally found above 1500 m. It is thus almost completely absent from the elevated continental plateaus of Turkey and Iran.

On the other hand, it is somewhat more xeric as compared with the wild wheat and penetrates relatively deep into warm steppes and deserts. Morphologically too, *H. spontaneum* is quite variable and several distinct races can be distinguished. Robust types with extremely large seeds and extraordinarily long awns occur in the catchment area of Upper Jordan Valley, often in close association with similarly robust *Triticum dicoccoides* form (Harlan, 1966). A much more slender desert type is found in the drier steppes and in desert dry water courses. This race is sporadically spread from the Negev to the steppic plateaux of Transjordan, northward to the Turkish border, and eastward to Iran and Afghanistan. It is a small, grassy type with kernels only half size of the robust races of the Eastern Galille.

All intermediate types between these extremes are widely spread in Palestine, Syria, Turkey and Iran (Harlan, 1966). Although the natural stands of *H. spontaneum* have been considered a natural part of the rich grass cover of open-woodland *Quercus brantii* belt in Zagros of Iran (Harlan, 1975), but recently, the species population densities have been increased in many other parts of our country and then may be considered as an invasive weed.

Characteristics of invasive species

Invasive weeds possess a variety of characteristics that enable them to disperse rapidly into new areas and out-compete crops and native or desirable non-native vegetation for light, water, nutrients, and space (Westbrooks, 2001). The success of a weed's biological pathway is dependent on several factors including the method or mechanism of dispersal, longevity of propagule, adaptiveness to varying environmental conditions, and competitive and reproductive ability.

Botanists, ecologists, and weed scientists have long been aware of the problem of establishment of non-native weed species and have gleaned knowledge of how some species reproduce, spread, and interact with crops and native and acceptable non-native species. To prevent economic and ecological diversity losses, it is necessary to prevent additional introductions and invasions of plant species that have the potential to become serious pests of agriculture, forest urban, and native areas. Understanding the basic biology and ecology of weeds is important to determine pathways of entry, spread, establishment, and persistence. Biology of pathways varies depending on the species and environmental surroundings.

Biological processes and characteristics that are most important for weeds to thrive are dependent on reproduction, dispersal, phenology, physiology, protection, habitat requirements, tolerances to environmental stress, and interspecific interactions (Brayson and Carter, 2004). Invader species affect the distribution, abundance, reproduction, and evolution of many native species (Sala et al., 1999). Humans are often directly or indirectly responsible for most introductions, whether intentional or unintentional, but animals and natural processes also disperse plants (Reed, 1977). The most common pathways of movement associated with humankind include contaminated soil, food, feed, fiber, ballast, and packing and bagging material. However, pathways for introduction and spread may be from ornamentals, forages, or plants used for erosion control that were once thought to be acceptable but have become weedy. Natural processes including wind, hurricanes, tornadoes, earthquakes, and floods are also responsible for plant dispersal but to a lesser extent than human activities (Brayson and Carter, 2004).

Natural barriers and restricted migration routes prevent many plant propagules from dispersing over great distances; however, the current speed and ease of world transportation by humans and their cargo has increased the rate and distance of dispersal of plant propagules. After introduction, a plant species may remain near the point of introduction without becoming a pest or the plant may continue dispersing from the initial point of entry. Unfortunately, newly introduced weeds are often unnoticed until after their numbers and range increase greatly. The period of time between introduction and invasion is termed the lag phase (Radosevich and Holt, 1984). Invasive species have an ability to undergo genetic changes due to selection pressure imposed by the alien environment and exhibit quick response anthropogenic disturbances (Sakai et al., 2001).

Factors affecting the invasiveness of *H. spontaneum*

A variety of both aboveground and belowground plant

traits can be used for identification of invasive species. Ehrenfeld (2004) accounted for some specific traits for invasive species including genetic variation, high allelopathic potential, high growth rate (measured as relative growth rate), high live tissue and litter nutrient contents, especially nitrogen, and root system size or morphology that contrasts strongly with native species. Based on Ehrenfeld's pointing out, wild barley has some considerable traits including high genetic variation (Brown et al., 1978; Nevo et al., 1986; Volis and Mendlinger, 1998; Volis et al., 2002b), high allelopathic potential (Hanson et al., 1981; Hanson et al., 1983; Barria et al., 1991; liu and Lovett, 1993; Liu et al., 2005; Hamidi et al., 2008a; Hamidi et al., 2008b; Hamidi et al., 2010b) high relative growth rate (Elberse et al., 2003; Poorter et al., 2005; Van Rijin et al., 2000), root characteristics (Ceccarelli et al., 1998; Invandic et al., 2000; Volis et al., 2001; Volis et al., 2002a; Volis et al., 2004), high competitive abilities (Hamidi et al., 2010a; Hamidi et al., 2011c; Hamidi et al., 2011b), the lack of effective herbicides caused increase in its resistance to used herbicides and consequently, changing in the flora and dominance of weeds that were not in the range of herbicides control, high tolerance to osmotic stress by reducing the cellular osmotic potential as a consequence of a net increase in solute accumulation and then, better uptake of water and nutrients from the soil under stress conditions (Matsuda and Riazi, 1981), sooner maturation than wheat in all climates of Iran, and then, can be an invasive species. Further, studies are thus essential to evaluate the other mechanism(s) of invasiveness of wild barley in alien environment especially regarding its colonization, expansion, establishment, and ecological impact as to take timely action for its management.

REFERENCES

Barria BN, Copaja SV, Niemeyer HM (1991). Occurrence of DIBOA in wild *Hordeum* species and its relation to aphid resistance. Phytochemistry, 31: 89-91.

Baskin JM, Baskin CC (1987). Temperature requirements for after-ripening in buried seeds of four summer annual weeds. Weed Res., 27: 385-388.

Brayson CT, Carter R (2004). Biology of pathway for invasive weeds. Weed Technol., 18: 1216-1220.

Brown AH, Nevo DE, Zohary D, Dagan O (1978). Genetic variation in natural population of wild barley (*Hordeum spontaneum*). Genetica, 49: 97-108.

Ceccarelli S, Grando G, Impiglia A (1998). Choices of selection strategy in breeding barley for strees environments. Euphytica, 103: 307-318.

Ehrenfeld JG (2004). Implications of invasive species for belowground community and nutrient processes. Weed Technol., 18: 1232-1235.

Elberse IAM, Van Damme JMM, Van Tienderen PH (2003). Plasticity of growth characteristics in wild barley (*Hordeum spontaneum* Koch) in response to nutrient limitation. J. Ecol., 91: 371-382.

Gutterman Y, Nevo E (1993). Germination comparison study of *Hordeum spontaneum* regionally and locally in Israel: A population in the Negev Desert Highlands and from two opposing slopes on Mediterranean Mount Carmel. Barley Genetics Newsletter, 22: 65-

71.

Hamidi R, Mazaheri D, Rahimian H, Alizadeh HM, Ghadiri H, Zeinaly H (2008a). Phytotoxicity effects of soil amended residues of wild barley (*Hordeum spontaneum* Koch) on growth and yield of wheat (*Triticum aestivum*). Desert, 13: 1-7.

Hamidi R, Mazaheri D, Rahimian H, Alizadeh HM, Ghadiri H, Zeinali H (2008b). Inhibitory effects of wild barley (*Hordeum spontaneum* Koch) residues on germination and seedling growth of wheat and its own plant. Desert, 11: 35-43.

Hamidi R, Mazaheri D, Rahimian H (2010a). Effects of nitrogen on *Hordeum spontaneum* (Koch) competition with winter wheat. Aust. J. Basic Apll. Sci., 4: 4695-4700.

Hamidi R, Mazaheri D, Rahimian R (2010b). Effect of wild barley (*Hordeum spontaneum* Koch) leaf and culm extracts on seed germination and seedling growth of winter wheat (*Triticum aestivum* L.). Iranian J. Weed Res., 2: 19-28.

Hamidi R, Mazaheri D, Rahimian R (2010c). Winter wheat nitrogen yields as affected by wild barley population densities and nitrogen. Aust. J. Basic Appl. Sci., pp. 4726- 4739.

Hamidi R, Mazaheri D, Rahimian R (2011a). Wild barley (*Hordeum spontaneum* Koch) seed germination as affected by dry storage periods, temperature regimes, and glumellae characteristics. Iranian J. Weed Sci., 15: (in press).

Hamidi R, Mazaheri D, Rahimian R (2011b) Wild barley (*Hordeum spontaneum* Koch) and winter wheat (*Triticum aestivum* L.) growth responses to nitrogen and population densities in a replacement series study. Can. J. Scient. Indust. Res., 2: 251-267.

Hanson AD, Traynor PL, Dits KM, Reicosky DA (1981). Gramine in barley forage – Effects of genotype and environment. Crop Sci., 21: 726 – 730.

Hanson AD, Dits KM, Singletary GW, Leland TJ (1983). Gramine accumulation in leaves of barley grown under high temperature stress. Plant Physiol., 71: 896 – 904.

Harlan JR (1966). Distributions of wild wheat and barley. Science, 153: 1074 - 1080.

Harlan JR (1975). Crops and Man. American Society of Agronomy, Crop Science Society of America, Madison, WI., p. 295.

Harlan JR, de Wet JMJ (1965). Some thought about weeds. Econ. Bot., 19: 16-24.

Invandic VC, Hackett A, Zhang AZ, Staub JE, Nevo E, Thomas WTB, Forster BP (2000). Phenotypic responses of wild barley to experimentally imposed water stress. J. Exp. Bot., 51: 2021-2029.

Klingman GC, Ashton FM (1982). Weed Science: Principles and Practices, 2nd edition. John Wiley and Sons, Inc. New York, p. 449.

Liu DW, Lovett JV (1993). Biologically active secondary metabolites of barley. I. Developing techniques and assessing allelopathy in barley. J. Chem. Ecol., 19: 2217-2230.

Liu DL, An M, Jhonson IR, Lovett JV (2005). Mathematical modeling of allelopathy: IV. Assessment of contributions of competition and allelopathy to interference by barley. Nonlinearity in Biology, Toxicology, Medicine, 3: 213-224.

Matsuda K, Riazi A (1981). Stress-induced osmotic adjustment in growing regions of barley leaves. Plant Physiol., 68: 571-576.

Nevo E, Kaplan D, Storch N, Zohary D (1986). Genetic diversity and environmental associations of wild barley, *Hordeum spontaneum* (Poaceae), in Iran. Plant Syst. Ecol., 153: 141-164.

Poorter H, van Rijin CP, Van Hola TK, Verhoeven KJF, de Jong YEM, Stam P, Lambers H (2005). A genetic analysis of relative growth rates and underlying components in *Hordeum spontaneum*. Oecologia, 142: 360- 377.

Radosevich SR, Holt JS (1984). Weed Ecology. John Wiley Publishing Co., p. 265.

Reed CF (1977). Economically Importance Foreign Weeds. Agriculture Handbook No. 498. Washington, DC: United States Department of Agriculture, p. 746 .

Sakai AK, Allendorf FW, Holt JS (2001). The population biology of invasive species. Ann. Rev. Ecol. Syst., 32: 305-332.

Sala OE, Chapin FS, Gardner RH, Lauenroth WK, Mooney HA, Ramakrishnan PS (1999). Global change, biodiversity and ecological complexity. In: The Terrestrial Biosphere and Global Change: Implications for Natural and Managed Ecosystems (Eds Walker B, Steffen W, Canadell J, Ingram J). Cambridge University Press, Cambridge, UK, pp. 304-328.

Van Rijin CPE, Heersche I, Van Berkel YEM, Nevo E, Lambers H, Poorter H (2000). Growth characteristics in *Hordeum spontaneum* populations from different habitats. New Phytologist, 146: 471-481.

Volis S, Mendlinger S (1998). Phenotypic variation and stress resistance in core and peripheral populations of *Hordeum spontaneum*. Biodiversity Conservation, 7: 799-813.

Volis S, Yakubov B, Shulgina I , Ward D, Zur V, Mendlinger S (2001). Tests for adaptive RAPD variation in population genetic structure of wild barley, *Hordeum spontaneum* Koch. Biological J. Linn. Soc., 74: 298 – 303.

Volis S, Mendlinger S, Ward D (2002a). Adaptive traits of wild barley plants of Mediterranean and desert origin. Oecologia, 133: 131 – 138.

Volis S, Mendlinger M, Ward D (2002b). Differentiation in populations of *Hordeum spontaneum* Koch along a gradient of environmental productivity and predictability: Plasticity in response to water and nutrient stress. Biol. J. Linn. Soc., 75: 301-312.

Volis S, Mendlinger M, Ward D (2004). Differentiation in populations of *Hordeum spontaneum* Koch along a gradient of environmental productivity and predictability: Intra-and interspecific competitive responses. Isr. J. Plant Sci., 52: 223-234.

Westbrooks RG (2001). Invasive species, coming to America: New strategies for biological protection through prescreening, early warning, and rapid response. Wild Land Weeds, 4: 5-11.

5

Characterization and impact of wood logging on plant formations in Ngaoundéré District, Adamawa Region, Cameroon

Tchobsala[1] and Mbolo, M.[2]

[1]Department of Biological Sciences, Faculty of Science, University of Ngaoundéré, P. O. Box 454, Cameroon.
[2]Department of Biology and Plant Physiology, Faculty of Science, University of Yaoundé I, P. O. Box 812, Yaoundé, Cameroon.

This study was conducted to characterize the different plant formations (shrubby, arborescent and woody savanna) and to ascertain the impact of wood logging on the floral diversity in the guinea savanna zone of Ngaoundéré District, Adamawa Region, Cameroon. The "Point-Centered Quarter (PCQ) Method" was used on 120 sites measuring 50 × 50 m. Results showed that according to the types of wood logging in the different plant formations, the species generally had an over-scattered distribution, and only the protected savannas had a gregarious distribution. The increased wood logging affects savannas' stability and the disappearance of the floral biodiversity which are consequently responsible for the accelerated degradation. This is an alarming situation which enhances the progress of desert and the loss of biodiversity in the guinea savanna of the Adamawa Region, Cameroon. It is advocated that a concerted effort between the government and the local population should be established to protect and save the biodiversity in the guinea savanna of the Adamawa Region, Cameroon.

Key words: Cameroon, impact, wood logging, distribution, biodiversity.

INTRODUCTION

The biodiversity is often used as a contracted shape of the biological diversity (Ndam, 1998; China et al., 2003). It groups together the generic and specific diversity, the populations and ecosystems and bases itself on the specific wealth and the relative abundance of the species. This biodiversity is actually endangered by the wood logging phenomenon which took an unequalled scale a quarter of century ago in the African savannas leading to the accelerated degradation of the natural resources which constitute the productive basic capital. In the Adamawa Region of Cameroon, several authors thought that the causes of the transformation and the degradation of the guinea savanna could be due to overgrazing, over-population led by human migration from the Far North region of Cameroon, agriculture and exploitation of wood by the local population (Rippstein, 1985; Yonkeu, 1993; Ndjidda, 2001; Tchotsoua, 2006). In the North-Cameroon, it has been asserted that the regression of the forest is due to the combined effects of wood logging, bush fire and overgrazing (Ntoupka, 1994, 1998). In Cameroon, the impact of wood logging on the distribution of the floral diversity is well known in the Southern part (Sonké, 1998; Guedje, 2002; Zacfack, 2005), but in the guinea savanna of Adamawa which is intermediate to the forested south and the Sahelian North, there is a paucity of information on the wood logging activities. Chouaibou (2006) reported the distribution of *Parkia biglobosa* in the district of Ngaoundéré which is the capital of the Adamawa Region, Cameroon; however, this region is among those that are threatened by the anthropological wood logging, bush fire and overgrazing.

Since the signing of the conventions on the presservation of the biological diversity and the use of biological resources in a long-lasting way in 1992 in Rio de Janeiro, Brazil, these conventions are yet to be implemented in our sub-Saharan guinea savannas, where many of the plant species are either cut or harvested by men for several uses. Thus, the present study was carried out to characterize the different plant formations and to estimate the impact of wood logging on the floral diversity in the guinea savanna zone of Ngaoundéré District, Adamawa Region, Cameroon.

MATERIALS AND METHODS

Study area

The study was undertaken in ten (10) villages namely: Béka Hooseré, Onaref, Wakwa, Tizon, Beskewal, Ngaohora, Borongo, Dang, Darang and Mban-Mboum all located in the Ngaoundéré district of the Adamawa Region, Cameroon (Figure 1). These villages are located at about 10 km for the shortest and 60 km for the farthest distance from Ngaoundéré the capital city of Adamawa Region, Cameroon. Ngaoundéré is located at latitude 7° 19' N and longitude 13° 34' E. Its population was estimated at about 230,000 inhabitants in 2001 (Tchotsoua, 2006) with an increase rate of 2.81% per annum. The main ethnic groups are the Fulbés, Mbororos, Gbayas, Mboums, Dourous, Yemyems, Hausas and the Koutinés. The economic activities of the local inhabitants are mainly animal husbandry and land farming. The soil of the area belongs to the geo-morphological domain of the plateau of Adamawa. They are characterized by sedimentary, volcanic, granitic and metamorphic rocks.

The vegetation of the Adamawa corresponds to a typical Sudano-guinea savanna constituted with shrubby, arborescent and woody savannas. These savannas are dominated by *Daniellia oliveri* and *Lophira lanceolata* (Letouzey, 1986). The precipitations are maximal in August and practically null from November to February. The hygrometric is maximal in August with a monthly average humidity of 81.38%.

Choice of the different wood logging zones in the guinea savannas of Adamawa Region, Cameroon

To choose the different wood logging, interviews were conducted with group of persons. The prospections with the population in the site were made. The types of wood logging in the savannas depended on the degree of accessibility to the site (absence or proximity to easy access road), the distance to the village (0 to 0.5, 0.5 to 1, 1 to 2, 2 to 4, 4 to 6, > 6 km) and the percentage of the wood cut. At the end of prospection, four types of wood logging were selected:

i) Pilot or witness logging (T_0): made up with natural formation where the estimated percentage of wood logging is less or equal to 10%. They are generally protected areas by the inhabitants;
ii) Weak logging (T_1): vegetation where the percentage of wood logging is between 11 and 25%;
iii) Average logging (T_2): vegetation where the percentage or wood logging is between 26 and 50%;
iv) Complete or total logging (T_3): vegetation where more than 50% of woods are cuts.

Experimental technique

The point-centered quarter (PCQ) technique described by Farid et al. (2006), Kevin (2007) and Tchobsala (2010, 2011) was used in this study. This technique consisted in choosing a direction at random in the savannas under study (Figure 2).

Experimental design

The study was a split-plot design with 3 factors (shrubby savanna, arborescent savanna and woody savannas) (Table 1). The pieces were numbered from 1 to 12, delimitated by numbered cement terminals or wood stakes. One hundred and twenty sites (3 types of savannas × 4 types of cuts × 10 villages) were selected with 30 sites for each treatment.

Data analysis

Calculations of relative frequency, relative density, relative dominance and relative important value of the species

Calculation of the relative frequency (Fr):

$Fr (\%) = \frac{A}{B} \times 100$, with A = number of the statements containing species; B = total number of the statements.

Calculation of the relative density (Dr):

$Dr (\%) = \frac{C}{D} \times 100$, with C = number of individuals of a species; D = total number of the individuals encountered on a considered surface.

Calculation of the relative dominance:

$Dre = \frac{Sb}{Sbt} \times 100$, with Sb = basal surfaces of a species and Sb = $\pi D^2/4$, D = the diameter of the stalk; Sbt = total sum of all the basal surfaces of all plant species considered in an upper diameter (≥ 5 cm) above the ground level by hectare (m²/ha).

Calculation of plant recovery rate:

The ligneous place setting (DC = canopy cover) of individual ligneous plants or of the population by hectare can be calculated as follows: DC = r^2, representing the area of projection of the foliage expressed, where m² (r = the averages horizontal distance of the trunk at the extremities of branches. DC (ha^{-1}) = the sum of surfaces (r^2) covered by the foliage of all the ligneous plants by ha, expressed in percentage (%) of total ligneous place setting on an hectare.

Calculation of the important value of curtis (IV):

IV (%) = Fr + Dr + Dre, with Fr = relative frequency, Dr = relative density and Dre = relative dominance. The (IV) varies between 0 and 300%.

Analysis of species distribution, species abundance and species dispersal

To analyze the distribution of species abundance, the models of

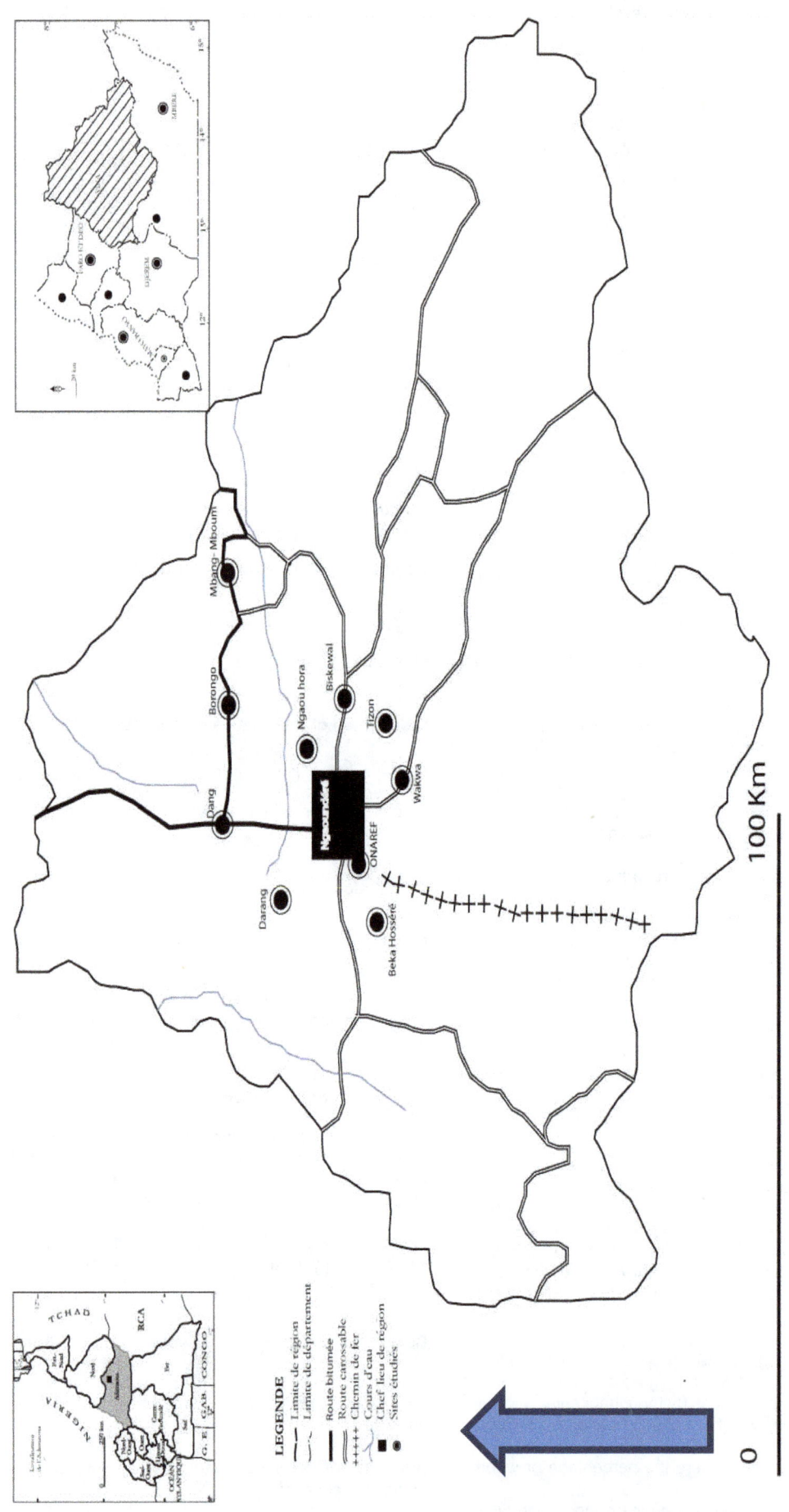

Figure 1. Map of study area.

Pichod-Viale, 1993), while the model of Pareto (Frontier and Pichod-Vilae, 1993) used to analyze the structure of the wet tropical dense forest was adopted to analyze the structure of the guinea savanna observed in the study which is a zone of transition between the wet tropical forest and the sahelian savanna. The species distribution is the distribution of the number of trees species by class diameter. To study the horizontal organization of the plant communities, we used the dispersal parameter of all the individuals of the community. The closest neighbors method described by Clark and Evan (1954) was used to analyze the dispersal of all the population. This method allows specifying the way and the degree of remoteness of the random distribution of individuals of a given population.

The ratio R is used in this case as:

$$R = r_{ob}/r_{at}\text{-}r_{ob},$$

with r_{ob}, as the average distance observed between the individuals of the community, r_{at}, as the distance waited between the individuals of density (d) and $r_{at} = \frac{1}{2}\,d^{1/2}$;

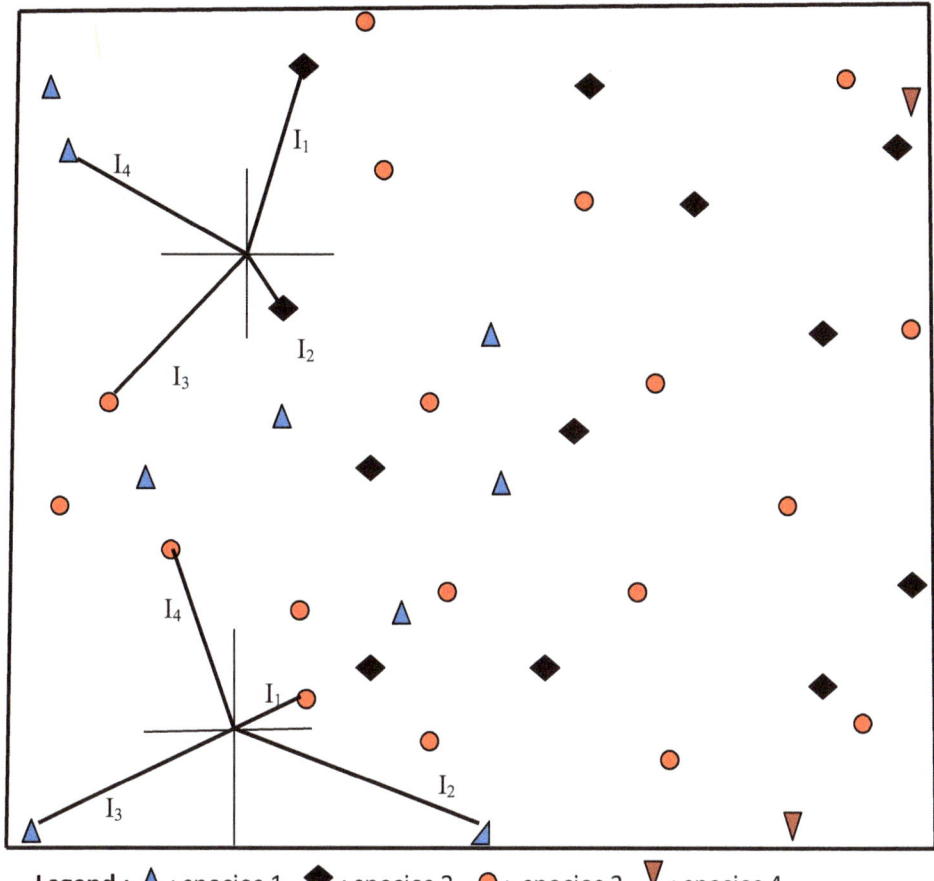

Legend : △ : species 1, ◆ : species 2 , ◯ : species 3, ▽ : species 4

Figure 2. Point-centered quarter (PCQ) technique.

Table 1. Experimental design.

					Village					
S/N	DAN	BEK	ONA	BOR	WAK	TIZ	NGA	BES	DAR	MBA
1	SbT2	SaT3	SbT0	SaT3	ScT1	SaT0	SaT0	SaT2	ScT1	SbT2
2	SbT3	SaT0	SbT1	SaT2	SbT0	ScT2	ScT0	ScT2	SbT0	SbT1
3	ScT3	SbT3	SaT2	ScT0	SaT0	ScT0	ScT3	ScT3	SbT2	SaT1
4	ScT1	ScT2	ScT1	ScT0	ScT3	SbT2	ScT3	SaT3	SbT2	ScT3
5	SaT2	ScT1	SaT3	ScT1	SaT1	SaT3	ScT2	SbT0	SbT1	SbT3
6	ScT3	SbT1	SaT0	SaT0	SaT1	ScT0	SaT0	ScT2	SbT0	SbT3
7	SaT1	SbT3	SaT1	SaT3	ScT2	SaT2	SbT1	ScT1	SbT1	SbT3
8	SbT1	SbT0	SbT1	SbT3	ScT1	SbT1	SbT3	ScT0	ScT3	SaT2
9	SaT2	SaT3	SaT1	SaT2	ScT0	ScT0	ScT2	ScT3	ScT2	ScT0
10	SbT2	ScT0	SaT1	ScT3	SbT1	SaT1	SbT0	SbT3	SaT0	ScT2
11	ScT2	SaT0	SbT3	SbT0	SbT2	SaT0	SaT2	SaT1	SbT0	SbT2
12	SaT2	SbT0	SbT1	SbT3	SaT0	SbT2	SaT1	SaT3	SaT3	ScT1

Key: BES, Beskewal; BEK, Beka; ONA, ONAREF; BOR, Borongo; WAK, Wakwa; TIZ, Tizon; NGA, Ngaouhoura; DAR, Darang; DAN, Dang; MBA, Mbang-Mboum; Sa, shrubby savannas; Sb, raised savannas; Sc: wooded savannas.

If R = 1, the distribution is random,
If R < 1, the distribution is grouped,
If R > 1, the distribution is over-scattered.

The statistical difference between r_{at} and r_{ob} can be appreciated by using the formula $C = (r_{ob} - r_{at}) / \partial r$, $t = 0.26136 N d^{1/2}$, with ∂r as the standard error on the average distance observed in the case of a

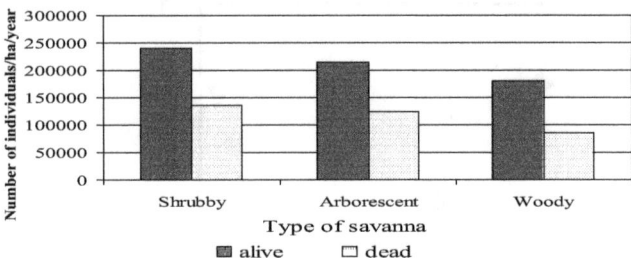

Figure 3. Biological state of the different plant formations in Ngaoundéré environs, Adamawa Region, Cameroon.

random distribution, d = the density, N = the number of individuals. The species dispersal concerns the individuals of the same species. The analysis of the distribution of the individuals of the same species within the inventories was made by the "run test" method (Siegel, 1956). We considered the value (1) for the presence and the value (0) for the absence. "Run test" examines the distribution of the species along the transect, allows to determine if we observe more or less sequences "run" than in the case of a random distribution. Significance levels were considered at $P \leq 0.05$, $P \leq 0.01$ and $P \leq 0.001$.

RESULTS

Characterisitics of the different plant formations in Ngaoundéré and environs

In whole, three types of plant formations were observed in the area: the shrubby, arborescent and woody savannas. The number of trees alive decreased from 241,860 trees/ha/year in the shrubby to 133,980 trees/ha/year in the woody savanna; however, the number of dead trees also decreased from 141,210 trees/ha/year in the shrubby savanna to 131,242 trees/ha/year in the arborescent savanna, but with a drastic decrease in the woody savanna. The mortality rate of these individuals is more important in shrubby and arborescent savannas. It is observed that in the three plant formations, the number of trees alive is almost twice the number of dead trees (Figure 3). This rate decreases in woody savannas. Indeed, shrubby and arborescent savannas undergo a strong pressure of wood cuttings and pastures.

Relative frequency, relative dominance, relative density and relative important values of the tree species in the different plant formations

Of all the 4,320 points of reading, *Hymenocardia acida* (58.01%) recorded the most important relative value. This species was the most important tree species in the savanna with a relative frequency of 18.54%, relative dominance of 8.56% and relative density of 30.96% (Table 3). *Lannea acida* (0.06%), *Carissa edulis* (0.05%) and *Mytragina inermis* (0.3%) had the least important values.

Relative frequency, relative dominance, relative density and relative important value of trees' genera in the different plant formations

Sixty two (62) genera were inventoried in the plant formations. The genera *Hymenocardia* (58.06%) had the most important relative value of all the genera encountered in the plant formations, while the genera *Antidesma* (0.07%) and *Mytragina* (0.03%) having the least values (Table 4).

Relative frequency, relative dominance, relative density and relative important value of the trees' families in the plant formations

Of the 34 families listed, Hymenocardiaceae (58.06%), Cesalpiniaceae (54.74%) and Annonaceae (28.38%) had the most relative important value, but Myrsinaceae (0.53%), Olacaceae (0.49%) and Tiliaceae (0.23%) families were sparsely found (Table 5).

Distribution of species abundance in the plant formations of Ngaoundéré environs

Figure 4 illustrates the rank- frequency of the distribution of the ligneous species in the plant formations of Ngaoundéré and environs. The most important species is *H. acida* (3134.4 plant individuals/ha/year) located on the vertical axis of the graph, while *C. edulis* and *M. inermis* were the least to be found in the plant formations of Ngaoundéré and environs located on the horizontal axis of the graph. The adjustment of this distribution in the model of Motumura gave a linear function of equation $\log_2 (y) = 8.6764 - 0.0895x$, with high significance ($P < 0.001$, $R^2 = 0.9515$). Where y is the frequency of the species (log N) and x its rank (log r).

Species distribution in relation to wood logging in the different plant formations

Table 2 describes the plant species distribution in relation to wood logging in the three types of plant formations studied. In shrubby and arborescent savannas, no significant difference was observed in both the number of live and dead trees between the different types of wood logging ($p \geq 0.05$), while in the woody savanna a significant difference was observed in the number of live trees between the different type of wood logging with an elevated number of trees (59070/ha/year) in T_2.

Distribution, density and recovery of the species according to the types of savannas and wood logging

Table 6 represents the results of distribution, density and recovery of the species in relation to the plant formations and wood logging. The distance between the individuals

Table 2. Distribution of trees species of the different plant formations in relation to wood logging/ha/year.

Plant formation	Type of Wood logging	Live trees	Dead trees
Shrubby Savanna	T_0	63,300	35,220
	T_1	63,360	34,980
	T_2	53,610	32,070
	T_3	61,590	31,710
	Total	241860	133980
Arborescent Savanna	T_0	46440	24360
	T_1	58020	35820
	T_2	62400	30660
	T_3	51990	35130
	Total	218850	125970
Woody Savanna	T_0	38010	19440
	T_1	43980	23280
	T_2	59070	23310
	T_3	45960	22560
	Total	187020	88590

Key: T_0 = Pilot or witness logging; T_1 = weak logging; T_2 = average logging; T_3 = total or complete logging.

varies from 3.77 to 5.54 m. There was no significant difference between the three types of plant formations and the various treatments ($p \leq 0.05$). The ratio R varies between 1.06 to 2.13 for shrubby savanna; 1.44 to 1.97 for the arborescent savanna and 1.15 to 2.18 for the woody savanna.

Distribution, recovery and density of the different types of plant formations in relation to distances from the villages surveyed

On the sites, the species were randomly distributed within 0.5 to 1 km (R = 0.99) (Table 7). They are grouped for the distances > 6 km (R = 0.91); 4 to 6 km (R = 0.93) and of 2 to 4 km (R = 0.71) and over-scattered for the distances of 0 to 0.5 km (R = 1.34) and of 1 to 2 km (R = 2.39) with regards to the villages.

Distribution, density and recovery of the species in the sites with easy access (T_3)

The recovery of trees is maximal in arborescent savanna with T_3 (1154.21 $m^2ha^{-1}year^{-1}$) from 2 to 4 km of the villages. In the arborescent savanna, the farthest distance from the villages was characterized by encountering the big trees such as *D. oliveri* and *Cesalpinia* sp (25 to 60 cm of diameter) that are not cut by lumberjacks. These species have a very important recovery in order of 40 to 60 m^2 by individual. In the whole of the sites and according to the distances with regard to the villages, trees have average distances varying between 1.55 and 6.29 m, with R oscillating between 0.31 and 2.92 (Table 8).

Specific dispersal of the species in the various plant formations in Ngaoundéré and environs

The distribution of the species was observed inside every type of savanna with treatments. The species which were present at least 10 times in 360 points of reading were held for the analysis of the specific dispersal. Among these species, only 15 on 102 species were inventoried by the method of distance between the species and were held for the analysis of the gregariousness. A total of 82 species represented scattered or over-scattered distributions. The species which have a significantly grouped distribution ($p < 0.05$) were *Annona senegalensis* in T_1, T_2 and T_3; *H. acida* in T_2 and T_3 and *Psorospermum febrifigum* in T_1 in shrubby savanna. In the arborescent savannas, only *H. acida* represented a grouped distribution (T_1 and T_2) (Table 9).

DISCUSSION

In the three types of plant formations observed in Ngaoundéré and environs, we found 102 species classified into 60 genera and 33 families. Three hundred and sixty one (371) stems.ha^{-1} were found alive in 2004 and 351 stems ha^{-1} in 2006. These results are similar to those reported by Thorgnang (2001) who listed 117 species into 80 genera and 37 families in the forest of Gawar. Our results are greater than those reported by Mahamat (1991) and Teicheugang (2000) who found 21 species and 11 botanical families in the National Park of Kalamaloué (4500 ha) and 75 species, 46 genera and 24 families in the forest reserve of Zamay, respectively. The majority of the species had a grouped distribution with high or very

Table 3. Relative frequency, relative dominance, relative density and relative important value of tree species in Ngaoundéré and environs.

S/N	Specie	FRe (%)	DRe (%)	Dr (%)	IV (%)	S/N	Specie	FRe (%)	DRe (%)	Dr (%)	IV (%)
1	Hymenocardia acida	18.54	8.56	30.96	58.1	51	Ficus sycomorus	0.09	0.62	0.01	0.72
2	Annona senegalensis	13.47	8.17	6.74	28.4	52	Flacourtia vogelii	0.3	0.11	0.26	0.67
3	Piliostigma thonningii	11.64	8.24	7.8	27.7	53	Nuxia congesta	0.16	0.11	0.35	0.62
4	Daniellia oliveri	4.4	17.4	3.09	24.9	54	Combretum sp,	0.14	0.36	0.09	0.59
5	Terminalia glaucescens	5.65	4.58	3.86	14.1	55	Erythrina senegalensis	0.16	0.19	0.22	0.57
6	Entada africana	4.14	6.61	1.9	12.7	56	Neoboutonia velutina	0.05	0.49	0.03	0.57
7	Harungana madagascariensis	4.47	2.32	4.18	11	57	Maesa lanceolata	0.07	0.1	0.36	0.53
8	Ficus sp.	0.07	0.03	7.06	7.16	58	Albizia coriaria	0.16	0.16	0.2	0.53
9	Lannea sp.	0.05	0.04	6.31	6.4	59	Malacantha alnifilia	0.25	0.03	0.23	0.51
10	Terminalia macroptera	2.38	2.44	0.76	5.57	60	Psidium guajava	0.21	0.07	0.22	0.5
11	Psorospermum febrifigum	2.22	1.38	1.51	5.11	61	Ximenia americana	0.12	0.27	0.1	0.49
12	Syzygium guineense var guineense	2.15	1.22	1.2	4.57	62	Combretum glutinosum	0.23	0.13	0.12	0.48
13	Croton macrostachyus	1.23	2.12	1.2	4.55	63	Indeterminate 1	0.07	0.32	0.08	0.47
14	Syzygium guineense var macrocarpum	2.01	1.05	1.12	4.18	64	Carissa spanrium	0.16	0.05	0.25	0.46
15	Cussonia barteri	1.62	2.07	0.43	4.12	65	Terminalia sp,	0.21	0.16	0.08	0.45
16	Erythrina sigmoidea	0.93	2.6	0.47	4	66	Ficus platyphylla	0.07	0.31	0.06	0.44
17	Lannea chimperi	1.2	1.89	0.48	3.57	67	Strichnos spinosa	0.14	0.22	0.05	0.41
18	Cinera macrostachys	0.37	0.14	2.57	3.09	68	Senna alata	0.37	0.02	0.1	0.4
19	Lophira lanceolata	1.09	0.7	1.29	3.08	69	Ochna schweinfurthiana	0.28	0.05	0.07	0.4
20	Bridelia ferruginea	1.13	1.1	0.82	3.04	70	Nauclea latifolia	0.14	0.03	0.21	0.38
21	Parkia biglobosa	0.44	2.43	0.16	3.03	71	Pavetta crassipes	0.05	0.27	0.03	0.34
22	Allophyllus africanus	1.44	0.65	0.79	2.88	72	Ficus capreaefolia	0.05	0.17	0.11	0.33
23	Zanthoxylum zanthoxyloides	0.14	2.34	0.07	2.55	73	Indeterminate 2	0.04	0	0.05	0.27
24	Terminalia laxiflora	1.02	0.67	0.8	2.49	74	Strichnos innocula	0.02	0.21	0.03	0.26
25	Gmelina arborea	0.21	2.1	0.04	2.35	75	Psychotria psychotrioides	0.02	0.05	0.17	0.25
26	Indeterminate 4	1.06	0.4	0.8	2.26	76	Albizia lebbeck	0.12	0.05	0.07	0.25
27	Vitex madiensis	0.28	1.72	0.22	2.22	77	Paulinia pinnata	0.05	0.16	0.02	0.23
28	Albizzia zygia	0.67	1.05	0.36	2.08	78	Lonchocarpus laxiflorus	0.02	0.05	0.15	0.22
29	Vitex doniana	0.76	0.94	0.38	2.07	79	Eugenia poliensis	0.14	0	0.08	0.22
30	Steganotaenia araliacea	1	0.5	0.52	2.02	80	Flacourtia indica	0.09	0.07	0.02	0.18
31	Lanha golungensis	0.32	1.42	0.27	2	81	Ficus sur	0.32	0.04	0.1	0.16
32	Maytenus senegalensis	0.81	0.79	0.35	1.95	82	Senna spectabilis	0.07	0.06	0.03	0.16
33	Trikilia rocka	0.86	0.63	0.41	1.9	83	Margaritaria discoidea	0.07	0.02	0.06	0.15
34	Bridelia ndellensis	1.06	0.28	0.52	1.86	84	Landolphia heudelotii	0.09	0.02	0.03	0.15
35	Protea madiensis	0.76	0.26	0.62	1.64	85	Vitex simplicifolia	0.05	0.09	0.01	0.15
36	Hyphaene thebarca	0.79	0.09	0.66	1.55	86	Vitex sp,	0.07	0.06	0.01	0.14

Table 3. Contd.

#	Species	Fre	DRe	Dr	IV
37	*Ekebergia senegalensis*	0.25	0.98	0.27	1.5
38	*Uapaca paludosa*	0.69	0.22	0.55	1.46
39	*Alchornea cordifolia*	0.14	0.96	0.15	1.25
40	*Vitellaria paradoxa*	0.39	0.52	0.33	1.24
41	*Oncoba spinosa*	0.07	0.79	0.34	1.2
42	*Securidaca longepedunculata*	0.49	0.25	0.44	1.18
43	*Sporospermum senegalensis*	0.63	0.22	0.33	1.18
44	*Allophyllus sp,*	0.21	0.18	1.14	1.14
45	*Trema orientalis*	0.28	0.24	0.47	1
46	*Ficus glumosa*	0.23	0.56	0.2	0.99
47	*Acacia siberiana*	0.32	0.27	0.3	0.89
48	*Phyllanthus muellerianus*	0.35	0.12	0.4	0.88
49	*Gardenia triacantha*	0.39	0.26	0.11	0.76
50	*Ochna afzeli*	0.35	0.14	0.25	0.74
87	*Grewia sp,*	0.07	0.05	0.01	0.13
88	*Terminalia micrantha*	0.09	0.02	0.01	0.12
89	*Jasmimum dichotomum*	0.07	0.05	0	0.12
90	*Ficus trichopoda*	0.05	0.01	0.05	0.11
91	*Grewia bicolor*	0.05	0.03	0.02	0.1
92	*Terminalia togoensis*	0.05	0.01	0.01	0.08
93	*Antidesma venosum*	0.02	0.06	0	0.08
94	*Lanea acida*	0.02	0	0.04	0.06
95	*Ficus thonningii*	0.05	0	0	0.06
96	*Gardenia ternifolia*	0.02	0	0	0.06
97	*Carissa edulis*	0.02	0.01	0.03	0.06
98	Indeterminate 3	0.02	0	0.02	0.04
99	*Mytragina inermis*	0.02	0.01	0	0.03
100	Total	100	100	100	300

Key words: Fre, Relative frequency; DRe, relative dominance; Dr, relative density; IV, Importance value of the species.

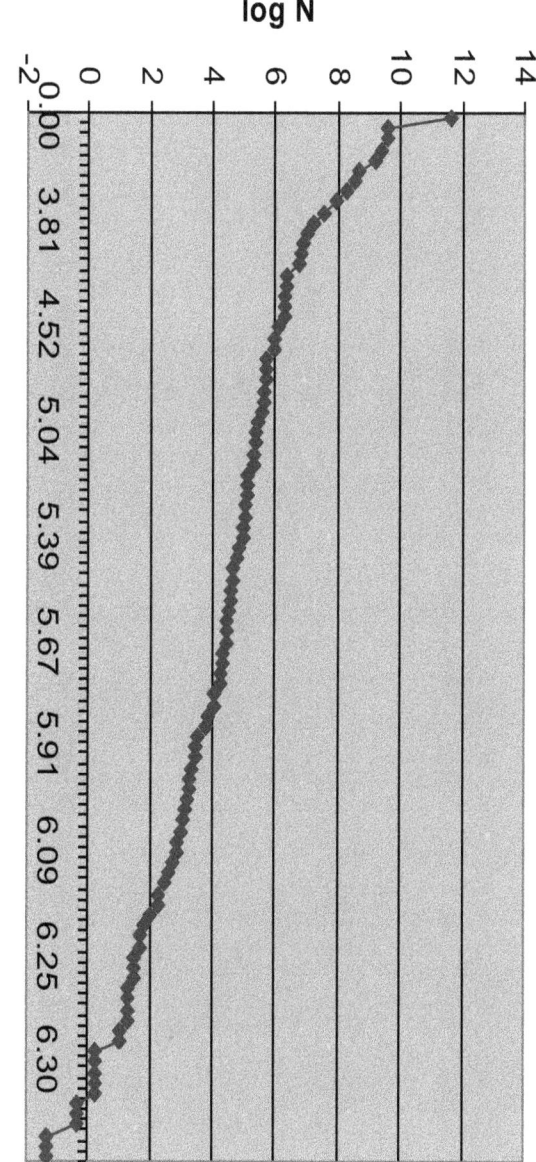

log N

log r

Figure 4. Diagram rank-frequency of the distribution of the woody species in the savannas.

Table 4. Relative frequency, relative dominance, relative density and relative important value of the genera in the different plant formations.

S/N	Genera	FRe (%)	DRe (%)	Dr (%)	IV (%)	S/N	Genera	FRe (%)	DRe (%)	Dr (%)	IV (%)
1	Hymenocardia	18.54	8.56	30.96	58.06	32	Indetermine 1	0.46	0.81	0.31	1.49
2	Annona	13.47	8.17	6.74	28.38	33	Uapaka	0.69	0.22	0.55	1.46
3	Piliostigma	11.64	8.24	7.8	27.68	34	Neoboutonia	0.02	1.36	0.01	1.39
4	Daniellia	4.42	17.4	3.11	25.11	35	Vitellaria	0.39	0.52	0.33	1.24
5	Terminalia	9.4	7.84	5.49	22.73	36	Securidaca	0.49	0.25	0.44	1.18
6	Entada	4.14	6.61	1.9	12.65	37	Indeternine 2	0.28	0.24	0.47	0.99
7	Harungana	4.47	2.32	4.18	10.97	38	Acacia	0.32	0.27	0.3	0.89
8	Lannea	1.27	1.93	6.83	10.03	39	Phyllanthus	0.35	0.12	0.4	0.87
9	Ficus	0.64	1.74	7.59	9.96	40	Flacourtia	0.39	0.18	0.28	0.85
10	Syzygium	4.16	2.27	2.32	8.75	41	Gardenia	0.41	0.29	0.11	0.81
11	Psorospermum	2.85	1.6	1.84	6.29	42	Ochna	0.63	0.19	0.32	0.78
12	Croton	1.28	2.61	1.23	5.12	43	Strichnos	0.16	0.43	0.08	0.67
13	Bridelia	2.19	1.38	1.34	4.91	44	Lanha	0.3	0.06	0.26	0.62
14	Vitex	1.16	2.81	0.62	4.59	45	Voacanga	0.16	0.11	0.35	0.62
15	Erythrina	1.01	2.79	0.69	4.57	46	Senna	0.35	0.08	0.13	0.56
16	Cussonia	1.62	2.07	0.43	4.12	47	Maesa	0.07	0.1	0.36	0.53
17	Allophyllus	1.6	0.82	1.52	3.94	48	Carissa	0.18	0.06	0.28	0.52
18	Cinera	0.37	0.14	2.57	3.08	49	Nauclea	0.23	0.05	0.24	0.52
19	Lophira	1.09	0.7	1.29	3.08	50	Burkea	0.25	0.03	0.23	0.51
20	Parkia	0.44	2.43	0.16	3.03	51	Psidium	0.21	0.07	0.22	0.5
21	Albizia	0.95	1.26	0.63	2.84	52	Ximenia	0.12	0.27	0.1	0.49
22	Eugenia	0.3	1.86	0.49	2.65	53	Malacantha	0.05	0.27	0.03	0.35
23	Zanthoxylum	0.16	2.34	0.07	2.57	54	Psychotria	0.02	0.05	0.17	0.24
24	Gmelina	0.21	2.1	0.04	2.35	55	Grewia	0.12	0.08	0.03	0.23
25	Pavetta	1.06	0.4	0.8	2.26	56	Paulinia	0.05	0.16	0.02	0.23
26	Steganotaenia	1	0.5	0.52	2.02	57	Jasmimum	0.02	0.05	0.15	0.22
27	Maytenus	0.81	0.79	0.35	1.95	58	Lonchocarpus	0.14	0	0.08	0.22
28	Trikilia	0.86	0.63	0.41	1.9	59	Margaritaria	0.07	0.02	0.06	0.15
29	Protea	0.76	0.26	0.62	1.64	60	Antidesma	0.05	0.01	0.01	0.07
30	Hyphaene	0.79	0.09	0.66	1.54	61	Indetermine 3	0.02	0	0.03	0.05
31	Ekebergia	0.25	0.98	0.27	1.5	62	Mytragina	0.02	0.01	0	0.03
							Total	100	100	100	300

Key words: Fre, Relative frequency; DRe, relative density; Dr, relative dominance; IV, relative important value of the genera.

Table 5. Relative frequency, relative dominance, relative density and relative important value of the families in the plant formations in Ngaoundéré and environs.

S/N	Families	Fred (%)	DRe (%)	Dr (%)	IV (%)	S/N	Families	Fred (%)	DRe (%)	Dr (%)	IV (%)
1	Hymenocardiaceae	18.54	8.56	30.96	58.06	18	Rutaceae	0.14	2.34	0.07	2.55
2	Cesalpinaceae	16.43	27.08	11.23	54.74	19	Apiaceae	1	0.5	0.52	2.02
3	Annonaceae	13.47	8.17	6.74	28.38	20	Celastraceae	0.81	0.79	0.35	1.95
4	Combretaceae	10.71	8.41	6.46	25.58	21	Meliaceae	0.86	0.63	0.41	1.9
5	Euphorbiaceae	4.1	4.81	3.2	12.11	22	Sapotaceae	0.69	0.58	0.59	1.85
6	Clusiaceae	4.47	2.32	4.18	10.96	23	Proteaceae	0.76	0.26	0.62	1.64
7	Anacardiaceae	1.27	1.93	6.83	10.03	24	Aricaceae	0.79	0.09	0.66	1.55
8	Moraceae	0.63	1.74	7.59	9.97	25	Méliantaceae	0.25	0.98	0.27	1.5
9	Myrtaceae	4.37	2.34	2.54	9.25	26	Rubiaceae	0.66	0.35	0.35	1.38
10	Verbenaceae	2.06	5.13	1.21	8.39	27	Polygalaceae	0.49	0.25	0.44	1.18
11	Indeterminate 1	6.22	10.71	5.58	7.51	28	Apocynaceae	0.34	0.17	0.63	1.13
12	Hypericaceae	2.85	1.6	1.84	6.29	29	Ulmaceae	0.28	0.24	0.47	1
13	Fabaceae	1.09	2.79	0.69	4.57	30	Flacourtiaceae	0.39	0.18	0.28	0.85
14	Sapindaceae	1.7	0.99	1.55	4.25	31	Loganiaceae	0.16	0.43	0.08	0.67
15	Indeterminate 2	0.8	2.26	1.18	4.24	32	Myrsinaceae	0.07	0.1	0.36	0.53
16	Ochnaceae	1.72	0.89	1.61	4.22	33	Olacaceae	0.12	0.27	0.1	0.49
17	Araliaceae	1.62	2.07	0.43	4.12	34	Tiliaceae	0.12	0.08	0.03	0.23
							Total	100	100	100	300

Table 6. Distribution, density and recovery of the species in relation to the plant formation and types of wood logging.

Wood logging	Shrubby savannas				Arborescent savannas				Woody savannas			
	T_0	T_1	T_2	T_3	T_0	T_1	T_2	T_3	T_0	T_1	T_2	T_3
D	3.92	5.34	5.3	5.3	4.89	4.47	4.78	4.8	4.32	4.37	5.54	3.77
R	1.06	1.96	2.1	2	1.92	1.44	1.58	1.75	1.66	1.57	2.18	1.15
Re	458	399.6	370	314	625	561	1098	653.3	1153	833	898	607

Keys: D, Mean distance between the individuals (m); Re, recovery of the stems $(m^2).ha^{-1}.year^{-1}$; R: ratio R of distance observed between the individuals of the community.

Table 7. Distribution, recovery and density of the three types of plant formations in relation to the distances from the villages surveyed.

Parameter	Distance from the village					
	0-0.5	0.5-1	1-2	2-4	4-6	>6
D	4.12	3.73	5.03	3	3.33	3.25
R	1.34	0.99	2.39	0.71	0.93	0.91
Re	587	560.8	956.2	659.2	723.7	540.2

Table 8. Distribution, density and recovery of the species in the sites to easy access.

Distance /village	Types of savannas	D	R	Re
0-0.5	A	4.87	1.57	176.12
0-0.5	B	1.55	0.31	73.74
0-0.5	C	2.59	0.45	351.43
0.5-1	A	4.60	1.50	171.1
0.5-1	B	1.51	0.30	71.04
0.5-1	C	2.59	0.45	345.41
1-2	A	4.59	1.21	692.4
1-2	B	6.29	2.92	357.93
1-2	C	4.78	2.18	561.66
2-4	A	2.66	0.47	770.56
2-4	B	4.79	1.85	275.71
2-4	C	4.48	1.58	435.13
4-6	A	5.37	2.61	1154.21
4-6	B	4.01	1.99	1354.32
4-6	C	3.68	1.64	352.73
>6	A	3.06	1.09	68.77
>6	B	2.18	0.73	172
>6	C	2.01	0.35	406.69

Key: A, shrubby savannas; B, arbores cent savannas; C, woody savannas.

Table 9. Specific dispersal of the species in the various plant formations.

Scientific names	Shrubby savanna				Arborescent savanna				Woody savanna			
	T_0	T_1	T_2	T_3	T_0	T_1	T_2	T_3	T_0	T_1	T_2	T_3
Allophyllus africanus	***	***										***
Annona senegalensis	**	*	*	*	**	**	**	***		***	***	***
Cussonia barteri	***	***										
Daniellia oliveri	***	**	***		***	***	***	***	***	***	***	***
Entada africana	***	**		***	***	***	***	***	***	***	***	***
Harungana madagascariensis		**	***	***	**	***	**	*	**	***	**	***
Hymenocardia acida	**	**	*	*	**	*	**	*		***	**	
Piliostigma thonningii	**	*	**	***	**	**	**	**	***	***	***	***
Psorospermum febrifugum		*	***	***		***			***		***	
Syzygium guineense var guineense		***	***		***	***			***	***	***	***
Syzygium guineense var macrocarpum		***	***		***	***			***	***		
Terminalia glaucescens	***	***		***	***	***	***	***	**	***	***	***
Terminalia laxiflora		***	***		***							
Terminalia macroptera		***			***					***		***
Soft butter tree		***					***					

Key: *significant (p < 0.05); **highly significant (p < 0.01); ***very highly significant (p < 0.001).

highly significance. These species are over-scattered in the outer-urban zone of Ngaoundéré because of the fragility of the ecosystems. However, our results are very different from those reported by Sonké (1998) who found 90 gregarious species in the reserve of the biosphere of Dja. The strong pressure of wood logging and pasture are at the origin of the low number of the gregarious species with regards to the closed circles.

In the peri-urban zone of Ngaoundéré, the women are the first to be responsible for wood logging activities which they use for energy and other culinary task. The organization of rural markets of the firewood constitutes a major threat to plant species in all the different plant formations in Ngaoundéré. In the peri-urban zone of Ngaoundéré, women are more involved in wood logging activities more than the loggers, local farmers and therapeutists. Mapongmetsem and Akagou (1997) showed that the situation of the firewood is already alarming in Adamawa and even worsened these recent years with the multiform economic crisis which Cameroon is passing through. This crisis involved people to have an increased quest for charcoal and firewood, thus increasing the rate of taking away significant quantities of wood from the natural formations. In addition to the firewood, peasants cut wood like non-woody forest products selectively.

The wild fruit-lofts like *Vitellaria paradoxa*, *Tamarindus indica*, *Syzygium guineense*, *Ximenia americana*, *Vitex doniana*, *S. guineense* and *Parkia biglobosa* are generally cut for human consumption. Gudjé (2002) reported that, the taking away of trees as not ligneous family products contributes to the destruction of vegetation cover. For the traditional pharmacopeia, *Piliostigma thonningii* and *Securidaca longepedunculata* are requested.

The local population appreciates *S. longepedunculata* for the treatment of rheumatisms; likewise pastoralists in the dry tropical zones use it to increase the availability of fodder at the end of the dry season and the rainy season. These pastoralists have a practice of cutting the highest branches, to lay them down on the ground and to place them at the disposal of their cattle and the smaller livestock (Ntoupka, 1999). The disappearance of the plant species mainly is due to the wood logging for heating and charcoal, the intensification of agriculture, the traditional pharmacopeia, the construction of the houses, the beekeeping through wood used in the hives. Species like *H. acida*, *S. guineense* spp, *D. oliveri*, *Terminalia* spp, *Strychnos spinosa* and *P. biglobosa* are over-exploited. The distribution of the species according to plant formations and the distances with regards to villages showed that the difficulty of access was one of the reasons. It can be noticed that the species are over-scattered in shrubby savannas (0 to 0.5 km), arborescent (> 6 km) and afforested (1 to 2 km).

With regards to recovery of the plant species, T_3 was observed having the highest activities of wood loggings which are very important; this consequently influenced the rate of regeneration which was also high. The sum of the recovery of the big trees and the regeneration of trees after logging is at the origin of the biggest recovery of trees in the outer-urban savanna of Ngaoundéré. However, the recovery of the species by hectare is more important in woody savanna.

CONCLUSION AND RECOMMENDATIONS

Wood logging activities represents a direct effect on the state of the individuals and consequently imposes an over-scattered distribution of the species in the different types of plant formations according to treatments. The number of the wood logging plant individuals was very important in shrubby savannas and afforested with the treatment T_3. *H. acida* has an important value in the Ngaoundéré savannas. It is the best regenerating plant in the vegetation and resists bushfire and wood logging in the guinea savanna of Adamawa. To manage our savannas, it is important to limit the wood logging around villages because of the over-dispersion and disappearrance of the endemic multi-purpose species in the zone. If the population does not become aware of wood logging activities and managing their forest heritage, we shall arrive at a total eradication of the gregarious endemic species of the region of Adamawa. It would be desirable that in the Sudano-Sahelian zone, victim of wood's logging begins to practise annual and seasonal wood logging rotation according to the types of plant formations. High practice of the wood logging should be sanctioned by the population itself and by the government. Other studies on wood logging activities should be carried out in the northern zone of Cameroon.

ACKNOWLEDGEMENTS

We are thankful to PRASAC/ARDESAC Project for support to this research work. We are grateful to Dr. R.S. Houmsou for the English language editing.

REFERENCES

China W, Kate D, Phyllida C (2003). La CDB pour les botanistes: une introduction à la Convention sur la Diversité Biologique pour les personnes qui travaillent avec des collections botaniques. Royal Botanic Gardens, Kew. Pp 94 .

Chouaibou N (2006). Structure de population et exploitation de *Parkia biglobosa* (Jacq.) Benth. (Mimosaceae) dans la zone périurbaine de Ngaoundéré. Mémoire de (D.E.S.S.). Université de Yaoundé I. 69pp.

Clark PJ, Evans FC (1954). Distance to nearest neighbour as a measure of spatial relationships in populations. Ecol. 35: 445-453.

Farid D, Nico K (2006). Empirical estimate of the reliability of the use of the Point-Centred Quarter Method (PCQM): Solutions to ambiguous field situations and description of the PCQM+ protocol, Biocomplexity Research Team c/o Laboratory of General Botany and Nature Management, Mangrove Management Group, Vrije Universiteit Brussel, Pleinlaan 2, B-1050 Brussel, Belgium, in Forest Ecology and Management 228: 1–18.

Frontier S, Pichot-Viale D (1993). Ecosystème, structure, fonctionnement, évolution. Masson, Collection d'écologie Paris. 21:392.

Guedje NM (2002). La gestion des populations d'arbres Comme outils pour une exploitation durable des Produits Forestiers Non Ligneux: l'exemple de *Garcinia lucida* (Sud -Cameroon). Thèse de Doctorat Université Libre de Bruxelles pp 221.

Kevin M (2007). Quantitative Analysis by the Point-Centered Quarter Method, Department of Mathematics and Computer Science Hobart and William Smith Colleges Geneva, NY 14456: 34.

Letouzey R (1986). Manual of forest botanic Tropical Africa. Nogen- sur -Marn C.T.F.T. France.1 :194.

Mahamat H (1991). Contribution à l'aménagement intégré des zones protégées de l'Extrême Nord-Cameroun: Cas du Parc National de kalamaloue. Mémoire de fin d'étude. COD/INADER, Dschang. Cameroun. pp 94.

Mapongmetsem PM, Akagou ZCH (1997). Situation des bois de feu dans les savanes humides de l'Adamawa. Flamboyant, 42: 29-33

Ndam N (1998). Tree regeneration, vegetation dynamics and the maintenance of biodiversity on Mount Cameroon: the relative impact of natural and human disturbance. Thesis, University of Wales, Bangor, UK. p 278.

Ndjidda (2001). Structures et dynamiques des espèces ligneuses dans les zones Sud Est du Parc National de Waza. Mémoire du Diplôme d'Ingénieur des Eaux et Forêts. Université de Dschang. p 62.

Ntoupka M (1994). Etude de la dynamique d'une savane arborée dans la zone soudano-sahélienne du Nord Cameroun sous les effets combinés du pâturage, du feu et de la coupe du bois. CNRS. CEPE. Louis Emberger de Monpellier. p. 97.

Ntoupka M (1998). Production utiles de bois sous perturbations anthropiques (pâturages et feux) dans la région soudano-sahélienne du Nord Cameroun. Actes du colloque. La foresterie de zone sèche. Ouagadoudou, p 12.

Ntoupka M (1999). Impacts des perturbations anthropiques (pâturages, feu et coupe de bois) sur la dynamique de la savane arbre en zone soudano-sahélienne Nord du Cameroun. p 226.

Rippstein G (1985). Etude sur la végétation de l'Adamawa ; évolution, conservation, régénération et amélioration d'un écosystème pâturé au Cameroon. Etudes et synthèses de l'IEMVT, 14 :233.

Siegel (1956). Nonparametric statistics for the behavioural sciences, 2è ed. Auckland, Bogota, Guatemela, Hambourg, Johannesbourg, Lisbon, London, Madrid, Mexico, New Delhi, Panama, Paris, San Juan, Sao Paolo, Sydney, Tokyo; McGraw-Will international Book Company. Dakar, 18: 145-234.

Sonké B (1998). Etude floristique et structurale de la forêt, de la réserve de faune de Dja (Cameroun). Thèse de Doctorat. Université Libre de Bruxelle. p. 267.

Tchobsala, Amougou, A., Mbolo, M. (2010). Impact of wood cuts on the structure and floristic diversity of vegetation in the peri-urban zone of Ngaoundere, Cameroon. Journal of Ecology, Nature and Environment, 2(11):235-258

Tchobsala (2011). Influence des coupes de bois sur la dynamique de la végétation naturelle de la zone peri-urbaine de Ngaoundéré (Adamaoua). Thèse de Docteur/Ph.D. Université de Yaoundé. 184 p.

Tchotsoua (2006). Evolution récente des territoires de l'Adamawa central: de la spatialisation à l'aide pour un développement maîtrisé. Université d'Orléans. Ecole doctorale sciences de l'homme et de la société. HDR. Discipline (Géographie-Aménagement-Environnement). p 267.

Teicheugang BP (2000). Etat et perspective de la Réserve Forestière de Zamay. Mémoire du Diplôme d'Ingénieur des Eaux et Forêts. Université de Dschang/FASA. p 82.

Thorgnang N (2001). Etat et perspective du boisé de Houbaré. Mémoire du Diplôme d'Ingénieur des Eaux et Forêts. Université de Dschang, p. 92.

Yonkeu S (1993). Végétations des pâturages de l'Adamawa (Cameroon). Ecologie et potentialités pastorales. Thèse Doctorat. Université de Rennes I, France. p. 207.

Zacfack L (2005). Impact de l'agriculture itinérante sur brûlis sur la biodiversité végétale et la séquestration du carbone. Thèse de Doctorat d'Etat. Université de Yaoundé I (Cameroon). p. 225.

6

Predictive modelling of the distribution of two critically endangered Dipterocarp trees: Implications for conservation of riparian forests in Borneo

Minerva Singh

School of Geography and Environment, University of Oxford, UK and Department of Plant Sciences, University of Cambridge, CB23EA, UK.

Riparian forests of Malaysian Borneo exhibit high tree species diversity. However, many of the tree species found in these riparian forests are conservation dependant, with their current conservation status (*sensu* IUCN Red List) varying from vulnerable to critically endangered. The present study had two objectives. Firstly, to identify the environmental factors associated with the distribution of two critically endangered tree species, *Shorea johorensis* and *Shorea inappendiculata* using a small number of occurrence records. Secondly, the research seeks to predict suitable habitat and distribution of these two species. The occurrence data and environmental variables are incorporated within a maximum entropy (MaxEnt) model to predict the distribution of the species and identify the environmental variables that influence the distribution. The research shows that for a small study area, the bioclimatic variables are relatively insignificant while factors such as, land use, tree cover play a prominent role in determining distribution of tree species.

Key words: MaxEnt, Borneo, riparian forests, land use changes, *Shorea johorensis*, IUCN, critically endangered

INTRODUCTION

Over the past few decades, the forests of South East Asia, especially those in Malaysian Borneo have undergone significant deforestation. The land use type comprises of a mosaic of human modified land uses- logged forests, forest fragments, riparian forests, oil palm plantations and pristine forests. Riparian forest corridors vary from natural riverine forest (such as gallery forests of tropical savannas) to remnant riparian buffer zones that have spared deforestation (Lees and Peres, 2008). The presence of riparian corridors has been associated with increasing regional species richness (Sabo et al., 2005). Riparian corridors of the study area in Malaysian Borneo contain a number of conservation dependant and vulnerable tree species notably the Dipterocarp species,

including IUCN red listed Critically Endangered species like *Shorea johorensis*, *Shorea inappendiculata, Dipterocarpus submellatus* and *Hopea nutans* (IUCN, 2008). Given the continuing levels of deforestation that the region faces, these riparian zones may be the last refuge for critically-endangered Dipterocarp tree species that have lost a significant portion of their original habitat. Hence, it is important to both evaluate the distribution of these species and identify the environmental factors which influence this distribution as a way of establishing conservation priorities.

Species distribution modelling has become a popular technique for the identification of suitable habitat and the evaluation of species' distribution for a wide variety of taxa. This modelling technique is applied in the present study in order to model the potential distribution and habitats of two critically-endangered Dipterocarp species, *Shorea johorensis* and *Shorea inappendiculata,* which are restricted to lowland forests of Borneo and Sumatra (Indonesia). Specifically, the Maxent modelling is based on a small dataset of presence-only records from riparian margins.

Maxent based SDMs have wide variety of applications. Applications range from estimation of species ranges (Moreno et al., 2011), identification of suitable habitats, establishing conservation priorities (Wilting et al., 2010) and predicting range shifts under future climate change scenarios (Thomas et al., 2004). Owing to its ability to produce useful results with a very small presence data, Maxent has proven itself to be useful for modelling the distribution of rare and endangered species.Tinoco et al. (2009) used Maxent to generate a species distribution model of the Violet throated metal tail hummingbird, a globally endangered bird species which is endemic to south-central Ecuador. The modelling was carried out using a limited species occurrence record. The Maxent model was able to identify that the species was restricted to small pockets in the Andes and has an extent of 2000 sq km. Further, the model helped identify the limiting factors of species distribution which included the presence of deep river canyons. The model also identified three distinct suitable habitats vital for species persistence. Thorn et al. (2009) modelled the distribution of a rare and nocturnal species of primates, the Asian slow loris in Maxent using 20 environmental variables along with information on protected areas to identify both the suitable habitat for the species and to prioritize different areas according to risk. SDMs can aid conservation planning of little known taxa or species with little survey data by highlighting unknown populations, suitable areas for reintroduction, key areas that could be studied in future and provide an assessment of potential risks (Thorn et al., 2009). Kumar and Stohlgren (2009) predicted the distribution and potential habitat of a critically endangered tree species in New Caledonia. Maxent modelling was used to carry out multi-species modelling to model the distribution of 56 endangered pinus tree species in Mexico. The modelling results were further used to evaluate if the pinus tree species are getting sufficient representation in the protected areas (Gutiérrez and Duivenvoorden, 2010).

This research will use species presence data of afore mentioned tree species within the Maxent modelling framework. The objectives of this research are to: (1) to predict suitable habitat and distribution for the recorded riparian tree species using a small number of occurrence records to inform conservation planning in a mixed landscape in the Malaysian Borneo; (2) to identify the environmental factors associated with species habitat distribution. This research uses species occurrence records, environmental layers (bioclimatic, land use and topogra-

phic data) within the using the Maxent model of maximization of entropy to identify suitable habitats and environmental factors which influence species distribution.

MATERIALS AND METHODS

Study species and occurrence data

Records of species occurrence were collected while carrying out fieldwork in the riparian forests of a mixed landscape comprising of forests that had undergone varying levels of logging, oil palm plantations and intact forests (117.5E, 4.5N) in Sabah, Malaysian Borneo. The entire study area has an approximate size of 75,000 hectares and the riparian zones are a small part of the landscape (Turner et al., 2011). During the course of the fieldwork, it was discovered that many of the tree species in the riparian margins such as *S. johorensis* are on the IUCN Red List of critically endangered species. These tree species (along with other tree species of the riparian forests) are now mostly restricted to patches and strips along the streams of low lying areas. The habitat and survival of these species is being threatened by deforestation and conversion of surrounding areas to oil palm plantations. This study considers two critically endangered tree species belonging to the Dipterocarpacae family- (a) *Shorea johorensis* (b) *Shorea inappendiculata* (IUCN, 2008). Maxent allows multiple species to be modelled simultaneously. Modelling of closely related species or species from the same family in this fashion may help identify the areas where the species occur and provide useful information on biogeographical patterns (Costa et al., 2010).

Environmental variables

Bioclimatic, topographic and land use related variables were used for modelling the distribution of the riparian tree species. A review of literature was carried out to identify which bioclimatic variables may explain the distribution of the tree species. The bioclimatic data included were: (a) Maximum temperature (b) minimum temperature (c) precipitation. These data were obtained from the WorldClim dataset (Hijmans et al., 2005). Digital Elevation Model (DEM) data was obtained from WorldClim dataset. DEM data was further used to calculate slope (in degrees) using the Spatial Analyst functionality of the ArcGIS 10. These data were further resampled to a 1 sq km resolution. In addition to the bioclimatic and topographic variables, the research uses the land use land change (LULC) information of the study area.

LULC map was generated using Landsat TM and ground truth data. An NDVI map of the region was also incorporated as one of the environmental variables. NDVI means Normalized Difference Vegetation Index and is a remote sensing based indicator of live green vegetation in the study area. NDVI can be used as a proxy for the health of the vegetation, plant growth and biomass production. In this research, NDVI was derived from data obtained from the SPOT satellite and was resampled to a 1 sq km resolution. Percent tree cover data was obtained from Moderate Resolution Imaging Spectroradiometer (MODIS) vegetation continuous field (VCF) data. This data has been included for the purpose of giving an indication of tree cover and deforestation in the entire landscape. Deforestation (and ensuing edge effects) in the surrounding landscape can have detrimental effect on the tree species isolated in the riparian margins. Since, the riparian forest zones are located near rivers, hydrological data were also included in the modelling process. A river drainage map and river flow map of the region was obtained from the HydroSHEDS database of the United States Geological Survey (USGS) data archive (USGS, 2012) and incorporated in the analysis. The latter was used to generate a layer representing Euclidian distance from the river. Distance from the rivers can be seen as a proxy to forest productivity which is higher at a river's edge (Cattau, 2010).

Modelling technique

A body of literature indicates that maximum entropy or Maxent modelling techniques perform better than many different modelling methods (Ortega-Huerta and Peterson, 2008) and may remain effective, despite small sample sizes (Benito et al., 2009). Further, it requires only species presence data and environmental variable (continuous or categorical) layers for the study area. This makes it a suitable choice for the species data and environmental variable data that has been collected in the present study.

In order to carry out the modelling of the riparian tree species, the research has used the freely available Maxent software, version 3.1 (http://www.cs.princeton.edu/~schapire/maxent/). Maxent is a maximum entropy based machine learning program that estimates the probability distribution for a species' occurrence based on environmental variables (Phillips et al., 2006). The environmental variable values at the presence localities impose constraints on the unknown distribution. The maximum entropy approach then approximates an unknown distribution using the known occurrences and background points (all points/grid cell values in the study region) that maximizes entropy, subject to the constraints imposed by the known occurrences. The result of Maxent shows a map where every grid has a value of 0 to 100 (if the result output format is set as cumulative) or 0-1 (if the result output format is selected as logistic); this represents the estimate of relative probability of species occurrence. Maxent is not strongly influenced by the number of environmental parameters used to build models because it ignores those that are non-informative, and uses regularization techniques to avoid over-parameterization (Phillips et al., 2006).

The study made use of presence records (collected from the field) and the afore-described environmental variables to model the potential habitat of the species under consideration and identification of suitable habitats. Significant sources of uncertainties exist in SDMs making it important to validate the results obtained from these models as a way of verifying the robustness of the model. The model generated by Maxent firstly was evaluated by their area under the curve generated for the model. The test data of models have a value of 0.767 (*S. johorensis*) and 0.864 (*S. inappendiculata*). Models that have an AUC value greater than 0.75 are considered to be useful (Elith, 2006). However, this is not the only method of validating the results of the Maxent model. Araujo et al. (2005) illustrate the use of independent validations and the feasibility of doing so by presenting the case of study of the observed distributional shifts among 116 British breeding bird species over a two decade period. However, it is difficult to obtain an independent datasets in many cases, especially when the species observation data maybe small. Redistribution is widely used in the assessment of accuracy. A part of data is used to calibrate/train the model and the other part is used to validate or test the model. Data partitioning techniques can be used to address the problem associated with redistribution methodologies. Some of these techniques include one-time data splitting of calibration and validation datasets. Although, no exact specifications exists, a review of the literature reveals that the models may be calibrated using 70% of the dataset sample obtained at a given point in time and the predictive accuracy of the model was evaluated using the 30% of the remaining data (Araujo et al., 2005).

However, this approach may not work with a small number of samples because the 'training' and 'test' datasets will be very small (Pearson et al., 2007). Hence it was decided to follow the jack-knife validation methodology developed by Pearson et al. (2007), which is shown to be effective for small sample sizes. Under the principle of these techniques, one locality/occurrence point is removed from the dataset, and the model built using the remaining '*n*–1' locality points. Thus, for a species with '*n*' localities, '*n*' individual models will be built for testing. Model accuracy and significance were evaluated based on the ability of each model to predict the one excluded test locality as present (Pearson et al., 2007).

RESULTS AND DISCUSSION

The Maxent model successfully predicted suitable habitats for both tree species. The predicted probability for the presence of the *S. inappendiculata* and *S. johorensis* is shown in Figure 1.

Probability of presence estimate has been defined on a 0-1 scale (logistic format selected for expressing the output data. This format is preferred over others are it expresses the estimates of probability of occurrence as predicted by included environmental variables, thus providing a comparatively more accurate interpretation of output data; Baldwin, 2009) and areas having a value greater than 0.5 could be considered to have suitable habitat for the persistence of the species (Stabach et al., 2009). However, areas most suitable for supporting the tree species are fragmented, especially in the case of *S. johorensis*. A visual examination of Figure 1 reveals that the distribution of species follows the pattern of river flow, while the presence of steeper slopes restricts the distribution of species; in addition, areas sited at higher altitudes were found to have the least suitable habitat for supporting the tree species under consideration. Further, the Maxent model also allows for performing an internal jack-knife test to quantify the importance of the variables in influencing the distribution of both tree species. The results are shown in Figure 2.

In Figure 2, altitude refers to elevation in metres (obtained from DEM); 'drainage' refers to the river drainage profile and pattern of the region; 'flow_corr' is the distance from the rivers; 'lulc_res' refers to the Landsat based land use land cover (LULC) map of the region which defines the different land use categories of the region; 'max_temp' and 'mean_temp' refer to the maximum and mean temperature of the study area; 'ndvi_spot' is the NDVI map of the study area; 'preceptn' is the annual rainfall; 'slope' is slope of the study area in degrees (obtained from DEM); 'tree_cov' is the percentage tree cover of the area, obtained from MODIS

Figure 2 indicates that in the case of *S. johorensis*, environmental variable with highest gain when used in isolation is 'flow_corr.' (or distance from the rivers), which therefore appears to have the most useful information by itself. The environmental variable that decreases the gain the most when it is omitted is also 'flow_corr.', which therefore appears to have the most information that is not present in the other variables. In the case of *S. inappendiculata*, the environmental variable with highest gain when used in isolation is altitude. This appears to have the most useful information by itself. The environ-

Shorea inappendiculata

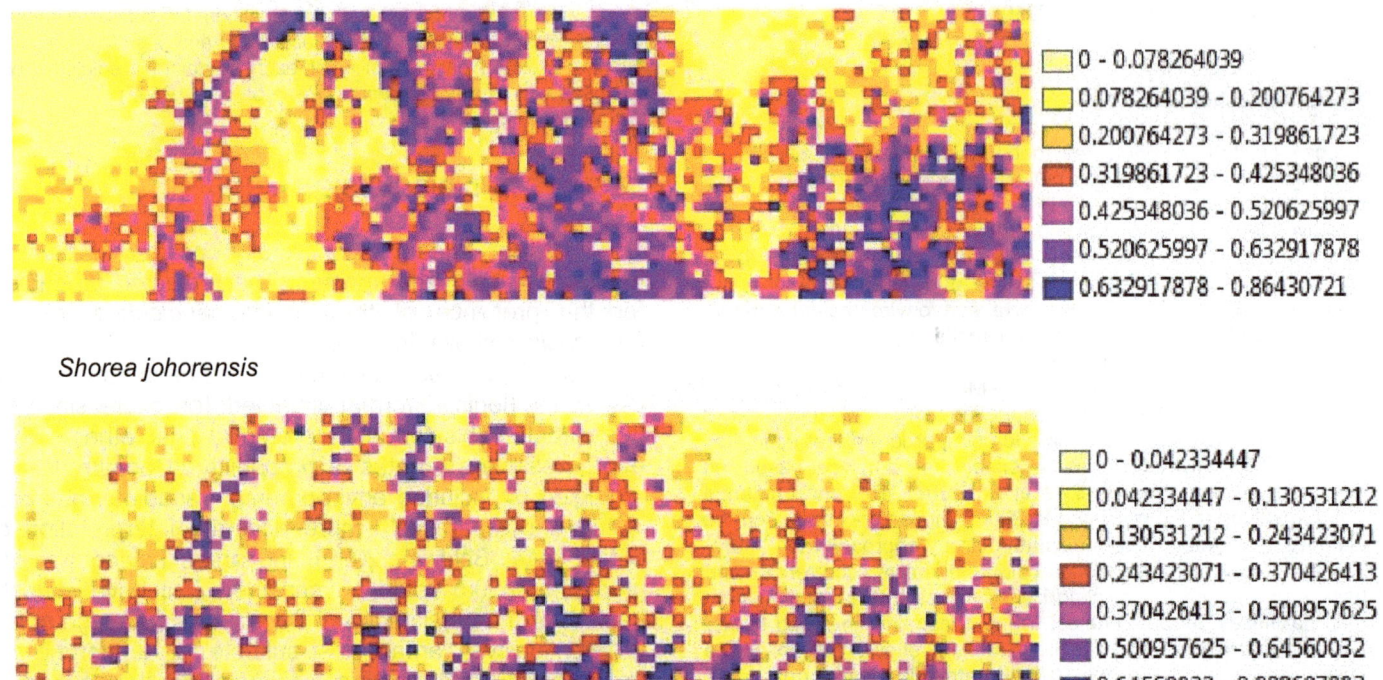

Shorea johorensis

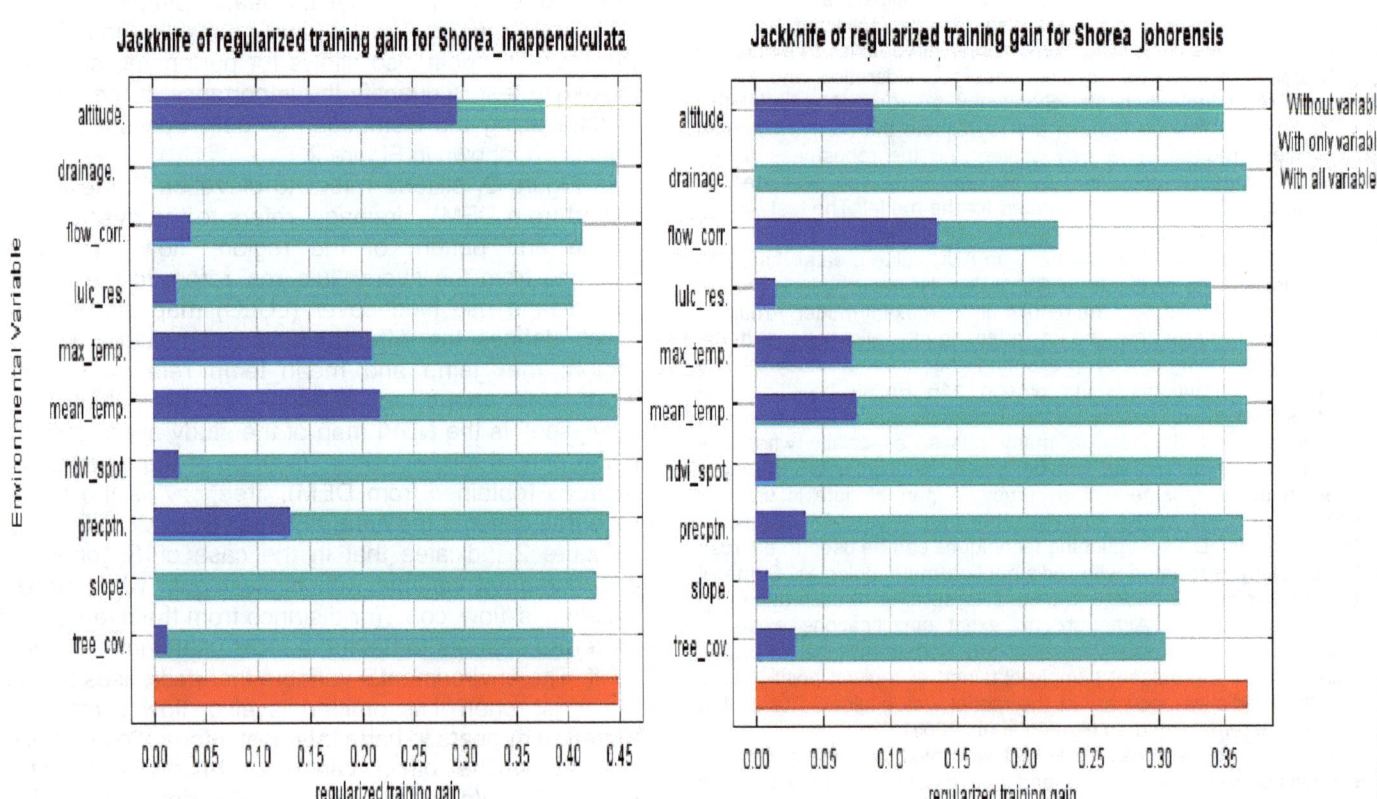

Figure 1. Predicted suitable habitat for *Shorea inappendiculata* and *Shorea johorensis*.

Figure 2. Results of jack-knife evaluations indicating the relative importance of predictor variables for *Shorea inappendiculata* and *Shorea johorensis* in the Maxent model.

Table 1. Selected environmental variables and their percentage contribution.

Variable	*Shorea johorensis*	*Shorea inappendiculata*
Altitude	24.4	53.3
Flow_corr	39.3	23.2
Tree_cov	15.4	6.5
Slope	9.2	3.3
NDVI_spot	3.5	2
LULC_res	5.8	6.7
Max_temp	1.5	0
Preception	0.8	0
Drainage	0	0

mental variable that decreases the gain the most when it is omitted is altitude. This therefore appears to have the most information that is not present in the other variables.

The Maxent model also quantifies the percentage contribution of the predictor variables in influencing the distribution of the tree species as shown in Table 1.

On the basis of this analysis, it may be argued that distance from the rivers and altitude are the most environmental variable that influence the presence and distribution of the riparian tree species under consideration. Both *S. johorensis* and *S. inappendiculata* are species of lowland forests and altitude can be seen as a limiting factor to their distribution. However, *S. inappendiculata* is less restrictive in its choice of habitat as compared to *S. johorensis* and is mainly dependant on topographic features. On the other hand, *S. johorensis* in addition to requiring suitable topographical conditions is also significantly dependant on eco-geographical characteristics such as distance from rivers and land use and quality characteristics such as percentage tree cover and NDVI.

This study has achieved its main objective of predicting the distribution and suitable habitats of the recorded riparian tree species on the basis of a relatively small number of presence records and environmental predictor variables. Further, the study has been able to identify the environmental factors that influence and limit the distribution of the species under consideration. The latter establishes the role played by topographic and land use patterns in influencing the tree species distribution in the study area. The research shows that for the small study area discussed in this paper, bioclimatic variables play a relatively insignificant role in determining the presence and distribution of tree species. However, other factors such as land use and vegetation quality factors such as tree cover, land use land change dynamics play an important role in influencing the presence and distribution of species. Additionally topographic factors such as altitude, distance from the rivers and slope also play an important role in influencing the presence and distribution of species.

These findings in turns have deep ramifications for conservation planning. Firstly, even though both the critically endangered species belong to the same family, their distributions are significantly different and environmental variables influenced their distribution differently. Hence, different conservation strategies may be required for conservation of species which belong to the same family and have the same conservation status. The results indicate that distance from rivers is an important determining factor for distribution of both species. For instance, it may be argued that in the immediate future it is important to focus on land use and vegetation quality factors as a way of ensuring the persistence of the tree species. One of the instances that can be noted is that of the Brazilian forestry legislation which requires the maintenance of riparian corridors on all private land holdings and these are required to have a pre-determined width (Lees and Peres, 2008). Given the role that distance from rivers plays in influencing the presence and distribution of tree species, laws similar to the aforementioned law may be considered as a way of protecting tree species in the study area.

In addition to land use changes, climate change is regarded as an important driver of biodiversity change. However, in this study, bioclimatic variables play a non-significant role in influencing the presence and distribution of the tree species. This may be attributed to two reasons. Firstly, the study area is small (covering only a small part of Malaysian Borneo).

Hence, it may be argued that while land use, topography and vegetation vary significantly over the landscape, climatic factors remain fairly constant throughout the study area. Secondly, bioclimatic variables are based on the interpolations of global climate data and thus have a coarse resolution. Hence, it is important to evaluate if the inclusion of finer scale climatic variables could improve the predictions of the SDM, especially when considering species response under future climate change scenarios.

Conclusions

The results of this study and the literature discussed previously indicates that Maxent can be useful in predicting species distribution and subsequently establishing conservation priorities both at local and regional scales. However, SDMs are fraught with significant uncertainties. Mainly uncertainties in the predictions from SDMs stem from the basic assumptions of the models, algorithms used, parameterization, the variables included for analysis or even the spatial scale of the variables (as demonstrated by Randin et al., 2009). Most SDMs are based on species data collected as a result of sampling carried out at a given point in time/space and the working postulate justifying the use of these data is that the species in question are in pseudo-equilibrium with their environment (Guisan and Thuiller, 2005). Further, sampling design too can introduce biases such as those stemming from incomplete sampling to focus on a particular geographic space as opposed to random sampling (Zimmerman et al., 2010). While Maxent offers the advantage of being able to use small samples, the accuracy of the models maybe compromised by sampling bias. Further, the transferability of Maxent results between sampled and unsampled areas needs to be interpreted with caution (Baldwin, 2009). It is important to minimize the inherent uncertainties in SDMs. This may be accomplished by the use of link different models, fine scale data (as opposed to coarse scale) and collection of detailed field records.

ACKNOWLEDGEMENTS

I would like to acknowledge and thank the following for their help in making this research possible: Sime Darby Foundation, Dr Rob Ewers, Dr Ed Turner, Hamzah Tangki, Dr Glenn Reynolds, Sabah Biodiversity Centre (SABC), Royal Society's South East Asia Rainforest Research Programme (SEARRP), Stability of Altered Forest Ecosystems (SAFE) project. A further thanks goes to my field research assistants Magad, Maria, Kiel, Was and Sabri.

REFRENCES

Araujo MB, Pearson RG, Thuiller Erhard WM (2005). Validation of species–climate impact models under climate change. Glob. Chang. Biol. 11: 1504-1513.

Baldwin RA (2009). Use of maximum entropy modelling in wildlife research. Entropy11:854–866.

Benito BM, Martınez-Ortega MM, Munoz LM, Lorite J, Penas J (2009). Assessing extinction-risk of endangered plants using species distribution models: a case study of habitat depletion caused by the spread of greenhouses. Biodiv. Conserv. 18:(9)2509-2520

Cattau M (2010). Using the Ecosystem Service Value of Habitat Areas For Wildlife Conservation: Implications of Carbon Rich Peat swamp Forests For the Bornean Orang-utan, http://www.outrop.com/uploads/7/2/4/9/7249041/cattau_megane_ma stersproject.pdf

Costa G, Nogueira C, Machado R, Colli G (2010). Sampling bias and the use of ecological niche modeling in conservation planning: a field evaluation in a biodiversity hotspot. Biodivers. Conserv. 19:883–899

Elith J, Graham CH, Anderson RP, Dudik M, Ferrier S, Guisan A, Hijmans RJ, Huettmann F, Leathwick JR, Lehmann A, Li J, Lohmann LG, Loiselle BA, Manion G, Moritz C, Nakamura M, Nakazawa Y, Overton JM, Peterson AT, Phillips SJ, Richardson K, Scachetti-Pereira R, Schapire RE, Soberon J, Williams S, Wisz MS, Zimmermann NE (2006). Novel methods improve prediction of species' distributions from occurrence data. Ecography 29: 129-151.

Guisan A, Thuiller W (2005). Predicting species distribution: offering more than simple habitat models. Ecol. Lett. 8: 993-1009.

Gutiérrez JA, Duivenvoorden JF (2010). Can we expect to protect threatened species in protected areas? A case study of the genus *Pinus* in Mexico Revista Mexicana De Biodiversidad 81: 875-882.

Hijmans RJ, Cameron SE, Parra JL, Jones PG, Jarvis A (2005). Very high resolution interpolated climate surfaces for global land areas. Int. J. Climatol. 25: 1965-1978

Kumar S, Stohlgren TJ (2009). Maxent modeling for predicting suitable habitat for threatened and endangered tree Canacomyrica monticola in New Caledonia. J. Ecol. Nat. Environ. 1(4): 94-98.

IUCN (2008). 2008 IUCN Red List of threatened species. <http://www.iucnredlist.org>

Lees A, Peres CA (2008). Conservation value of remnant riparian forest corridors of varying quality for Amazonian birds and mammals. Conserv. Biol. 22: 439-449.

Moreno R, Zamora R, Molina JR, Vasquez A, Herrera MA (2011). Predictive Modelling of Microhabitats For Endemic Birds in South Chilean Temperate Forests Using Maximum Entropy (MaxEnt). Ecol. Inform. 6: 364-370

Ortega-Huerta MA, Peterson AT (2008). Modeling ecological niches and predicting geographic distributions: a test of six presence-only methods. Revista Mexicana De Biodiversidad 79: 205-216.

Pearson RG, Raxworthy CJ, Nakamura M, Peterson AT 2007. Predicting species distributions from small numbers of occurrence records: a test case using cryptic geckos in Madagascar. J. Biogeo. 34: 102- 117.

Phillips SJ, Anderson RP, Schapire RE (2006). Maximum entropy modelling of species geographic distributions. Ecol. Model. 190: 231-259.

Randin CF, Engler R, Normand S, Zappa M, Zimmermann, NE, Pearman, PB, Vittoz P Thuiller W. and Guisan A. 2009: Climate change and plant distribution: local models predict high-elevation persistence. Chang. Biol. 15:1557-1569.

Sabo JL, Sponseller R, Dixon M, Gade K, Harms T, Heffernan J, Jani A, Katz G, Soykan C, Watts J, Welter J (2005). Riparian Zones Increase Regional Species Richness by Harbouring Different, Not More Species. Ecology 86(1)56-62

Stabach JA, Laporte N, Olupot W (2009). International Journal of Biodivers. Conserv. 1(5):177-186

Tinoco BA, Pedro X, Latta SC, Graham CG (2009). Distribution, ecology and conservation of an endangered Andean humming bird Violet throated Metaltail. Bird Conserv. Int. 19:63-76

Thorn JS, Nijman V, Smith D, Nekaris KAI (2009). Ecological niche modelling as a technique for assessing threats and setting conservation priorities for Asian slow lorises (Primates: Nycticebus). Divers. Distrib. 15: 289-298.

Thomas CD, Cameron A, Green RE et al. 2004. Extinction risk from climate change. Nature 427:145–147.

Turner EC, Snaddon JL, Ewers RM, Fayle TM, Foster WA (2011). The impact of oil palm expansion on environmental change: putting conservation research into context.in I. M. A. dos Santos Bernardes, editor. Environmental Impact of Biofuels. InTech

United States Geological Survey (USGS) 2012. <http://gisdata.usgs.gov/website/HydroSHEDS/viewer.php>

7

The woodland tree *Brachystegia floribunda* facilitates the encroachment of forest tree species into miombo woodlands in northern Malawi

Tomohiro Fujita

Graduate School of Asian and African Area Studies, Kyoto University, Japan.

Although, human activities have caused major losses of tropical forest, there are also reports of forest encroachment and expansion into surrounding sparse vegetation types (savannah and woodland). Large savannah-woodland trees may promote forest encroachment into their surroundings by facilitating seed deposition and subsequent establishment of forest species. This work sought to determine whether large individuals of the woodland tree *Brachystegia floribunda* facilitate encroachment of montane forest tree species. To this end, environmental parameters, seed rain and seedling distribution of montane forest tree species were examined in south-east Africa. The seed rain and the numbers of tree seedlings were both higher below large *B. floribunda* trees than below: (i) small specimens of this woodland tree and (ii) in open grassland. Moreover, large individuals of *B. floribunda* modified microhabitat conditions more than small trees. Thus, large *B. floribunda* trees probably facilitate the encroachment of montane forest trees into surrounding woodlands.

Key words: *Brachystegia floribunda*, tropical forest, grassland, woodlands.

INTRODUCTION

Savannah and/or woodland habitats interspersed with closed-canopy forest are common in many tropical regions (Hennenberg et al., 2006). Although, there has been a major loss of forest area in these regions due to factors such as logging (Nepstad et al., 1999); studies have also documented the encroachment and expansion of forest into more open vegetation types (Banfai and Bowman, 2007; Mitchard et al., 2009; Bowman et al., 2010). Generally, forest trees seldom establish adjacent to these sparsely vegetated areas, because the open vegetation is often more resistant to fire and water stress and competition from grass species (Shararm et al., 2009); however, these limiting factors may not apply under a savannah-woodland tree (Hoffmann, 2000). By providing shade, savannah-woodland trees can suppress the growth of grass, leading to less frequent occurrence

of fires and less competitive conditions (Shararm et al., 2009). A tree can also improve water status by reducing solar radiation (Shararm et al., 2009). In addition, seed dispersal can be greater below trees than in a treeless grassland, because bats and birds may be more likely to visit tree crowns than grassland. Thus, savannah-woodland trees adjacent to forest may play an important role in forest encroachment (Rolhauser et al., 2011).

The effect of savannah-woodland trees on promoting forest tree encroachment may vary depending on tree size. Seed rain may be greater below large trees than small trees, because bats and birds are more likely to perch on taller, large-crowned trees than on the smaller crowns of short trees (Aukema and Martínez del Rio, 2002; Lasky and Keitt, 2012). Seed deposition below large trees may be important in establishing forest tree seedlings, as their crowns may create a better environment for seedling germination and survival than the smaller crowns of short trees. The present study examined the role played by woodland trees in the encroachment of forest trees in northern Malawi (southeastern Africa). In south-central Africa, about 2.7 million km^2 are covered with miombo woodland made up of leguminous species in three closely related genera: *Brachystegia*, *Julbernardia*, and *Isoberlinia* [Fabaceae subfamily Caesalpinioideae (Campbell et al., 1996)]. These areas also contain patches of lowland rain, montane rain, and riparian forests, all of which differ from miombo woodland in floristic composition and structure (White et al., 2001). The vegetation dynamics between these forests and miombo woodlands are unclear (White, 1983). Experiments have shown that the forests expand into miombo woodland when fires are prevented (Trapnell, 1959; White et al., 2001); miombo woodland, especially in wetter regions, can be an alternative stable state to forest.

The present study focused on the role of *Brachystegia floribunda* as a facilitator of montane forest tree encroachment. This species was chosen because large numbers of forest tree saplings occur under the dense canopy of *B. floribunda* (Fujita submitted), which is generally the dominant tree in miombo woodland (White, 1983) and in the study area. The dry pods of the species are explosively dispersed (Chidumayo and Frost, 1996), and the seed itself does not attract frugivores. However, non-fruiting trees may attract as much outside seed rain as do fruiting trees by providing a calling perch and shelter from predators (Carriere et al., 2002). In this context, the following specific postulates were tested. First, tall *B. floribunda* individuals modify microhabitat conditions differently from short trees or grasses in treeless vegetation openings. Second, the seed rain of forest trees is greater below tall *B. floribunda* trees than below short trees or grasses in treeless opening. Third, the number of forest tree seedlings is highest below tall *B. floribunda* trees, reflecting greater seed input and beneficial environmental modifications.

MATERIALS AND METHODS

Study area

The study was conducted in the periods of January to March 2012 and August to September 2012 on private land located in a rural zone managed by the villagers of Ntchuka (10° 58' S, 34° 04' E) on the north Vipya Plateau in northern Malawi. The region is vegetated predominantly by miombo woodland, although, there are stands of montane rain forest on mountain crests (above ~1800 m elevation), in valleys, and on mountainous slopes (1700 to 1800 m elevation). The size of these patches varies from about 10 m^2 to ~ 1 ha. The mean annual rainfall on the north Vipya Plateau exceeds 1270 mm, which is distributed across a wet season from December to April, with a dry season from May to November (Chapman, 1970). The Vipya Plateau is underlain by undifferentiated basement complex rocks that are mainly gneisses (Chapman, 1970). The soil of the study area was a well-drained red and sandy clay loam. Most of the

private land area other than montane forest and other forest patches is burnt by the local inhabitants at intervals of 2 to 3 years during the late dry season (September to December). Fires are set to clear footpaths, as paths overgrown by tall grass become difficult to traverse and dangerous due to the presence of snakes.

Montane forests are typically less flammable due to the dense canopy that excludes grasses and maintains a more humid understorey. Therefore, fire is unlikely to penetrate far into the forest. Few trees are harvested from this area of private land, as it is located far from local villages.

Measurement of standing trees

To describe the vegetation structure in the study area, all standing trees were measured in a 50 × 50 m plot established on a mountainous slope bearing miombo woodland. The plot was located about 100 m from a montane forest. In the plot, all mature trees of more than 1 cm in Diameter at Breast Height (DBH) were identified, and their height and crown radius were measured. The DBH of trees greater than 5 cm was also measured.

Experimental design

Environmental parameters, seed rain and seedling densities of montane forest trees were measured in three microhabitats: a treeless opening (hereafter, open microsite), a site beneath short *B. floribunda* trees (<3 m tall; hereafter, short-tree microsite) and a site beneath tall *B. floribunda* trees (>5 m tall; hereafter, tall-tree microsite). The open microsite lacked both mature trees and a canopy cover. The microsites were generally round in shape with radius of 2 m. The sizes of tree microsites (short-tree or tall-tree) exactly matched the crown areas above them. Three microsites of each type were located in miombo woodland located at least ~50 m from the montane forest (Figure 1). Short-tree and tall-tree microsites were established within 20 m of one another. Open microsites were located at least 10 m from short-tree and tall-tree microsites. Five replicates of each microsite type were established.

Measurement of environmental conditions

Environmental parameters were measured in each replicate of the three microsite types (n = 15). Parameters selected included aboveground microclimate variables, soil water content, and the proportion of grass cover. To estimate canopy openness, four hemispherical canopy images were taken at each microsite 1-m above ground using a fish-eye wide-angle lens (Raynox DCR-CF; Yoshida Industry Co., Ltd., Tokyo, Japan). The images were captured at the midpoint of the crown radius in each of the cardinal directions from trunks of short and tall trees. At the open microsite, the photographs were taken at 1-m in each of the cardinal directions from its centre. The hemispherical photographs were taken during the rainy season (February 2012). Gap light analyser software (Frazer et al., 1999) was used to calculate canopy openness. The overall canopy openness at each microsite was calculated as the mean of the four values. Air temperature and relative humidity (RH) were measured with a data logger (T and D, Ondotori RH TR72U, Matsumoto, Japan) 1 m from the trunks of short and tall trees and at the centre of the open microsite. The measurements were made on 1 to 2 September 2012. Soil water content at 12-cm depth was measured using time domain reflectometry (TDR) probes (Campbell Scientific, Hydrosense, Townsville, Australia) during the rainy season of March 2012 and the dry season of September 2012. The soil water content measurements were taken at the same points at which the canopy openness photographs were taken.

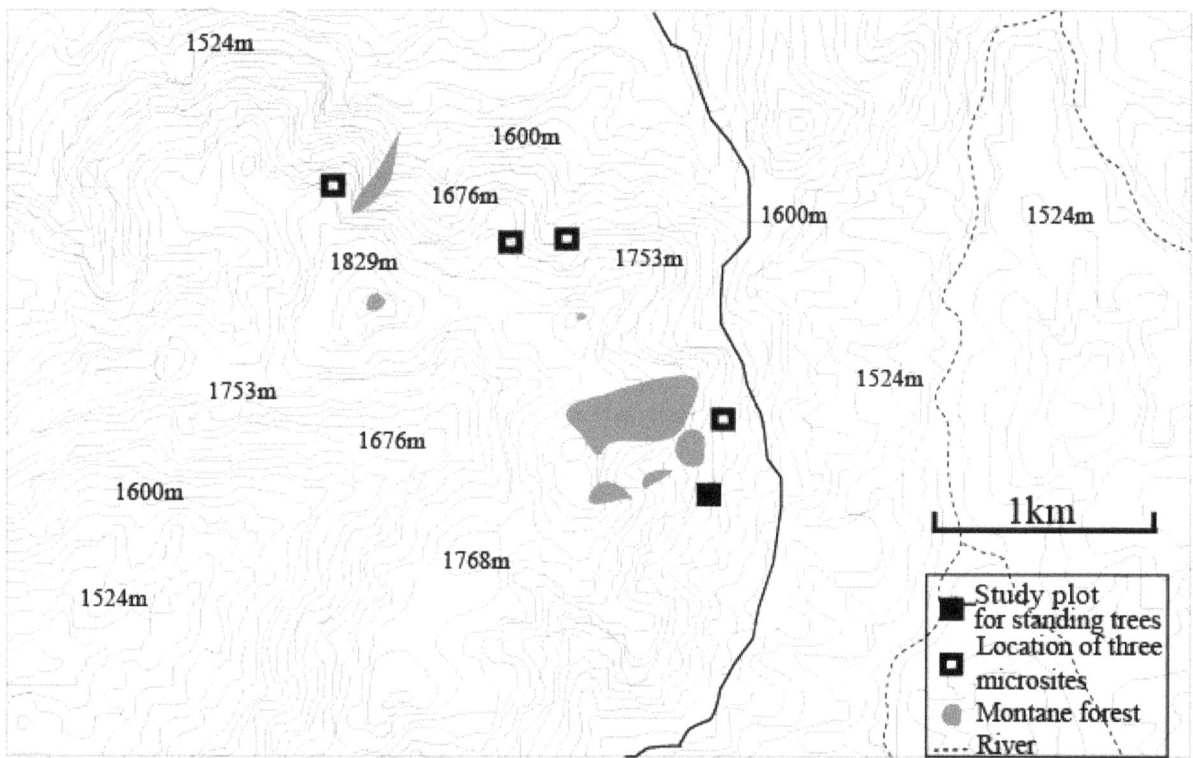

Figure 1. Map of the study plots, microsites and montane forest patch. One of the replicates of microsites (open, short-tree, tall-tree) exists within study plot for standing trees. Thus, only four replicates of microsites are shown in the map. Miombo woodland surrounds the montane forest patch.

The overall soil water content at each microsite was calculated as the mean of the four values. To estimate the proportions of grass cover, three quadrats (each 1 × 1 m) were established at 1 m from each tree bole (short-tree and tall-tree microsites) and at the centres of open microsites. The direction of the first quadrat was determined haphazardly, and the other quadrats were placed on bearings 120 and 240° from the direction of the first quadrat. In each quadrat, the proportion of grass cover was visually estimated using 10% cover-class intervals. The overall proportion of grass cover at each microsite was calculated as the mean of the three values.

Measurement of seed rain and seedling density

Seed rain was measured between January 2012 and March 2012 (rainy season). During the rainy season, many montane forest tree species bear fruit (Dowsett-Lemaire, 1985). Seed traps were installed where the quadrats had been located. The traps each comprised a 70 × 70 cm sheet of fine-mesh net secured to the ground (14,700 cm²/microsite). Each net had sides 5 cm high that prevented seed from washing away, but allowed for entry of seed predators. Thus, the seed rain values may be underestimated. Initially, a conventional method was used to measure seed rain (traps were inverted cones of polyethylene cloth with circular mouths of polyethylene pipe; each was supported by three vertical PVC plastic pipes adjusted so that the receiving face was 1 m above the ground). However, these traps were easily visible and were damaged by local villagers. Each microsite was visited and collected from twice a week; seeds were identified and counted. The seed rains in replicate traps at each microsite were combined for analysis. Only the number of montane forest tree seeds is reported here, as the focus of this study was on encroachment of

forest trees into miombo woodland. Montane forest trees were defined following Friis (1992) and White et al. (2001), as tree species that occur mainly in montane rain forests.

Before setting the seed traps, all montane forest tree seedlings (30 cm ≤ H ≤ 130 cm) were counted in quadrats in which the proportions of grass cover had been estimated. The seedlings were identified by an expert from the national herbarium and botanical gardens of Malawi.

Data analysis

Data were checked for normality using the Kolmogorov-Smirnov test and for homoscedasticity using the Bartlett test. The data for the proportion of grass cover, density of seed rain, and seedlings of montane forest tree were analysed using nonparametric methods, as the data violated the assumption of homoscedasticity. Differences in these values among microsites were compared using the nonparametric Steel-Dwass multiple comparisons test. Tukey-Kramer multiple comparison tests were used to detect differences in canopy openness, soil water content, mean air temperature and RH among microsites. All statistical analyses were conducted with R software ver. 2.12.2 (R Development Core Team, 2011).

RESULTS

Composition and structure of standing trees

In total, 198 individuals from 13 species of standing tree (DBH > 1 cm) were recorded. The basal area (BA) was

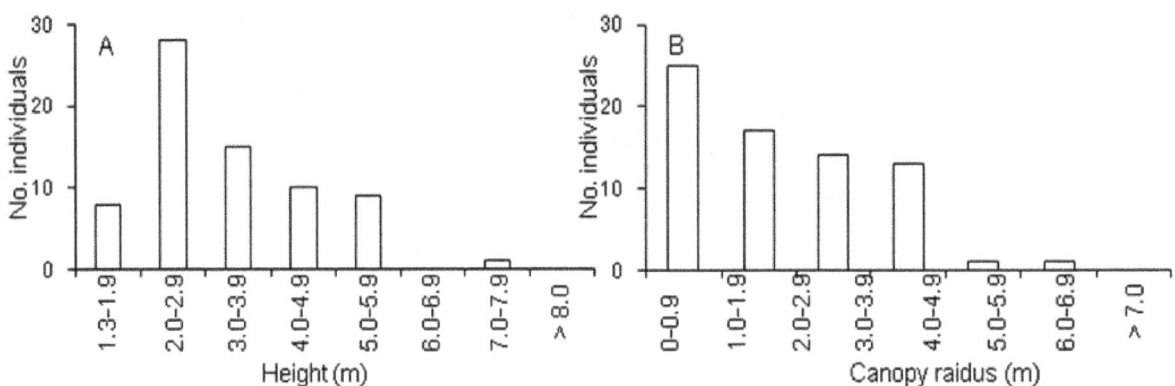

Figure 2. Height (A) and canopy radius (B) distribution of *Brachystegia floribunda* in study plot (0.25 ha).

Table 1. Environmental conditions (n = 5 per microsite) measured among three microsites (means ± SE).

Variable	Open	Short-tree	Tall-tree
Canopy openness (%)	71.1±9.3a	29.9±6.9b	18.9±4.7b
Ta (°C)	25.2±0.8a	21.9±0.4b	21.4±0.4b
RH (%)	42.1±2.5a	49.9±1.8b	51.9±1.3b
SWC (rainy season) (%)	27.6±1.2	31.3±1.4	30.4±1.6
SWC (dry season) (%)	5.3±0.8	4.8±0.2	4.9±0.3
Grass cover (%)	58.0±4.5	48.0±6.5	22.7±9.0

See text for microsite definitions. Variables: Ta, mean air temparature; RH, relative humidity; SWC, soil water content. Variables for which significant differences between microsite were found are in bold. Different letters show significant differences at α < 0.05 for the same variable.

6.6 m²/ha, tree density was 0.08 individuals/m² and the median DBH was 8.4 cm. *B. floribunda* made up the vast majority (> 90%) of the basal area. The analyses presented here were conducted exclusively on *B. floribunda*. The height distribution and crown radius of *B. floribunda* had a right-skewed distribution (skewness = 3.02, P<0.01, height; skewness = 2.16, P<0.05, crown radius). Although, most trees were fairly short (half were < 3 m) and small-crowned (half were < 2 m), a few were tall and large-crowned (Figure 2).

Microhabitat characterisation

Table 1 shows the environmental variables. Canopy openness and mean air temperature were significantly lower in the short- and tall-tree microsites than in the open microsite, but similar between the short- and tall-tree microsites (Tukey-Kramer test; p = 0.55, canopy openness; p = 0.85, mean air temperature). RH was higher in the short- and tall-tree microsites than in the open microsite, but similar between the short- and tall-tree microsites (Tukey-Kramer test; p = 0.75). Soil water content did not differ among microhabitats in either the rainy (one-way ANOVA; p = 0.17) or dry seasons (one-way ANOVA; p = 0.79). The proportion of grass cover in the tall-tree microsite tended to be lower than that in the

open and short-tree microsites, although, there were no differences between these latter sites (Steel-Dwass test; p = 0.09) or between the short- and tall-tree microsites (Steel-Dwass test; p = 0.14).

Seed and seedling density of montane forest trees

In all, 40 seeds of montane forest trees from three species were found: *Parinari excelsa* Sabine (Chrysobalanaceae), *Apodytes dimidiata* E. Mey. (Icacinaceae) and *Syzygium guineense* ssp. *afromontanum* (Willd.) DC. (Myrtaceae). The seeds of these species are dispersed by animals. The seeds were found in all three microsites, but most (83%) were found in the tall-tree microsite (Figure 3). The seed density was significantly higher in the tall-tree microsite than in the short-tree microsite (Steel-Dwass test; p < 0.05). Furthermore, there was a tendency towards higher seed densities in the tall-tree microsite (in comparison to open microsites), but this between-site difference was not significant (Steel-Dwass test; p = 0.10). Thirty-two seedlings of montane forest tree from five species were found: *Cussonia spicata* Thunb. (Araliaceae), *Diospyros whyteana* (Hiern) F. (Ebenaceae), *A. dimidiata*, *Rapanea melanophloeos* (L.) Mez (Myrsinaceae) and *S. guineense* ssp. *afromontanum*. The seeds of all of these trees are also

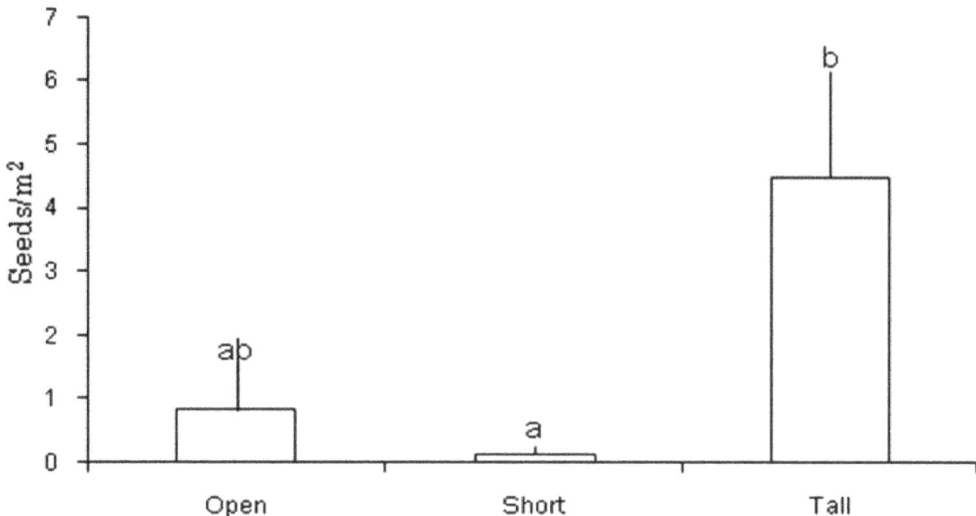

Figure 3. Mean seed density of montane forest in the three microsites. Error bars = 1 SE. Different letters indicate statistically significant differences between microsites at the *P* < 0.05 level.

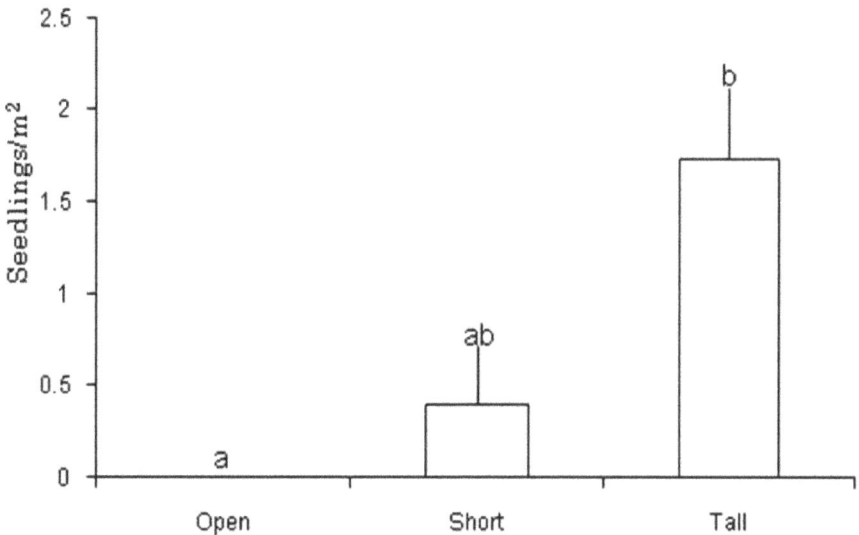

Figure 4. Mean seedling (30 cm ≤ H ≤ 130 cm) density of montane forest in the three microsites. Error bars = 1 SE. Different letters indicate statistically significant differences between microsites at the *P* < 0.05 level.

dispersed by animals. The seedlings were found in short- and tall-tree microsites, but not in the open microsite. There was a tendency towards higher seedling density in the tall-tree microsite (in comparison to the short-tree microsite) (Figure 4), but this between-site difference was not significant (Steel-Dwass test; P = 0.09).

DISCUSSION

Seeds of montane forest trees in miombo woodland

The deposition of seeds of montane forest trees was confirmed in all three microsites, but most were found in the tall-tree site (Figure 3). The greater seed rain at this site was probably related to the number of birds that preferentially perch on tall and large-crowned trees (Aukema and Martínez del Rio, 2002; Lasky and Keitt, 2012). Bird preferences are most probably related to better calling sites in large trees, or better protection from predators (McDonnell, 1986; Roxburgh and Nicolson, 2008). This outcome is consistent with previous research showing that there are large seed rains beneath tall tree individuals located in abandoned fields (McDonnell, 1986; Slocum and Horvitz, 2000). McDonnell (1986) suggested that perch height relative to neighbouring perches, not absolute height, has the most influence on attracting birds.

This may be the case in this study as well. Tall trees (5 m < height) were the tallest structures in the vici-nity and were relatively scarce in the study area (Figure 2A). The results indicate that tall *B. floribunda* trees may act as dispersal foci (Clark et al., 2004) for montane forest trees as they encroach into miombo woodland.

The present results suggest that tree height may partly influence the seed deposition of montane forest trees into miombo woodland. However, other factors such as canopy architecture (Slocum and Horvitz, 2000), forage density (Howe, 1979) and mistletoe infection (their fruits attract frugivores) may also influence seed-deposition patterns of montane forest trees. Future studies should include these factors and examine the spatial pattern of seed deposition.

Seedling establishment of montane forest trees in miombo woodland

Seed arrival is the first step in forest tree encroachment into other vegetation types. The establishment of forest trees after seed arrival is limited by many factors such as repeated fires (Hoffmann, 2000), competition with grasses (Holl, 2002), water stress (Bowman and Panton, 1993), low soil fertility (Kellman and Miyanishi, 1982), seed predation by rodents (Holl, 2002) and browsing by antelope (Shararm et al., 2009). In this miombo wood-land, the establishment success of montane forest trees seems to be facilitated by the presence of *B. floribunda* (Figure 4). Potential mechanisms that may allow montane forest trees to establish in miombo woodlands are presented as follows:

Amelioration of water stress can be important in esta-blishing montane forest trees in miombo woodland (Holl, 1999). Lower canopy openness and air temperature and higher RH in tall-tree microsites, compared to the open microsite (Table 1) can reduce leaf temperature and transpiration loss, consequently, inducing a more favoura-ble water balance in tree seedlings. These modifications could be critical for seedlings to withstand drought, even without improving soil water conditions, as found by Gomez-Aparicio et al. (2005). Reduced grass cover in the tall-tree microsite (Table 1) may also have a positive effect on the establishment of montane forest trees. For example, less grass cover may decrease the occurrence of fire.

Hennenberg et al. (2006) showed that fires occur frequently with high grass cover, but rarely with low grass cover. Suppressed fire occurrence is crucial for the survi-val of montane forest trees, which have a low resistance to fire (Lawton, 1978; Kikula, 1986). In Africa, including this region, fire-exclusion experiments have resulted in closed forest stands (Swaine et al., 1992; White et al., 2001). Low grass cover may also reduce competition for seedlings. Several studies in tropical regions have shown that competition with grasses reduces seedling survival and growth (Holl, 2002; Sharam et al., 2009). The present

study suggests that tall *B. floribunda* trees facilitate not only seed arrival, but also the establishment of montane forest trees in miombo woodland.

Unexpectedly, aboveground microclimatic factors and soil water content did not differ significantly between the tall- and short-tree microsites (Table 1). However, the proportion of grass cover tended to be higher in the short-tree microsite (Table 1). The difference in the proportion of grass cover can critically influence the establishment of montane forest trees, as high grass cover induces some negative effects, as described earlier.

ACKNOWLEDGMENTS

I thank Dr. Kazuharu Mizuno for a critical review of this manuscript, the Forestry Research Institute of Malawi for their helpful advice in the field. This research was funded by the Japan Society for the Promotion of Science Global COE Program (E-04): In Search of Sustainable Humanosphere in Asia and Africa.

REFERENCES

Aukema JE, Martínez del Rio C (2002). Where does a fruit-eating bird deposit mistletoe seeds? seed deposition patterns and an experiment. Ecology 83:3489-3496.

Banfai DS, Bowman D (2007). Drivers of rain-forest boundary dynamics in Kakadu National Park, northern Australia: a field assessment. J. Trop. Ecol. 23:73-86.

Bowman D, Murphy BP, Banfai DS (2010). Has global environmental change caused monsoon rainforests to expand in the Australian monsoon tropics? Landscape Ecol. 25:1247-1260.

Bowman D, Panton WJ (1993). Factors that control monsoon-rain-forest seedling establishment and growth in north Australian Eucalyptus savanna. J. Ecol. 81:297-304.

Campbell B, Frost P, Byron N (1996). Miombo woodlands and their use: overview and key issues. In: Campbell B (ed.), The Miombo in transition: woodlands and welfare in Africa. Center for International Forestry Research, Bogor. pp. 1-5.

Carriere SM, Andre M, Letourmy P, Olivier I, McKey DB (2002). Seed rain beneath remnant trees in a slash-and-burn agricultural system in southern Cameroon. J. Trop. Ecol. 18:353-374.

Chapman JD (1970). PART II Description of the forest. In: Chapman JD, White F (eds.), The evergreen forests of Malawi. Commonwealth Forestry Institute, Oxford. pp. 113-180.

Chidumayo E, Frost P (1996). Population biology of miombo trees. in: Campbell B (ed.), The Miombo in transition: woodlands and welfare in Africa. Center for International Forestry Research, Bogor. pp. 59-72

Clark CJ, Poulsen JR, Conno EF, Parker VT (2004). Fruiting trees as dispersal foci in a semi-deciduous tropical forest. Oecologia 139:66-75.

Dowsett-Lemaire F (1988). The forest vegetation of the Nyika Plateau (Malawi-Zambia): ecological and phenological studies. Bull. Jard. Bot. Nat. Belg. Bull. Nat. Plantentuin Belg. 55:301-392.

Frazer GW, Canham CD, Lertzman KP (1999). Gap Light Analyzer (GLA), Version 2.0: Imaging software to extract canopy structure and gap light transmission indices from true-colour fisheye photographs, users manual and program documentation, Simon Fraser University, Burnaby, British Columbia, and the Institute of Ecosystem Studies, Millbrook, New York.

Friis I (1992). Forests types and forest trees of northeast tropical Africa, Royal Botanic Garden, London. p.396.

Gomez-Aparicio L, Gomez JM, Zamora R, Boettinger JL (2005). Canopy vs. soil effects of shrubs facilitating tree seedlings in

Mediterranean montane ecosystems. J. Veg. Sci. 16:191-198.

Hennenberg KJ, Fischer F, Kouadio K, Goetze D, Orthmann B, Linsenmair KE, Jeltsch F, Porembski S (2006). Phytomass and fire occurrence along forest-savanna transects in the Comoe National Park. Ivory Coast. J. Trop. Ecol. 22:303-311.

Hoffmann WA (2000). Post-establishment seedling success in the Brazilian Cerrado: A comparison of savanna and forest species. Biotropica 32:62-69.

Holl KD (1999). Factors limiting tropical rain forest regeneration in abandoned pasture: Seed rain, seed germination, microclimate, and soill. Biotropica 31:229-242.

Holl KD (2002). Effect of shrubs on tree seedling establishment in an abandoned tropical pasture. J. Ecol. 90:179-187.

Howe HF (1979). Fear and frugivory. Am. Nat. 114:925-931.

Kellman M, Miyanishi K (1982). Forest seedling establishment in neotropical savannas - observations and experiments in the mountain pine ridge savanna, Belize. J. Biogeo. 9:193-206.

Kikula IS (1986). The influence of fire on the composition of Miombo woodland of SW Tanzania. Oikos 46:317-324.

Lasky JR, Keitt TH (2012). The effect of spatial structure of pasture tree cover on avian frugivores in eastern Amazonia. Biotropica 44:489-497.

Lawton RM (1978). A study of the dynamic ecology of zambian vegetation. J. Ecol. 66:175-198.

McDonnell MJ (1986). Old field vegetation height and the dispersal pattern of bird-disseminated woody-plants. Bull. Torrey Bot. Club 113:6-11.

Mitchard ETA, Saatchi SS, Gerard FF, Lewis SL, Meir P (2009). Measuring woody encroachment along a forest-savanna boundary in central Africa. Earth Interact. 13:1-29.

Nepstad DC, Verissimo A, Alencar A, Nobre C, Lima E, Lefebvre P, Schlesinger P, Potter C, Moutinho P, Mendoza E, Cochrane M, Brooks V (1999). Large-scale impoverishment of Amazonian forests by logging and fire. Nature 398:505-508.

R Development Core Team (2011). R: A language and environment for statistical computing. R foundation for Statistical Computing, Vienna, Austria (http:// www.R-project.org.).

Rolhauser A, Chaneton E, Batista W (2011). Influence of conspecific and heterospecific adults on riparian tree species establishment during encroachment of a humid palm savanna. Oecologia 167:141-148.

Roxburgh L, Nicolson SW (2008). Differential dispersal and survival of an African mistletoe: does host size matter? Plant Ecol. 195:21-31.

Shararm GJ, Sinclair ARE, Turkington R, Jacob AL (2009). The savanna tree Acacia polyacantha facilitates the establishment of riparian forests in Serengeti National Park, Tanzania. J. Trop. Ecol. 25:31-40.

Slocum MG, Horvitz CC (2000). Seed arrival under different genera of trees in a neotropical pasture. Plant Ecol. 149:51-62.

Swaine MD, Hawthorne WD, Orgle TK (1992). The effects of fire exclusion on savanna vegetation at Kpong, Ghana. Biotropica 24:166-172.

Trapnell CG (1959). Ecological results of woodland burning experiments in Northern Rhodesia. J. Ecol. 47:129-168.

White F (1983). The vegetation of Africa, Unesco, Paris. p. 356.

White F, Dowsett-Lemaire F, Chapman JD (2001). Evergreen forest flora of Malawi, Royal Botanic Gardens, London. p. 697.

8

Habitat structure of flat-headed cusimanse (*Crossarchus platycephalus*) in Futa Wildlife Park, Ondo state, Nigeria

Oguntuase B. G. and Agbelusi E. A.

Department of Ecotourism and Wildlife Management, Federal University of Technology Akure, Ondo State, Nigeria.

The study explored the habitat structure of flat-headed cusimanse (*Crossarchus platycephalus*) in Federal University of Technology, Akure (FUTA) Wildlife Park, Ondo State. It was undertaken during both wet and dry seasons of the year 2012. In assessing the habitat, ten plots (10 x 10 m^2 each) were marked out of the total land area of the Park. A total of fifteen (15) different species were identified with diameter at breast height values ranging from 4 to 130 cm. The frequency of tree species contained in the plots revealed *Funtumia elastica* to be of the highest frequency. *Elaeis guineensis* emerged as the highest dominating woody plant. The habitat evaluation revealed that, the species of the flat-headed cusimanse are adapted to fairly dense vegetation and preferred to utilize fallen logs, thicket of bush, environments with considerable wetness with pond or stream and crevices formed by fallen logs and rock outcrops as habitats. The study has given a base line information on where flat-headed cusimanses are likely to be seen, places and features they utilize as habitat in order to ensure easier ecological studies on them and to see these features not as useless and nuisance especially the fallen, dead logs, the thickets and the crevices, but as principal components of some animal's habitat.

Key words: Crevice, dominance, diversity, dwarf mongoose, requirements, patrol.

INTRODUCTION

Very little information is given on flat-headed cusimanse in Nigeria. Nigeria is one of the African countries to which this specie is endemic and FUTA Wildlife Park is one of the areas where this animal is found in about three different colonies with no documentation, whatsoever, on their habitat in this park. Therefore, baseline knowledge is very much needed on the habitat of the animal for further ecological studies.

The flat-headed cusimanse (*Crossarchus platycephalus*) is a dwarf mongoose endemic to Benin, Cameroon and Nigeria. This species was once regarded as a subspecies of the common cusimanse (*Crossarchus*

obscurus) (Goldman and Hoffmann, 2008). It occurs in rainforests of Benin, Nigeria, Cameroon, Equatorial Guinea, Congo Republic and Central African Republic (Wilson and Reeder, 2005). The specific locality of a specimen collected in Benin is uncertain and given as "Agouagou" which is well beyond the rainforest zone (Goldman in press). Habitat ecology is an important aspect of ecology, dealing with the study of organism in relation to its environment. Habitat is defined as an area that provides the food, water, cover and space that a living thing needs to survive and reproduce. The quality and quantity of a particular type of habitat determines the

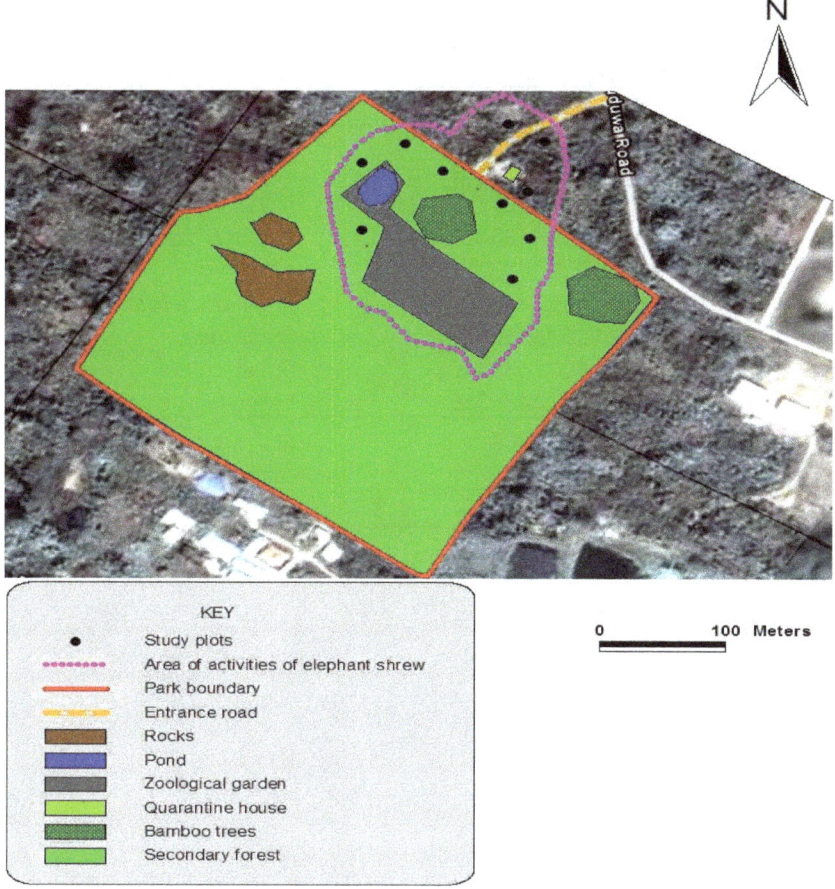

Figure 1. Map of FUTA Wildlife Park, showing the studied plots and the distribution of the animal.

number and variety of its inhabitants (Esbjorn-Hargens, 2005).

Every wildlife species requires a general environment to live. To properly manage land for the benefit of wildlife, there is a need to understand those things in the environment that wild animal need to survive and reproduce, the environment or natural home where a wild animal lives is called its habitat, just like humans, wild animals have specific requirements that they have at home. Habitat for any wild animal must provide: cover (shelter) from weather and predators; food and water for nourishment and space to obtain food, water and to attract a mate (Greg, 2009). While shelter, food and water are basic requirements, how wildlife obtain these requirements varies (Greg, 2009). The habitat of a species describes the environment over which a species is known to occur and the type of community that is formed as a result (Whittaker et al., 1973).

More specifically, habitats can be defined as regions in environmental space that are composed of multiple dimensions, each representing a biotic or abiotic environmental variable; that is, any component or

characteristic of the environment related directly (forage biomass and quality) or indirectly (elevation) to the use of a location by the animal (Beyer, 2010).

MATERIALS AND METHODS

Study area

This study was carried out in the Wildlife Park of The Federal University of Technology, Akure. The Wildlife Park is located at Akure, the state capital of Ondo State, Nigeria. The state lies between latitudes 50 45' and 70 52'N and longitudes 4020' and 60 05'E. Its land area is about 15,500 km^2 (UNAAB-IFSERAR, 2010).

FUTA Wildlife Park

The park covers a total area of 89,100 m^2 (8.91 ha), it is a lowland tropical rainforest found on longitude 7°C, 21 min north and latitude 5°C, 20 min with an average rainfall of 1650 to 1700 mm annually. Study area is located on elevation 1200 m above sea level (Afolayan and Agbelusi, 1987). The study area is under laid with crystalline basement rock which imposes a partially rugged topographic relief on the area (Figure 1). The lower elevation is

Table 1. The GPS readings and other features of each plot.

Plot	GPS reading	Other features
Plot 1	07.29627°N, 005.14332°E	Thicket of bush, fallen logs
Plot 2	07. 29628°N, 005.14350°E	Fallen logs
Plot 3	07.29584°N, 005.14366°E	Small rocks, fallen log
Plot 4	07.29551°N, 005.14281°E	Fallen logs, termite hill and small rocks
Plot 5	07.29607°N, 005.14234°E	*Aspilia africana*, stream
Plot 6	07.29627°N, 005.14332°E	-
Plot 7	07.29638°N, 005.14305°E	Small scattered rocks
Plot 8	7.29695°N, 005.14330°E	Small rocks
Plot 9	07.29693°N, 005.14335°E	Small rocks
Plot 10	07.29656°N, 005.14353°E	Fallen logs, small rocks

Table 2. Tree species composition of the randomly selected plots.

Scientific name	Common name	Local name (Yoruba)	Average DBH (cm)
Ficus capensis	African Mustard tree	Opoto	25
Antiaris africana	Antiaris	Oriro	14
Funtumia elastica	Wild rubber	Ire	17
Newbouldia laevis	Tree of life	Akoko	6.75
Elaeis guineensis	Red oil Palm	Igi ope	72
Alchornea laxiflora	Three-veined bead string	Pepe	8
Zanthoxylum zanthoxyloides	Lecaniodiscus	Ata	56
Lecaniodiscus cupanioides	Sapindaceae	Aka	6.8
Ficus exasperata	Sand paper tree	Epin	15.5
Milicia excelsa	Iroko	Iroko	14
Ceiba pentandra	White silk cotton	Araba	130
Alchornea cordiflora	Christmas bush	Esin	5.25
Piper guineensis	Piper	Ayere	51
Spondias mombin	Hog plum/yellow mombin	Ekikan	11.8

about 95 m above sea level while the higher elevation is above 140 m above sea level. The terrain of Federal University of Technology Akure, Wildlife Park can be described as undulating with small outcrop scattered about. The elevation of the park varies between 215 and 320 m (Afolayan and Agbelusi, 1987).

Flora and fauna

The study site is one of the vegetation typical of secondary forest with herbaceous undergrowth. The vegetation is a combination of tropical trees, shrubs and herbaceous plants in great diversity such as: *Tetrapleura tetraptera, Trichilia emetic, Newbouldia laevis, Jatropha gossypifolia, Aframomum melegueta, Elaeis guineensis, Diospyros* spp., *Khaya ivorensis, Milicia excelsia, Aspilia Africana* among others (Abu, 2010).

The park is dominated with a large diversity of rodents, though other families of animal do exist. The fauna resources in general include: bush buck, duicker, grasscutter, giant rat, squirrel and rock python (Idowu, 2010).

Selection of site and sampling

In assessing the habitat of flat-headed cusimanse in FUTA Wildlife

Park, ten different points were marked out, the GPS readings and other features of the selected plots are given in Table 1. The ten locations were sampled following the procedure of Barbour et al. (1999) using area sampling after it has been modified by designing the quadrat in such a way that it considered the scenes of activities of the animal, for this study, 10 × 10 m square quadrat was used, in this way, the activities of the animal was noticed, and possible routes of considerable distances were well accommodated within the quadrats.

Total enumeration of the tree species in each location was assessed. The diameter at breast height (DBH) of individual tree was taken using measuring tape and recorded; as well as the occurrence of each species. From the data obtained, density, relative density, frequency and dominance were calculated. Other features of importance in the plots such as: rock outcrops, fallen logs, termite hills and bush thickets were carefully observed. The plants utilized by the animal in each plot were assessed.

Direct/indirect observation of the animal

Direct and indirect observation of the animal was done concurrently. The animals were directly observed thrice in a week for a period of 4 months each during the wet and dry season of the year. This was done by quietly moving around the selected plots, and intermittently waiting and listening, for any available sign, this was done to determine their presence and activities. The animals were observed from 08:00 to 12:00 h in the morning and 16:00 to 19:00 h in the evening. Flat-headed cusimanse were indirectly observed in FUTA Wildlife Park by looking out for activities of the animals in the plots earlier marked out. Traces of movement, habitation, foot prints and droppings among others were carefully looked out for and observed.

Data analysis

The analysis of data was done using the methods described by Brower et al. (1998) as follows:

Density (Individual/M^2): The number of individuals within a chosen area (m^2 ha):

$$\text{Density} = \frac{\text{Number of individuals of a species}}{\text{Area sampled}}$$

Relative density: The density of one species as a percentage of total density:

$$\text{R.D} = \frac{\text{Density for a species}}{\text{Total density for all species}}$$

Frequency: The percentage of total quadrats or points that contains at least one individual of a given species:

$$\text{Frequency} = \frac{\text{No. of quadrats in which species occur}}{\text{Total number of quadrat sampled}}$$

Relative frequency: The frequency of one species as a percentage of total frequency:

$$\text{R.F} = \frac{\text{Frequency value for a species}}{\text{Total frequency value for all species}}$$

$$\text{Dominance} = \frac{\text{Total basal area value for a species}}{\text{Area sampled}}$$

RESULTS

The studied plots contained fifteen different species of plant as shown in Table 2. In all the plots studied, *Funtumia elastica* had the highest occurrence, while *Ceiba pentandra, Ficus exasperata* and *Milicia exelsa* had the lowest occurrence as shown in Table 3.

Table 1 gives the global positioning system readings of each plot and shows other features of importance being utilized by the animal such as: thick blanket of bush, fallen logs of wood, decaying logs, small rock outcrops, *A. africana* vegetation, small pond/stream and termite hill. The density and the relative density of the plots studied as shown in Figure 2 revealed plot 8 as the most dense with the value of 0.12 tree/m^2, followed by plot 4 with the

value of 0.10 tree/m^2 and plot 7 with the value of 0.08 tree/m^2. The plots were sparsely vegetated with tree species, though in some plots, the vegetation cover was enough to give limited light penetration. Apart from the tree cover, herbaceous plants also gave the kind of dense vegetation required by the animal most especially for dashing away from threat. Table 3 also shows the dominant species of each plot. *F. elastica* and *E. guineensis* are the most dominant species of the total plots studied.

DISCUSSION

The habitat of flat-headed cusimanse as described by Angelici et al. (1999) is similar to the habitat of the animal in FUTA Wildlife Park which is characterized by certain plant species which are typical of rain forest zone such as: *Ficus capensis, Antiaris africana, F. elastica, Newbouldia laevis, Milicia excelsa* and *C. pentandra*. These tree species have their DBH ranging from 1.5 to about 115 cm with *C. pentandra* being of the highest DBH. Most of which have good percentage of canopy cover and give a dense and cool environment as required by the animal. The habitat utilized by the animal also include: thickets, rocky outcrops, fallen and dead logs, hollows, burrows and crevices for protection, resting and mating; and fruiting plants such as *E. guineensis, Alchornea cordiflora* and non- fruiting plant, such as *A. africana* utilized as food. The density and relative density of the studied plots showed and described the preferred habitat of the animal, which means that the animal utilizes areas with considerable wood density in the primary forest region as studied in FUTA Wildlife Park. These areas are so preferred by the animal probably because some of the woody plants apart from providing cool environment also possess fruits on which the animal feeds, and the non-woody vegetation protects against predators, so also are crevices and fallen logs of wood which make the animal prefers these selected areas to the rest of the park. Their activities are limited to these areas as shown in Figure 1.

In all the ten plots selected for study, as the animal's habitat, *E. guineensis* dominated 3 plots thereby emerging as the best indicator of the animal's habitat. This animal has its activities pronounced in those plots where *E. guineensis* are present. The distribution of the animal is restricted to the areas where their requirement for food, water and protection are met.

Conclusion

Flat-headed cusimanse based on this study are known to utilize areas with considerable wetness, cool dense environment with limited light penetration as habitat.

Table 3. Tree species contained in individual studied plot.

Tree species	Plot 1	Plot 2	Plot 3	Plot 4	Plot 5	Plot 6	Plot 7	Plot 8	Plot 9	Plot 10
Ficus capensis	+	−	+	−	−	−	−	+	+	−
Antiaris africana	+	−	−	−	−	+	−	−	+	+
Funtumia elastica	++(60)	+	+	+	−	−	++(64)	+	++(59)	+
Newbouldia laevis	−	+	−	+	−	−	−	−	−	−
Elaeis guineensis	−	++(98)	−	+	++(98)	−	−	++(82)	−	+
Alchornea laxiflora	−	+	−	−	−	−	+	−	−	+
Zanthoxylum zanthoxyloides	−	+	−	−	−	−	+	−	−	−
Lecaniodiscus cupanioides	−	−	++(78)	−	−	−	−	−	−	++(53)
Ficus exasperata	−	−	−	+	−	−	−	−	−	−
Melicia excels	−	−	−	+	−	−	−	−	−	−
Ceiba pentandra	−	−	−	++(52)	−	−	−	−	−	−
Alchornea cordiflora	−	−	−	−	+	+	−	−	−	−
Piper guineenseis	−	−	−	−	−	++(84)	+	+	−	−
Spondias mombin	−	−	−	−	−	−	+	+	−	+

+: Present; ++: dominant with rate (%) in parenthesis; -: absent.

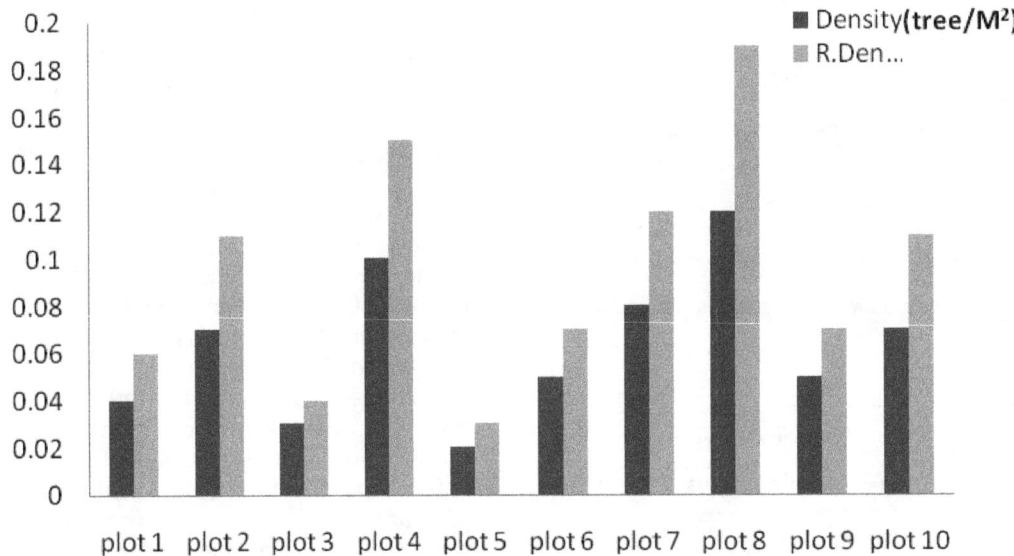

Figure 2. Density (tree/m^2) and relative density of the studied plots.

Certain species of tree that might probably be indices of the animal's habitat include: *E. guineensis, A. laxiflora, F. elastica, N. laevis, F. exasperata, M. excelsa, C. pentandra, Piper guineensis, Spondias mombin, Ficus* spp. and *A. Africana*, while the presence of *E. guineensis* will most probably indicate the presence of the animal anywhere. Some of these tree species are fruiting plants desired by the animal while some provide the canopy cover and escape route required by the animal.

This study has given base line information on the habitat of flat-headed cusimanse in a secondary forest of FUTA Wildlife Park.

REFERENCES

Abu MI (2010). Ethnobotanical study of Federal University of Technology, Akure (FUTA) Wildlife Park. A thesis in the Department of Ecotourism and Wildlife Management, pp. 31-35.

Afolayan TA, Agbelusi EA (1987). A feasibility report on FUTA and its botanical garden (A paper presented to FUTA).

Angelici FM, Grimod I, Politano E (1999). Mammals of the Eastern Niger Delta (Rivers and Bayelsa States, Nigeria): An environment affected by a gas-pipeline. *Folia Zoologica* 48: 249-264.

Barbour MG, Burk JH, Pitts WD, Gilliam FS, Schwartz MW (1999). Terrestrial plant ecology, 3rd edition. Benjamin Cummings.

Beyer HL (2010). The interpretation of habitat preference metrics under use–availability design, Phil. Trans. R. Soc. B 365 (1550): 2245–54. doi: 10.1098/rstb.2010.0083.

Brower JE, Zar JH, von Ende CN (1998). Field and laboratory methods for general ecology, 4th edition. Wm. C. Brown Co., Publishers, Dubuque, Iowa.

Esbjorn-Hargens S (2005). "Integral Ecology: An Ecology of Perspectives" (PDF). J. Integral Theory Pract. 1(1): 2–37. http://www.vancouver.wsu.edu/fac/tissot/IU_Ecology_Intro.pdf.

Greg KY (2009). Habitat Requirements of Wildlife: Food, *Water, Cover and Space. Sheet 14.* Clemson University Cooperative Extension Service.

Goldman C, Hoffmann M (2008). *Crossarchus platycephalus*. In: IUCN 2008. IUCN Red List of Threatened Species. Downloaded on 22 March 2009. Database entry includes a brief justification of why this species is of least concern.

Idowu DS (2010). Assessment of fauna composition in the Federal University of Technology, Akure, Wildlife Park'. A thesis in the Department of Ecotourism and Wildlife Management. Pp. 21-26.

UNAAB-IFSERAR (2010). Institute of food security, Environmental Resources and Agricultural Research. 'Ondo State Diagnostic Survey Report, December, 2009.

Whittaker RH, Levin SA, Root RB (1973). Niche, Habitat, and Ecotope. Am. Nat. 107 (955): 321–338. doi:10.1086/282837. JSTOR 2459534.

Wilson DE, Reeder DM (2005). Mammal Species of the World, A Taxonomic and Geographic Reference (Eds.). (3rd ed.). Johns Hopkins University Press, Baltimore, Maryland.

9

Composition of understory vegetation in tree species of Cholistan desert, Pakistan

Muhammad Farrukh Nisar[1] , Farrukh Jaleel[2], Muhammad Waseem[3], Sajil Ismail[4] and Muhammad Arfan[5]

[1]College of Bioengineering, Chongqing University, Chongqing 400044, China.
[2]College of Chemistry and Chemical Engineering, Chongqing University, Chongqing 400044, China.
[3]Department of Biology, Allama Iqbal Open University (AIOU), Islambad (44000), Pakistan.
[4]Department of Botany, Govt. Sadiq Egerton (SE) College, Bahawalpur (63100), Pakistan.
[5]Department of Biology, Lund University, Sweden.

In the present study the understory vegetation or communities under the canopy of tree species of Cholistan desert (27° 42' and 29° 45' North and longitudes 69° 52' and 75° 24' East) and the effect of canopy cover on the soil chemistry were studied. Quadrats of 1 m^2 were laid to record the different plant species underside the canopy cover of tree species and then frequency, density and plant cover were recorded, whereas relative frequency, relative density, relative cover and IVI for each of the plant species were computed following the standard methods. The study was repeated for three times and field guides were used to identify the plant species. Soil analysis showed that tree species greatly modified the soil chemistry beneath the canopy cover. *Stipagrostis plumosa*, *Salsola baryosma* and *Cenchrus cilaris* were the major plant species that form the understory communities of tree species due to their close association with each other in the Cholistan desert. Other plant species also take part in forming the understory vegetation of tree species but to some less extent as they fail to form any type of association with each other.

Key words: Understory, vegetation, Cholistan desert, canopy, quadrat, IVI, soil.

INTRODUCTION

The Cholistan desert, a stretch of about 26,000 km^2, situated in the Southern part of the Punjab province, Pakistan. The vegetation is a typical of arid regions and comprises of xerophytic plant species, adapted to extreme seasonal temperature, moisture fluctuation and a wide variety of edaphic conditions (Arshad et al., 2008). Vegetation cover is comparatively better in eastern region (200 mm rainfall zone) than the hyper arid southern region (100 mm rainfall zone) (Nisar et al., 2010). The soil topography and chemical composition is playing an important role in plant distribution in the area. The association of certain plant species to certain soils at different places is very common (Arshad et al., 2008).

It has been recognized that understory biomass increases with an increase in incident solar energy (Zavitkovski, 1976). In fact, the understory layer is affected by a reduction of light, when canopy is closed and stratified and by nutrient availability (Gilliam and Turrill, 1993). The different nutrient requirements of the species in a stand and their relative contributions to the chemistry litter affect soil nutrient availability and other soil properties such as pH and net nitrification (Ste-Marie and Pare, 1999). The relationship between overstory and understory is complex and also dependant on permanent site factors such as surface deposit. It has been demonstrated that canopy type exert its influence on nutrient availability, affects

Table 1. Soil analysis collected from underside the canopies of tree species of Cholistan desert, Pakistan.

Soil property	*Acacia jacquemontii*	*Prosopis juliflora*	*Prosopis cineraria*	*Tamarix aphylla*	*Acacia nilotica*
pH	8.6	8.2	8.3	8.6	8.4
E.C. (ds/m)	1.26	1.62	1.31	4.57	1.21
P (ppm)	5.5	6.2	4.9	5.03	5.2
K (ppm)	109.3	123.3	105.6	117.3	113.0
Ca (meq/L)	0.60	0.98	0.69	1.02	0.69
Mg (meq/L)	0.27	0.45	0.38	3.09	0.47
Na (meq/L)	0.39	0.19	0.23	0.596	0.05
O.M. (%)	0.28	0.34	0.25	0.28	0.30
Texture	Sandy loam	Sandy loam	Sandy loam	Sandy loam	Sandy loam

understory composition (Legare et al., 2001). Communities in fact are the mirrors of landmass or indicators of land's biological resources and based upon this information. The plant communities are considered as reliable indicators of environments and economic potentialities of the area (Arshad and Akbar, 2002). The association of certain plant species to certain soils at different places is very common and different plant communities at different soil types have been identified (Arshad et al., 2007, 2008). Arshad (2003) has reported the major parameters that, are responsible for vegetation distribution in Cholistan desert are the salinity and pH. While Rao et al. (1989) are of the view that phytosociological groups are determinant of the soil types as the edaphic factors influence the vegetation more than any other factor. By reviewing the above reports the present study was conducted. The principal objective of this study was to determine the impacts of tree species on composition of understory vegetation and the chemistry of soil under it.

MATERIALS AND METHODS

The study was conducted in Cholistan desert (27° 42' and 29° 45' North and longitudes 69° 52' and 75° 24' East) on 10 different sites to determine the understory vegetation composition of five plant species namely *Acacia jacquemontii*, *Prosopis cineraria*, *Prosopis juliflora*, *Tamarix aphylla* and *Acacia nilotica*. Quadrat method was used to record understory vegetation (Oosting, 1957; Chul and Moody, 1983). Three quadrats measuring 1 x 1 m^2 were randomly taken from underside of the canopy of each plant species. The study was repeated for three times. The plant species were identified in the field with field guide (Rao et al., 1989). Frequency, density and plant cover were recorded, whereas relative frequency, relative density and relative cover for each of the plant species was computed following the methods described by Hussain (1989). Importance value index (IVI) was calculated by the direct summation of relative frequency, relative density and relative cover of each plant species (Curtis and McIntosh, 1951; Muller-Dombois and Ellenberg, 1974; Chul and Moody, 1983). The species having the highest IVI were considered as the leading dominant plant species of understory vegetation.

Soil samples from underside of each plant were also taken and were transferred to Soil Testing Laboratory, for complete soil assay to judge the effect of plant canopy on the soil enrichment. Chemical analysis conducted included soil pH, organic matter (Walkley-Black method), Phosphorus (Bray-2 method), and exchangeable cations,

namely: sodium (Na), potassium (K), calcium (Ca) and magnesium (Mg). The exchangeable cations were extracted by leaching the sample soil with normal ammonium acetate and analyzed spectrophotometrically. The soil chemical analyses were done in triplicate to verify the results.

RESULTS AND DISCUSSION

The results of soil samples collected from underside the canopies of *P. cineraria*, *P. juliflora*, *T. aphylla* and *A. nilotica* are presented in Table 1. The texture of the soil was sandy and the soil was found to be moderately alkaline with the highest pH 8.6 in the soil collected from underside the canopy of *T. aphylla*, while minimum pH 8.2 was recorded in the soil samples collected from underside the canopy cover of *P. juliflora* (Table 1). The pH of the soil samples collected from underside the canopies of *A. nilotica* and *P. cineraria* were 8.4 and 8.3, respectively.

The electrical conductivity (E.C.) was maximum (4.57 ds/m) in the soil samples from underside the canopy of *T. aphylla*. This increase of E.C. may be due to addition of salt accumulated leaves of this plant. Minimum E.C. 1.21 ds/m was recorded in the soils from underside the canopy of *A. nilotica*. The concentration of E.C. in the soils collected from underside the canopy *P. juliflora* and *P. cineraria* was 1.31 ds/m and 1.26 ds/m.

The soil samples collected from underside the canopy of trees in Cholistan desert were analyzed for soluble ions (Table 1). The maximum phosphorus contents (6.2 ppm) were recorded in the soil collected from underside the canopy cover of *P. juliflora*, while minimum phosphorus contents (4.9 ppm) was noted in the soil sample from underside the canopy of *P. cineraria*. Phosphorus contents in the soil underside the canopy of *T. aphylla* and *A. nilotica* was 5.03 ppm and 5.2 ppm, respectively. Similar results were observed for Potassium contents in soil samples. Maximum potassium contents (123.3 ppm) were recorded in soil samples underside the canopy of *P. juliflora* and minimum 105.6 ppm underside the canopy of *P. cineraria*. While potassium contents in soil samples collected from underside the canopy cover of *T. aphylla* and *A. nilotica* was 117.3 ppm and 113.0 ppm.

The results of the exchangeable cations (Na, Mg, Ca) measured in the different soil samples collected from underside the canopy cover of *P. juliflora*, *P. cineraria*, *T. aphylla* and *A. nilotica* varied greatly (Table 1). The maximum amount of these exchangeable cations was noted in the soil underside the canopy of *T. aphylla* namely: Ca (1.02 meq/L), Mg (3.09 meq/L) and Na (0.596 meq/L), while minimum amount of Ca (0.69 meq/L) was recorded in the soils from *A. nilotica* and *P. cineraria*. Calcium contents in the soils from the underside canopy cover of *P. juliflora* was 0.98 meq/L. Magnesium ions in soil samples from underside the canopy cover of *P. cineraria* were 0.38 meq/L, *P. juliflora* 0.45 meq/L and *A. nilotica* 0.47 meq/L. Minimum amount of Na (0.05 meq/L) was recorded in soil sample from underside the canopy of *A. nilotica* while soil samples from underside the canopy cover of *P. juliflora* and *P. cineraria* showed 0.19 and 0.23 meq/L of Na ions, respectively.

The accumulation of organic matter was meager in all the soil samples collected underside the canopies of studied tree species (Table 1). Highest organic matter (0.34 %) was recorded in the soil from underside the canopy cover of *P. juliflora* and minimum (0.25%) under *P. cineraria*. In the soil from *A. nilotica* and *T. aphylla* organic matter was 0.30 and 0.28%, respectively.

Plant density

Number of plant species recorded under the canopy of *P. juliflora*, *P. cineraria*, *T. aphylla* and *A. nilotica* varied greatly from one plant species to other species (Figure 1). Under the canopy cover of *P. juliflora* nine plant spe-cies were recorded. *S. plumosa* appeared as the domi-nant plant species having maximum plant density (23.55), whereas minimum plant density (0.11) was recorded by *Panicum antidotale*, *Lasiurus scindicus* and *Leptadenia pyrotechnica*. The density of *Cenchrus ciliaris* was 2.22, *Haloxylon salicornicum* 2.11, *Salsola baryosma* 1.88, *Tribulus terristris* 0.44 and *Suaeda fruiticosa* 0.22. Eleven plant species were recorded under the canopy of *P. cineraria*. *S. plumosa* again appeared as dominant plant having maximum density (15.00), while the lowest plant density was attained by *Launea procumbens* (0.11). Plant density of *C. ciliaris* and *S. baryosma* was 10.77 and 9.33. However the density in other plant species ranged between 0.22 to 1.00.

Thirteen plant species were recorded growing under the canopy of *T. aphylla*. Maximum density (12.77) was scored by *S. plumosa* closely followed by *S. baryosma* (11.20). Minimum plant density was attained by *P. cineraria* (0.11). Density of other plant species including *Chenopodium album*, *Sonchus asper*, *L. scindicus*, *P. antidotale*, *Alhaji morarum*, *Launea nudicaulis*, *C. ciliaris*, *Ochthocloa compressa*, *L. procumbens* and *Cynodon dactylon* ranged from 0.22-8.11. Under the canopy cover of *A. nilotica* thirteen plant species were noted. Out of which *S. plumosa* attained the highest density (6.66). Both *Calligonum poly-*

gonoides and *H. salicornicum* scored minimum plant density (0.11). The range of plant density in *P. juliflora*, *L. pyrotechnica*, *S. fruiticosa*, *L. nudicaulis*, *L. scindicus*, *A. nilotica*, *O. compressa*, *S. baryosma*, *L. procumbens* and *C. ciliaris* was 0.22 to 5.88.

Plant frequency percentage

Plant frequency recorded under the canopy of *P. juliflora*, *P. cineraria*, *T. aphylla* and *A. nilotica* varied from one species to the other (Figure 2). Under the canopy of *P. juliflora*, *S. baryosma* and *C. ciliaris* appeared as commonly occurring plants having maximum plant frequency (55.55%). Whereas minimum plant frequency (11.11%) was recorded by *L. scindicus*, *L. pyrotechnica*, *P. antidotale*, *S. fruiticosa* and *T. terristris*. Frequency of *Stipagrostis plumosa* and *H. salicornicum* was 33.33 and 44.44%. Underside the canopy cover of *P. cineraria* highest plant frequency was noted by *S. baryosma* and minimum plant frequency (11.11%) was recorded by *L. procumbens*, *Citrulus colocynthis*, *C. setigerus*, *Sesuvium sesuvioides* and *O. compressa*. *C. polygonoides* and *C. ciliaris* scored plant frequency 22.22% and 77.78%. Plant frequency percentage varied under the canopy cover of *T. aphylla*. Maximum plant frequency (55.55%) was shown by *L. procumbens* and minimum (11.11%) was recorded by *S. plumosa*, *Chenopodium album*, *P. cineraria* and *Sonchus asper*. The range of plant frequency in other plant species was between 22.22 to 44.44%. Underside the canopy of *A. nilotica* highest plant frequency (88.89) was recorded by *C. ciliaris* and minimum (11.11%) was scored by *H. salicornicum*. *C. polygonoides*, *A. nilotica*, *L. pyrotechnica* and *P. cineraria*. Plant frequency in other plants ranged between 22.22 and 55.55%.

Importance value

Figure 3 shows the pattern of change in the importance value of understory plant species growing as common associates under the canopy cover of tree species. According to the importance value (92.75) *S. plumosa* appeared as dominant plant species, whereas *T. terristris* appeared as the rare plant with importance value 8.56. Importance value of associated plants such as *S. baryosma*, *H. salicornicum*, *C. ciliaris*, *P. antidotale*, *L. pyrotechnica*, *S. fruiticosa* and *L. scindicus* was 45.61, 41.78, 36.39, 21.64, 21.64, 15.99 and 15.63, respectively.

Under the canopy of *P. cineraria*, *S. baryosma* and *S. plumosa* appeared as the dominant plant species, scoring importance value 63.81 and 60.16. While the rare plant species under the canopy of *P. cineraria* was *L. procumbens* having importance value 3.82. The importance value of other associated plant species such as *C. ciliaris*, *P. cineraria*, *C. polygonoides*, *O. compressa*, *T. terristris*, *Ctrulus colocynthis*, *Cenchrus setigerus* and *Sessuvium sessuvioides* was 50.03, 42.88, 32.58, 13.44, 10.58, 8.59, 7.44 and 4.69, respectively.

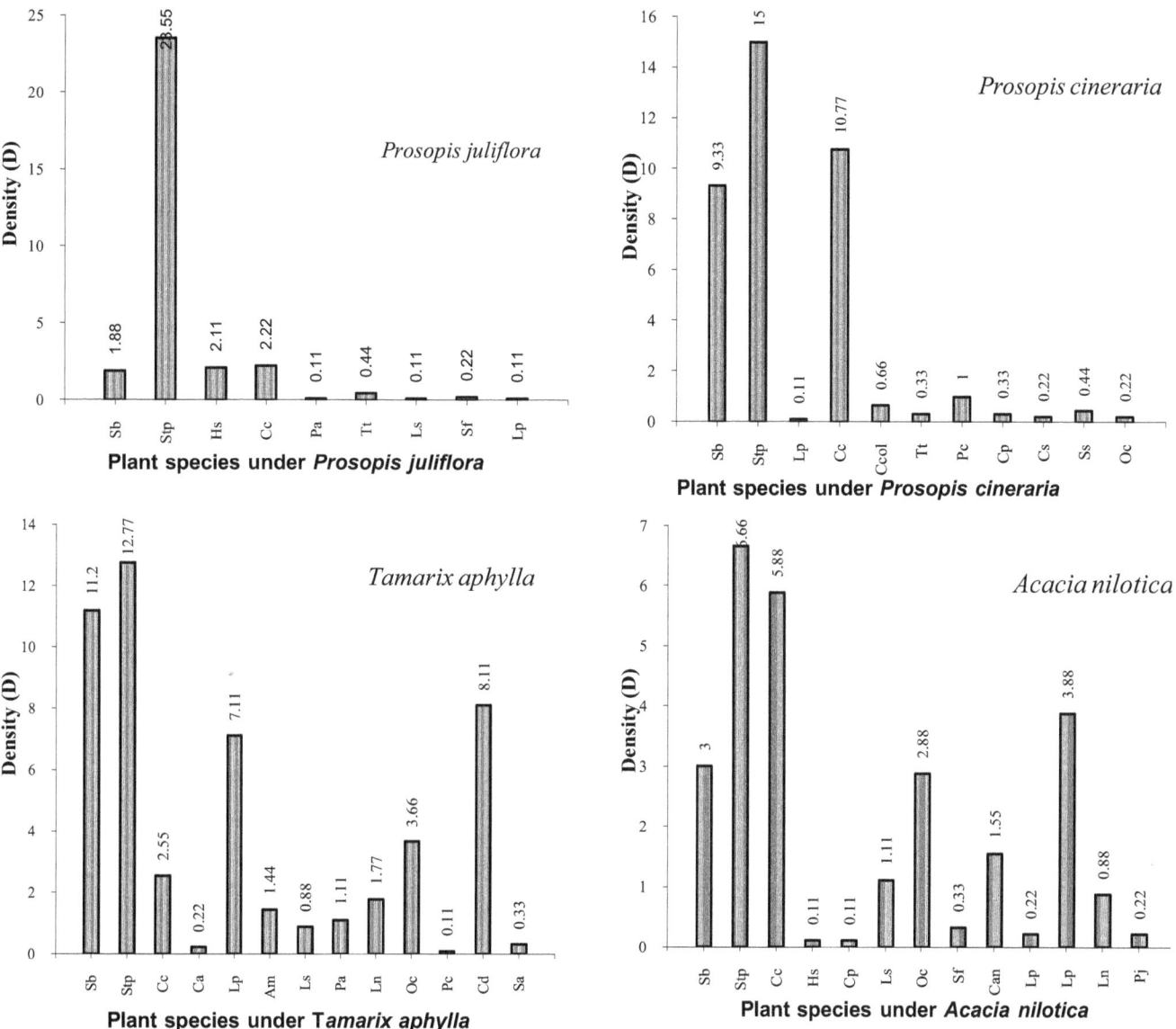

Figure 1. Density of the plant species forming the under-story vegetation of major tree species of Cholistan desert. Sb, Salsola baryosma; Stp, Stipagrostis; Hs, Haloxylon salicornicum; Cc, Cenchrus ciliaris; Aj, Aerva javanica; Tt, Tribulus terristris; Acj, Acacia jacquemontii; Cp, Calligonum polygonoides; Dg, Dipterygium glaucum; Pa, Panicum antidotale; Ls, Lasiurus scindicus; Sf, Suaeda fruiticosa; Lp, Leptadenia pyrotechnica; Ccol, Citrulus colocynthis; Pc, Prosopis cineraria; Ss, Sessuvium sesuvioides; Oc, Ochthocloa compressa; Ca, Chenopodium album; Lp, Launea procumbens; Ln, Launea nudicaulis; Am, Alhagi morarum; Cd, Cynodon dactylon; Sa, Sonchus asper; Can, Acacia nilotica; Pj, Prosopis juliflora; Cs, Cenchrus setigerus.

Under the canopy of *T. aphylla*, *S. baryosma* was dominant plant with maximum importance value 63.81 while minimum (6.86) importance value was scored by *P. cineraria*. Under the canopy of *A. nilotica*, *C. ciliaris* was recorded as the dominant plant with importance value 66.19. While minimum importance value (6.84) was scored by *H. salicornicum* and *C. polygonoides*. The associated plants under the canopy of *A. nilotica* were *S. baryosma*, *S. plumosa*, *S. fruiticosa*, *L. procumbens*, *O. compressa*, *L. scindicus*, *L. nudicaulis*, *A. nilotica*, *L. pyrotechnica* and *P. juliflora* having importance value 44.02, 42.64, 28.38, 26.60, 25.74, 13.41, 12.58, 12.22, 7.24 and 7.26, respectively.

The tree species in the present study were of the same height and age but this introduces a bias while comparing the effect of the tree canopies on the understory vegetation. The composition of understory vegetation underside the canopy cover may differ from one tree species to another one as the canopy of tree species had a great influence on shrub and herb cover underside it (Klinka et al., 1996). This may be due the fact that there occurs competition among understory plant species for soil moisture, soil nutrients or may be to escape high solar radiation and heat stress to avoid excessive transpiration as there is already scarcity of water in deserts.

Soil characteristics were the most striking factors which

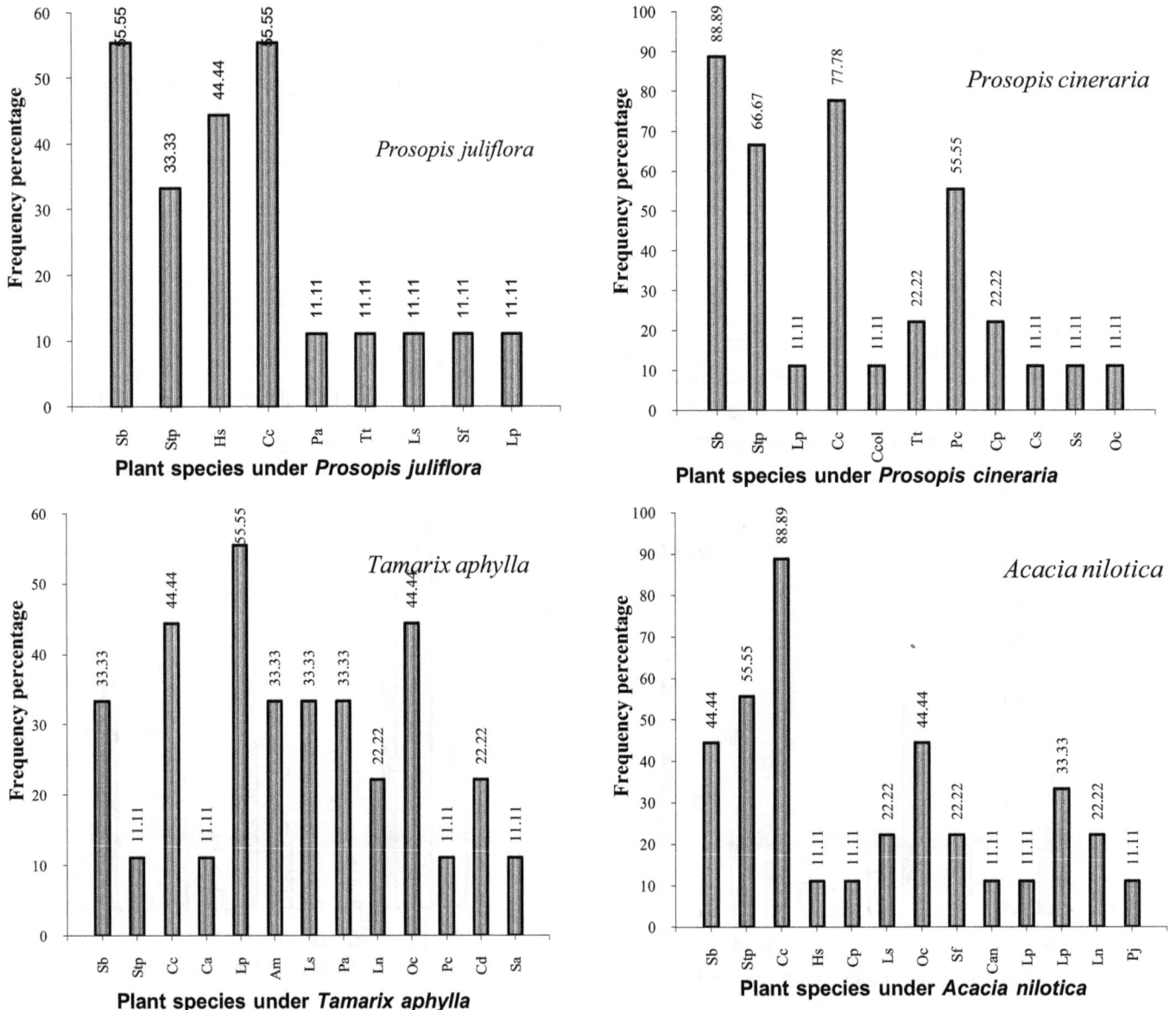

Figure 2. Frequency percentage of plant species forming the under-story vegetation of major tree species of Cholistan desert. Sb, Salsola baryosma; Stp, Stipagrostis; Hs, Haloxylon salicornicum; Cc, Cenchrus ciliaris; Aj, Aerva javanica; Tt, Tribulus terristris; Acj, Acacia jacquemontii; Cp, Calligonum polygonoides; Dg, Dipterygium glaucum; Pa, Panicum antidotale; Ls, Lasiurus scindicus; Sf, Suaeda fruiticosa; Lp, Leptadenia pyrotechnica; Ccol, Citrulus colocynthis; Pc, Prosopis cineraria; Ss, Sessuvium sesuvioides; Oc, Ochthocloa compressa; Ca, Chenopodium album; Lp, Launea procumbens; Ln, Launea nudicaulis; Am, Alhagi morarum; Cd, Cynodon dactylon; Sa, Sonchus asper; Can, Acacia nilotica; Pj, Prosopis juliflora; Cs, Cenchrus setigerus.

best explained the composition of ground flora underside the tree canopy. More precisely the organic matter availability of soil best explained the vegetation composition (Missin et al., 2001; Hutchinson et al., 1999). *P. cineraria* had high amount of soil organic matter under its canopy as compared to other tree species and this was due the reason that the leaves of *P. cineraria* are smaller and can easily be decomposed, as the result much litter was produced (Gower and Son, 1992). Top soil pH and other earth-alkaline cations (Na, Mg, Ca) were highest in the soil samples underside the canopy of *T. aphylla*. It was

explained that *T. aphylla* was a highly salt tolerant and incorporate these salts in its metabolism and finally these salts ooze out through salt glands present in the leaves. The highest phosphorus and potassium contents were observed under *P. cineraria* while all other tree species were intermediate.

The species richness and diversity was much higher underside the canopy cover of *A. nilotica* (13 plant species) and *T. aphylla* (13 plant species) very closely followed by *P. cineraria* (11 plant spp.) and 9 plant species were recorded underside the canopy of *P. juliflora*.

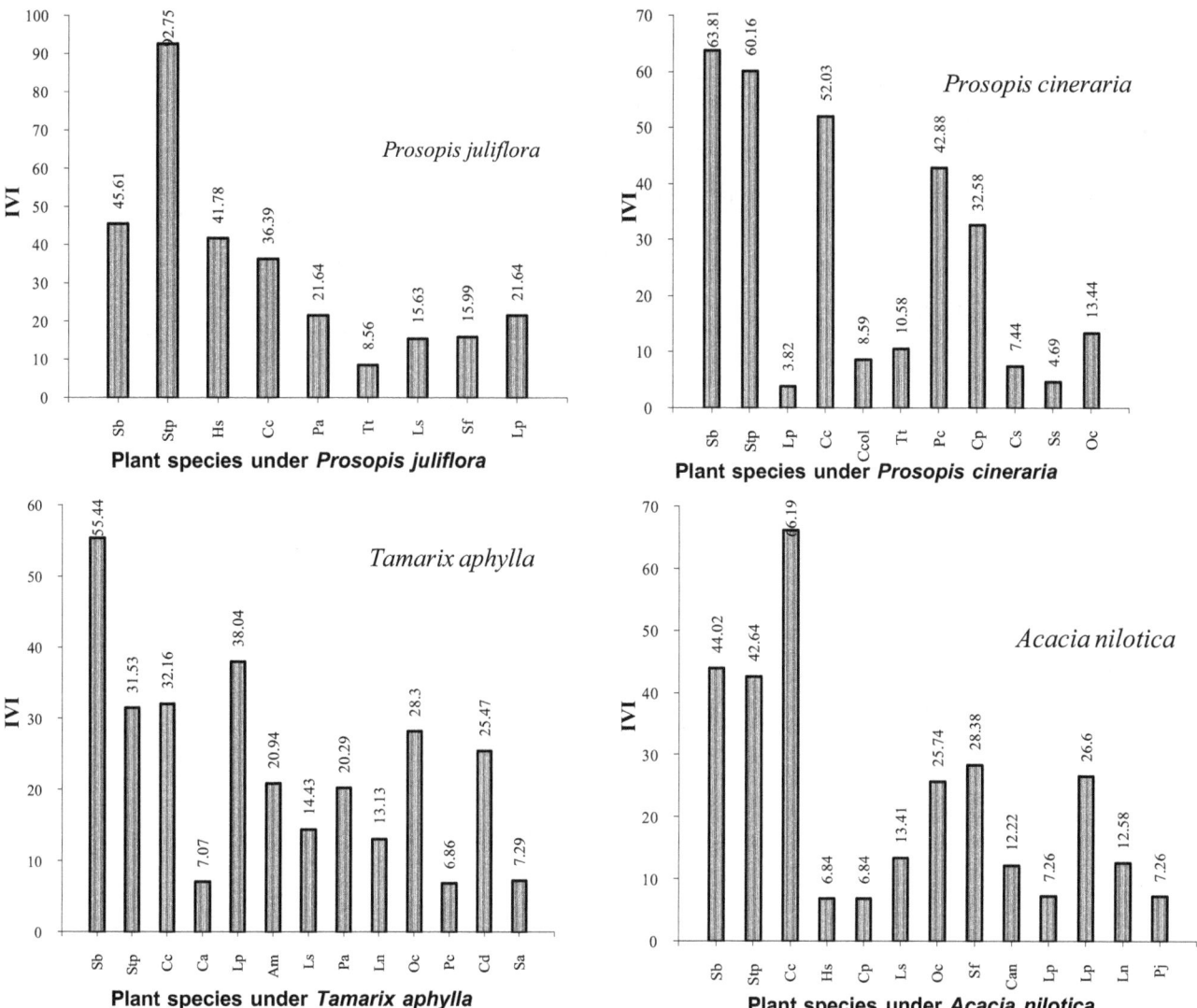

Figure 3. IVI of the plant species forming the under-story vegetation of major tree species of Cholistan desert. Sb, *Salsola baryosma*; Stp, *Stipagrostis*; Hs, *Haloxylon salicornicum*; Cc, *Cenchrus ciliaris*; Aj, *Aerva javanica*; Tt, *Tribulus terristris*; Acj, *Acacia jacquemontii*; Cp, *Calligonum polygonoides*; Dg, *Dipterygium glaucum*; Pa, *Panicum antidotale*; Ls, *Lasiurus scindicus*; Sf, *Suaeda fruiticosa*; Lp, *Leptadenia pyrotechnica*; Ccol, *Citrulus colocynthis*; Pc, *Prosopis cineraria*; Ss, *Sessuvium sesuvioides*; Oc, *Ochthocloa compressa*; Ca, *Chenopodium album*; Lp, *Launea procumbens*; Ln, *Launea nudicaulis*; Am, *Alhagi morarum*; Cd, *Cynodon dactylon*; Sa, *Sonchus asper*; Can, *Acacia nilotica*; Pj, *Prosopis juliflora*; Cs, *Cenchrus setigerus*.

S. plumosa, S. baryosma and *C. ciliaris* appeared as the most frequently and abundantly occurring plant species forming the understory composition of vegetation. Some scientists reported the same canopy cover effects on understory vegetation (Callaway et al., 1991) while many others reported that the effect of tree canopy cover on understory vegetation diversity was much low (Whitney and Foster, 1988; Hong et al., 1997). The variation in the understory flora was may be due to the presence of thick litter layers or some plant species were sensitive to thick litter layers (Sydes and Grime, 1981 a, b; Holderegger, 1996).

It is concluded from the present study that tree species greatly modified the soil chemistry beneath the canopy cover. *S. plumosa, S. baryosma* and *C. cilaris* were the major plant species that form the understory flora of tree species due to their close association with each other in the Cholistan desert. Other plant species also take part in forming the understory vegetation of tree species but to some less extent as they fail to form any type of association with each other.

REFRENCES

Arshad M, Akbar G (2002). Benchmark of plant communities of Cholistan desert. Pak. J. Biol. Sci. 5: 1110-1113.

Arshad M (2003). Cholistan desert ecosystem monitoring for future management. Annual Technical report of a project sponsored by WWF-Pakistan. (Unpublished)

Arshad M, Hussain A, Ashraf MY, Naureen S, Moazzam M (2008). Edaphic factors and distribution of vegetation in the Cholistan desert, Pakistan. Pak. J. Bot. 40(5):1923-1931.

Arshad M, Ashraf MY, Ahmad M, Zaman F (2007). Morpho-genetic variability potential of *Cenchrus ciliaris* L., from Cholistan desert, Pakistan. Pak. J. Bot. 39: 1481-1488.

Callaway RM, Cipollini D, Barto K, Thelen GC, Hallett SG, Prati D, Stinson K, Klironomos J (1991). Novel Weapons: Invasive plant suppresses fungal mutualists in America but not in its native Europe. Ecology 89(4):1043-1055.

Chul KS, Moody K (1983). Comparison of some methodologies for vegetation analysis in transplanted rice. Korean J. Crop Sci. 28: 310-318.

Curtis JT, McIntosh RP (1951). An upland forests continues in the prairie forest border region of Wisconsin. Ecology 32: 476-449.

Gilliam FS, Turrill NL (1993). Herbaceous layer cover and biomass in a young versus a mature stand of a central Appalachian hardwood forest. Bulletin of the Torrey Botanical Club 120:445-450.

Gower ST, Son Y (1992). Differences in soil and leaf litterfall nitrogen dynamics for five forest plantations. Soil Sci. Soc. Am. J. 56: 1959-1966.

Holderegger R (1996). Effects of litter removal on the germination of *Anemone nemorosa* L. *Flora* 191: 175-178.

Hong Q, Klinka K, Sivak B (1997). Diversity of the understory vascular vegetation in 40 year old and old growth forest stands on Vancover Island, British Columbia. J. Veg. Sci. 8: 778-780.

Hussain F (1989). Field and Laboratory Manual of Plant Ecology. National Academy of Higher Education, University Grants Commission, Islamabad, Pakistan.

Hutchinson TF, Boerner EJ, Iverson LR, Sutherland S, Sutherland EK (1999). Landscape patterns of understory composition and richness across a moisture and nitrogen mineralization gradient in Ohio (USA) Quercus forests. Plant Ecol. 144:177-189.

Klinka K, Chen HYH, Wang Q, de Montigny L (1996). Forest canopies and their influence on understory vegetation in early seral stands on West Vancouver Island. Northwest Sci. 70:193-200.

Legare S, Bergeron Y, Pare D (2001). Comparison of the understory vegetation in boreal forest types of southwest Quebec. Can. J. Bot. 79:1019-1027.

Missin L, Du Bus de Warnaffe G, Jonard M (2001). Effects of fertilization on the vascular ground vegetation of European beech (*Fagus sylvatica* L.) and sessile oak (*Quercus petraea* (Matt.) Lieb.) stands. Ann. For. Sci. 58: 829-842.

Muller-Dombois D, Ellenberg H (1974). Aims and methods of vegetation ecology. John Willey and Sons. Inc., New York.

Nisar MF, Iram S, Akhtar Y, Majeed A, Ismail S, Lin F (2010). AFLP based analysis of genetic diversity in buffle grass. World Appl. Sci. J. 10(5):560-567.

Oosting HJ (1957). Study of plant community. Freeman Co., 2nd Ed. 62-64, 3050.

Rao AR, Arshad M, Shafiq M (1989). Perennial grass germplasm of Cholistan desert and their phytosociology. Cholistan Institute of Desert Studies, Islamia University of Bahawalpur. pp. 84.

Ste-Marie C, Pare D (1999). Soil, pH, and N availability effects on net nitrification in the forest floors of a range of boreal forest stands. Soil Biol. Biochem. 31:1579-1589.

Sydes C, Grime JP (1981a). Effects of tree leaf litter on herbaceous vegetation in deciduous woodland. I- Field Investigation. J. Ecol. 69: 237-248.

Sydes C, Grime JP (1981b). Effects of tree leaf litter on herbaceous vegetation in deciduous woodland. II – An Experimental Investigation. J. Ecol. 69:249-262.

Whitney GG, Foster DR (1988). Overstory composition and age as determinants of the understory flora of woods of central New England. J. Ecol. 76: 867-876.

Zavitkovski J (1976). Ground vegetation biomass, production and efficiency of energy utilization in some northern Wisconsin forest ecosystems. Ecology 57:694-706.

10

Reproductive aspects of common carp (*Cyprinus carpio* L, 1758) in a tropical reservoir (Amerti: Ethiopia)

Mathewos Hailu

Ziway Fisheries Resources Research Center, Ethiopia.

The reproductive aspect of common carp (*Cyprinus carpio* L, 1758) in the Amerti reservoir (9°.63' N, 37°. 23' E) was studied monthly between August 2011 and July 2012. A total of 496 specimens, comprising 231 males and 265 females were captured during the sampling period. There were no significant differences between the sex ratio (χ^2 = 2.33; P= 0.126). The mean ± SD of Fulton's condition factor was 1.22 ± 0.14 for both sexes. The size at first sexual maturity (L_{50}) for male *Cyprinus carpio* was 27.2 cm fork length (F_L) while the females attained L_{50} at 28.3 cm F_L. Absolute fecundity (F) varied between 36955 and 318584 with a mean ± SD of 170937 ± 13084 for the length group (270-470 mm F_L). The relationship between F and F_L were significant (F=0.224*$F_L^{3.708}$, R^2 = 0.933; p<0.001). The mean monthly water temperature that ranged from 18.9 to 23.1°C during the study period appears to favor year round spawning of common carp in the reservoir.

Key words: Breading season, fecundity, maturity, species introduction.

INTRODUCTION

The rising demand for electric power, potable water and irrigation has led to the construction of different dams in Ethiopia (Kloos and Legesse, 2010). Furthermore, these reservoirs are sources of income and enhanced nutritional health by supplying fish protein for the adjacent community. Most of these reservoirs are stocked with *Oreochromis niloticus* and *Cyprinus carpio* or sustained themselves with the riverine fish (Tariku et al., 2009; Tigabu, 2010).

The common carp (*C. carpio* L, 1758) is native to coastal areas of the Caspian and Aral Seas (Balon, 1995). The ability of common carp to withstand various environments has made it one of the major exotic species to spread throughout the world (Britton et al., 2007; Sivakumaran et al., 2003). Despite its benefit in commercial fishery *C. carpio* is regarded as a pest fish because of its tendency to destroy vegetation (Miller and Crowl, 2006; Zambrano et al., 2001) increase water turbidity (Lougheed et al., 1998) and decrease habitat heterogeneity for native species (Perrow et al., 1999). Natural spawning of *C. carpio* varies according to environmental factors. In tropical and subtropical regions carp usually mature during their first year and may spawn several times within a given year (Sivakumaran et al., 2003). Female gonad development is continuous when temperatures are above 16°C (Crivelli, 1981).

C. carpio was introduced in Ethiopia in 1936 for aquaculture (Welcome, 1988); it has been stoked in various reservoirs and natural lakes to enhance fish yield by filling the available niche (Dagne and Degefu, 2007; Tedla and Haile-Meskel, 1981). Even though *C. carpio* is wide spread in Ethiopia little is known about its reproductive biology. The aim of this study was to assess the adaptability status and reproductive pattern of *C. carpio* in a tropical highland reservoir, including its size at sexual maturation and seasons of spawning to enable sustainable fishery exploitation.

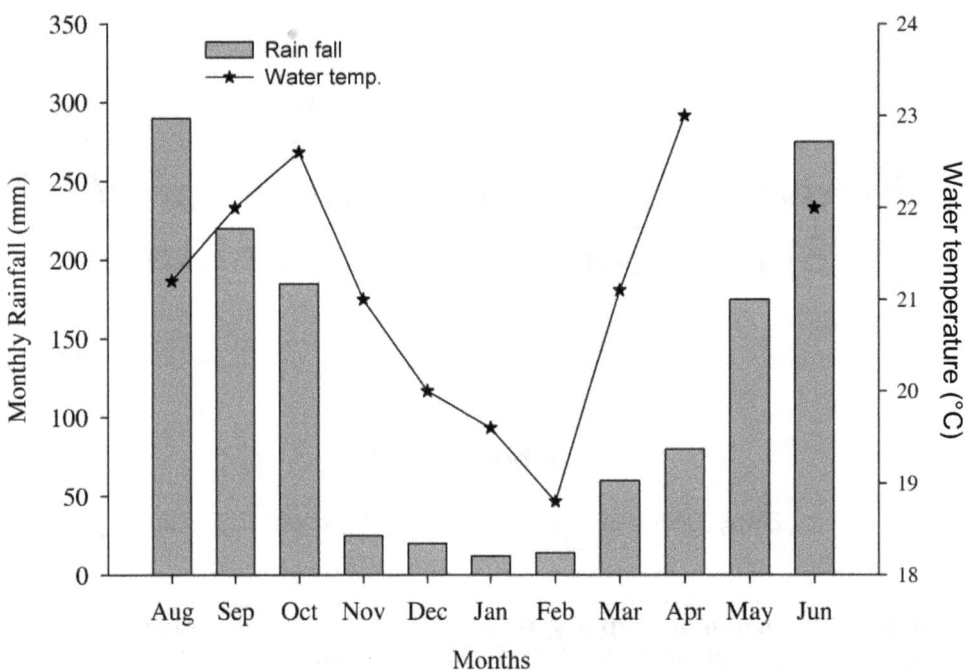

Figure 1. Monthly variation in water temperature and rainfall around Amerti Reservoir (2011- 2012).

MATERIALS AND METHODS

Study area

Amerti reservoir (9°.63' N, 37°.23' E) is located at an altitude of 2243 m above sea level. The reservoir was built to supply water to the adjacent Fincha reservoir through a tunnel for hydroelectric power generation (OADB, 1996). The reservoir was stocked with *C. carpio, O. niloticus* and *Tilapia zilli*. The water temperature of the reservoir ranges between 18.9 and 23.1°C (Figure 1). The area has annual rainfall of 1823 mm with long rains occurring from May to August and the short rains from November to February (Figure 1).

Sample collection

Fish samples were collected monthly between August 2011 and June 2012 using multifilament gill net having 60, 80, 100 and 120 mm stretched mesh size. The panel length of each mesh size was 25 m and the depth was 3 m. Immediately after capture, the fork length (F_L) and total weight (T_W) were measured and weighted to the nearest 1 mm and 0.1 g, respectively. Sex was determined from gonads of the specimens. The sex ratio was computed and Chi-square (χ^2) test was used to determine if it varied from 1:1 (Zar, 1999).

The length-weight relationship was determined using the following equation:

$$W = aF_L{}^b$$

Where, W: weight in grams; F_L: fork length in centimeter.

While, Fulton's condition factor (F_C) (King, 1995) was computed using the following equation:

$$F_C = W/T_L{}^3$$

Where, W: total weight in g; T_L: total length in cm

Maturity estimation

Length at which 50% of both sexes reached maturity (L_{50}) was determined from the percentages of mature fish that were grouped in 1 cm length classes and fitted to the logistic equation described by Echeverria (1987). The breeding season was determined from the percentages of fish with ripe gonads taken each month. The absolute fecundity (F) of individual females was determined gravimetrically (Bagenal and Braum, 1987), with the number of ripe oocytes counted from triplicates of 1 g sub-sample of the ovary. The relative fecundity was also estimated from the number of oocytes per unit body weight of a matured *C. carpio* (Sivakumaran et al., 2003). The relationship between absolute fecundity and fork length was determined using least squares regression.

RESULTS

A total of 496 *C. carpio* comprising 231 (47%) males and 265 (53%) females were captured during the sampling period. The sex ratio between females and males 1:1.15 did not differ significantly from 1:1 (χ^2 = 2.33; P = 0.126). The fork length of the specimens ranged from 14.0 to 45 cm (Figure 2), while the weight ranged between 56.9 and 1535.8 g. The length-weight relationships was curvilinear (Figure 3) and statistically significant with W = 0.022 $L_F{}^{2.923}$ (R^2 = 0.977, P < 0.001).

The Fulton's condition factor (Mean ± SD) of *C. carpio* for both sexes was 1.22 ± 0.14. The value varied from

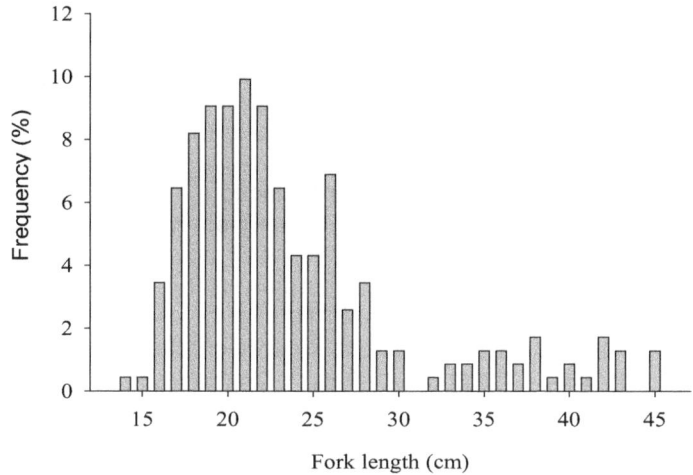

Figure 2. Length frequency distribution of sampled *C. carpio* in Amerti reservoir.

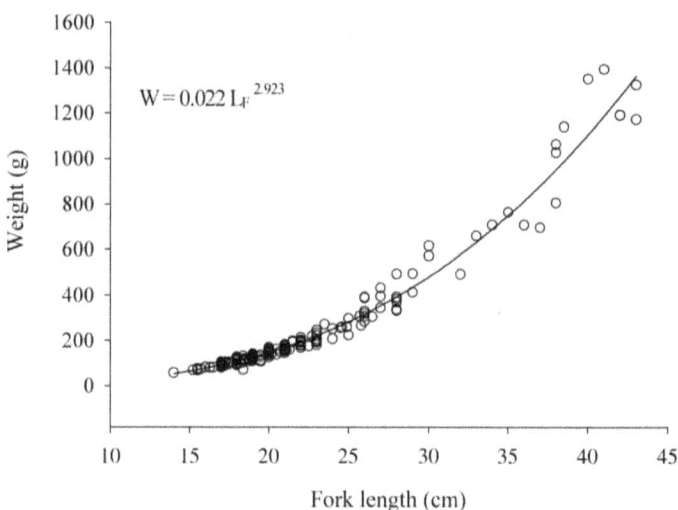

Figure 3. Length-weight relationship of *C. carpio* in Amerti Reservoir.

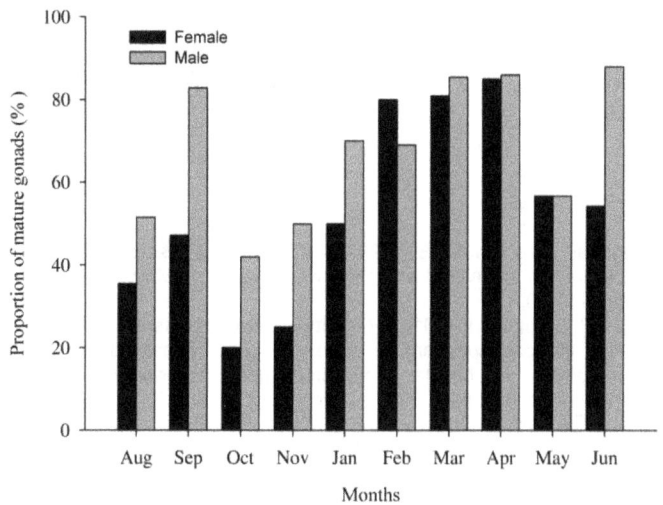

Figure 4. Breading pattern of *C. carpio* in Amerti Reservoir (August 2011 to June 2012).

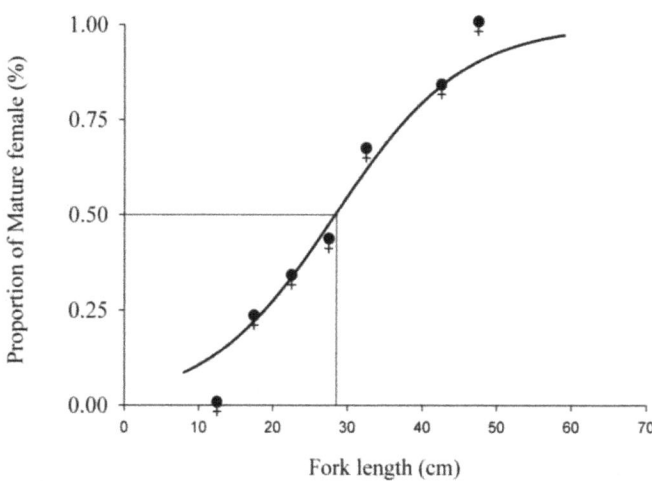

Figure 5. Size at first sexual maturity (L_{50}) for female *C. carpio* in Amerti Reservoir.

1.22 to 1.46 for males and 1.23 to 1.77 for females.

The breeding season of *C. carpio* was determined from percentages of fish with ripe gonads taken monthly from August 2011 to June 2012. Breeding was continuous throughout the year; however the most intense breeding activity was from February to April 2012 (Figure 4). During high breeding time, 69 - 86% of the males and 80 - 85% of the females were found with ripe gonads (Figure 4). The proportion of fish with ripe gonads was relatively low during the months of October and November.

The size at first sexual maturity for male *C. carpio* was 27.2 cm F_L while the females attained first sexual maturity at 28.3 cm F_L (Figure 5). The smallest male found with ripe gonads was 15.5 cm F_L and weighed 74.3 g while the corresponding female was 17 cm F_L and weighed

86.9 g.

The estimated absolute fecundity of female *C. carpio* varied between 36955 and 318584 oocytes per specimen with a mean ± SD fecundity of 170937 ± 13084. The relationship between F and F_L were significant (F = 0.224*$F_L^{3.708}$, R^2 = 0.933; p<0.001). Relative fecundity varied between 98 284 and 310 964 oocytes kg^{-1} total body weight, with a mean of 177 786 ± 48 427 oocytes kg^{-1} total body weight.

DISCUSSION

Reservoirs can result in productive fisheries based on

their location and management status (Marmulla, 2001). Stocking of exotic species can enhance yields, as long as they are not influencing the indigenous species which are preferred by the community. C. carpio has adapted well in the Amerti reservoir with its reproductive trait.

The reproductive cycle and pattern of gonad development of C. carpio in natural ecosystems greatly depends on the ambient water temperature (Smith and Walker, 2004; Tempero et al., 2006). Spawning in C. carpio occurs at a water temperature of around 18°C (Fernández-Delgado, 1990). The mean monthly water temperature that ranged from 18.9 to 23.1°C during the study period appears to favor year round spawning of common carp in the reservoir. Year round breeding pattern for C. carpio has been reported in other tropical lake (Oyugi et al., 2011).

Length at first maturity is variable, ranging from 90 to 430 mm T_L for both males and females in wild C. carpio population (Fernández-Delgado, 1990; Prochelle and Campos, 1985). The first maturity for male and female which were 155 mm F_L 170 F_L respectively did not vary from other studies. However, the length at first maturity (27.2 cm F_L for female and 28.3 cm F_L for male) is lower than Lake Naivasha were it is 340 mm for male and 420 mm (L_T) for females (Oyugi et al., 2011).

Length and age at maturity in common carp are related to latitude and sex; males often mature before females, and fish mature earlier at low latitudes when compared with higher latitudes (Tempero et al., 2006). In Amerti Reservoir, males matured at smaller sizes (27.2.6 cm F_L) than females (28.3 cm F_L). This may be related to the preparation of females to sustain large number of eggs. Males attaining maturity at a smaller size than females are also reported in both temperate and tropical aquatic ecosystems (Britton et al., 2007; Tempero et al., 2006).

Determination of fecundity and spawning season are indispensable in understanding the population dynamics of fish species (Sivakumaran et al., 2003). Fecundity also explains the degree of invasiveness and ecosystem impacts (Bajer and Sorensen, 2010). The mean relative fecundity in Amerti reservoir 177 786 kg^{-1} is larger than 97 200 oocytes kg^{-1} reported in Newzeland (Tempero et al., 2006); but comparable with 163 000 eggs kg^{-1} total body weight (Sivakumaran et al., 2003). Females C. carpio can carry >1000000 oocytes for length groups >60 cm (Bajer and Sorensen, 2010).

Maturing at small size with large number of oocytes is one of the physiological advantages of the traits of the carp which appear to have provided their population with resilience to the exploitation by providing rapid growth to maturity and the opportunity for early life reproduction prior to their capture.

The establishment and year round reproduction of C. carpio in a tropical environment with high fecundity has shown that introducing the species to other natural lakes with prior indigenous fish can threaten their ecology. Indigenous species are expected to have a competitive advantage over newly arriving species adapted to different species and resource availability (Vermeij and Dudley, 2000). However, physiological advantages of displacement by C. carpio have been well documented (Jia et al., 2008). Once C. carpio is established in a system, eradication is extremely expensive and in many cases impossible (Zambrano et al., 2006).

In conclusion, C. carpio introduced in Amerti Reservoir has adapted successfully. The long spawning period and multiple spawning characteristics of common carp were favored by the ambient temperature of the reservoir. Care should be taken in introduction of C. carpio to natural lakes in similar habitats, as it can disrupt native species. Successful fisheries in Amerti reservoir from common carp could be established with a minimum catch size of 28 cm F_L, which could be achieved by using gillnet with a mesh size of 80 mm and above.

ACKNOWLEDGEMENTS

My sincere gratitude goes to the technical staff of Ziway Fisheries Resources Research Center, particularly Mr Getachew Senbete, Mitiku Bonxa and Dawit Imiru for their assistance during the fieldwork. The study was financed by the Oromia Agricultural Research Institute (OARI).

REFERENCES

Bagenal TB, Braum E (1987). Methods for assessment of fish production in freshwaters. Blackwell Scientific Publications, London. pp. 165-201.

Bajer P, Sorensen P (2010). Recruitment and abundance of an invasive fish, the common carp, is driven by its propensity to invade and reproduce in basins that experience winter-time hypoxia in interconnected lakes. Biol. Invasions 12:1101-1112.

Balon EK (1995). Origin and domestication of the wild carp, Cyprinus carpio: from Roman gourmets to the swimming flowers. Aquaculture 129:3-48.

Britton JR, Boar RR, Grey J, Foster J, Lugonzo J, Harper DM (2007). From introduction to fishery dominance: the initial impacts of the invasive carp Cyprinus carpio in Lake Naivasha, Kenya, 1999 to 2006. J. Fish Biol. 71:239-257.

Crivelli AJ (1981). The biology of the common carp, Cyprinus carpio L., in the Camargue, southern France. J. Fish Biol. 18:271-290.

Dagne A, Degefu F (2007). Proceedings of the 15 the Annual conference of the Ethiopian Society of AnimalProduction (ESAP), Addis Ababa, Ethiopia. pp.163-171.

Echeverria TW (1987). Thirty-four species of California rockfishes: maturity and seasonality of reproduction. US. Fish. Bull. 85:229-250.

Fernández-Delgado C (1990). Life history patterns of the common carp, Cyprinus carpio, in the estuary of the Guadalquivir river in south-west Spain. Hydrobiologia 206:19-28.

Jia Y, Chen Y, Xie S and Yang Y (2008). Physiological advantages may contribute to successful invasion of the exotic Cyprinus carpio into the Xingyun Lake, China. Environ. Biol. Fish 81:457–463.

King M (1995). Fisheries biology assessment and management. Blackwell Science Ltd, Oxford. p.341.

Kloos H, Legesse W (2010). Water Resources Management in Ethiopia: Implications for the Nile Basin. Cambria Press, New York. p.415.

Lougheed VL, Crosbie B and Chow-Fraser P (1998). Predictions on the effect of common carp (Cyprinus carpio) exclusion on water quality, zooplankton, and submergent macrophytes in a Great Lakes wetland. Can. J. Fish. Aquat. Sci. 55:1189-1197.

Marmulla G (2001). Dams, fish and fisheries: Opportunities, challenges and conflict resolution. FAO Fisheries Technical Paper. No. 419 Rome.

Miller SA, Crowl TA (2006). Effects of common carp (*Cyprinus carpio*) on macrophytes and invertebrate communities in a shallow lake. Freshw. Biol. 51:85-94.

OADB (Oromia Agriculture Development Bureau) (1996). Land resource and socioeconomic survey report of common watershed. OADB, Addis Ababa, Ethiopia.

Oyugi DO, Cucherousset J, Ntiba MJ, Kisia SM, Harper DM, Britton JR (2011). Life history traits of an equatorial common carp *Cyprinus carpio* population in relation to thermal influences on invasive populations. Fish. Res. 110:92–97.

Perrow MR, Jowit AJD, Leigh SAC, Hindes AM, Rhodes JD (1999). The stability of fish communities in shallow lakes undergoing restoration: expectations and experiences from the Norfok Broads (U.K.). Hydrobiologia 408: 85–100.

Prochelle O, Campos H (1985). The biology of the introduced carp *Cyprinus carpio* L., in the river Cayumapu, Valdivia, Chile. Stud. Neotrop. Fauna Environ. 20:65–82.

Sivakumaran K, Brown P, Stoessel D, Giles A (2003). Maturation and Reproductive Biology of Female Wild Carp, *Cyprinus carpio*, in Victoria, Australia. Environ. Biol. Fishes 68:321-332.

Smith BB, Walker KF (2004). Spawning dynamics of common carp in the River Murray, South Australia, shown by macroscopic and histological staging of gonads. J. Fish Biol. 64:336-354.

Tariku E, Mengistu S, Tussa D, Tugie D, Tadesse Z (2009). Proceedings of the 16th Annual conference of the Ethiopian Society of Animal Production (ESAP), Addis Ababa. Ethiopia. pp. 95-102.

Tedla S, Haile-Meskel F (1981). Introduction and transplantation of freshwater fish species in Ethiopia. SINET. J. Sci. 4: 69-72.

Tempero GW, Ling N, Hicks BJ, Osborne MW (2006). Age composition, growth, and reproduction of koi carp (*Cyprinus carpio*) in the lower Waikato region. N. Z. J. Mar. Freshw. Res. 40:571-583.

Tigabu Y (2010) Stocking based fishery enhancement programs in Ethiopia. Ecohydrol. Hydrobiol. 10: 241-246.

Vermeij GJ, Dudley R (2000). Why are there so few evolutionary transitions between aquatic and terrestrial eco-systems? Biol. J. Linn. Soc. 70:541–554.

Welcome RL (1988). International introductions of inland aquatic species. FAO Fish Tech Pap 294:31.

Zambrano L, Martínez-Meyer E, Menezes N, Peterson AT (2006). Invasive potential of common carp (*Cyprinus carpio*) and Nile tilapia (*Oreochromis niloticus*) in American freshwater systems. Can. J. Fish. Aquat. Sci. 63:1903–1910.

Zambrano L, Scheffer M, Martínez-Ramos M (2001). Catastrophic response of lakes to benthivorous fish introduction. Oikos 94:344-350.

Zar JH (1999). Biostatistical analysis. Prentice-Hall, New Jersey. pp. 663.

11

A new model: Herbaceous species diversity along the environmental gradient in the typical hilly areas of Henan Province

Bing-Hua Liao[1,2,3]*, Pei-Song Liu[1], Zhen-Zhong Wen[1], Sheng-Yan Ding[3], Hai-Long Yu[2], Zhi-Chao Wang[2], Zhong-Kai Li[2], Huan-Xin Chu[2], Wen-Liang Li[2] and Yi Shen[2]

[1]The Key Laboratory of Ecological Restoration in Hilly Areas, Forestry Department of Henan Province, Ping-ding-shan University, Ping-ding-shan, Henan Province,China,467000.
[2]Department of Environment and Geography, Ping-ding-shan, Henan Province, China, 467000.
[3]Institute of Ecological Science and Technology, College of Life Sciences, Henan University, Kaifeng, China, 475001.

Ecological model is important. The relationship between biodiversity and disturbance gradient, however, makes it difficult to compare data from different studies and draw general conclusions. These results assessed the relationship between disturbance gradients and herbaceous species diversity in the typical hilly areas on varying spatial scales along the environmental gradient in Henan Province. Using community ecology techniques and quantitative measurements of disturbance, we detected a linear relationship between weighting values of disturbance intensity and herbaceous species diversity (Simpson), which were significantly correlated ($P<0.01$) in the differential ecosystem types along different disturbance gradient. Understanding a linear relationship between herbaceous species dynamics and their causes in the differential type's landscapes is essential for further research of local ecosystem functions reaching the goal of ecosystem sustainable development in the context of biodiversity conservation. These results indicate that weighting values of disturbance are the most important environmental factors affecting the herbaceous species diversity in building a model of the ecosystem. This model may help policy makers formulate better ecological conservation and restoration plans with ecosystem.

Key words: Model, herbaceous species diversity, disturbance gradient, significantly, the typical hilly areas.

INTRODUCTION

Ecosystems are typically filled with large numbers of plant species along environmental gradient, making species-centered studies of systemic processes and functions extremely difficult, if not outright impossible, to carry out (Liao et al., 2010; Liao et al., 2011a, 2011b). The intermediate disturbance Hypothesis (IDH) suggests that species diversity will be maximal at intermediate levels of disturbance (Connell, 1978). Moreover, many experiments have assessed the relationship between plant species biodiversity and disturbance intensity from IDH perspective along different disturbance gradients in theoretical ecology (Biswas et al., 2010; Boutin et al., 2008; Leis et al., 2005). For example, Leis et al. (2005) found that disturbance up to intermediate levels can be used to maintain biodiversity by enriching the plant species pool. Further, Biswas et al. (2010) suggested that plant species richness and diversity, functional richness and diversity reached peaks at moderate disturbance intensity in riparian and upland plant communities.

Therefore, the relationship between plant species diversity and intermediate disturbance intensity has many

---

*Corresponding author. E-mail: lbh@henu.edu.cn.

Abbreviation: IDH, Intermediate disturbance hypothesis.

Table 1. The relationship between IDH and biodiversity from the different perspectives.

The relationship between IDH and biodiversity from the different perspectives	Author
Water level fluctuations, fire and grazing are essential for maintaining plant diversity.	Keddy, 2005.
Increased diversity at intermediate disturbance was due primarily to increased evenness.	Aronson, 1995.
Diversity and soil properties were best at intermediate disturbance levels.	Zhang et al., 2010.
The IDH pattern was obtained for low frequency dependence and low immigration.	Cordonnier et al., 2006.
Generate a period-bubbling bifurcation structure and population dynamics that are most variable at intermediate disturbance frequencies.	Reluga, 2004.

real world applications in the field of ecological conservation and restoration (Table 1). Here, using plant community ecology techniques, this study analysis asked three key questions regarding the relationship between herbaceous species diversity and intermediate disturbance levels in the differential ecosystem types. First, this study asked if the IDH facilitate human understanding of the contribution of the relationship between herbaceous species diversity dynamics and disturbance gradient to different ecosystem types, and if these relationships are more useful than others. Next, this study asked if the large set of the relationship between herbaceous species diversity dynamics and disturbance gradient from IDH perspective can be compressed into a general model that is universally applicable to ecosystem studies. Finally, this study asked how the links between herbaceous species diversity and disturbance along disturbance gradient can be applied to real world scenarios in order to prevent biodiversity loss and ecosystem manage degrading or degraded ecosystems.

MATERIALS AND METHODS

The ecosystems of the typical hilly areas are results of the historical natural and anthropogenic activities in the Henan Province. Quantitative assessments mainly depended on disturbance of human-driven and season driven changes in herb biodiversity and were based on a careful choice of landscape scenarios (Table 2).

Using community ecology techniques, we examined the influences of different disturbance on dynamics of herbaceous species diversity and disturbance interactions along the different disturbance gradient along elevations gradient on the northern slope of the *Fu-Niu* Mountain Natural Reserve in August, 2011. Several plots (e.g. typical hilly ridge areas, typical hilly boundary areas and typical hilly center areas) were established per 1 × 1 km area. A total of 20 plots were set. Each study plot, consisted of one 20 × 20 m tree layer quadrate, five (the center and four corners of the study plot) 2 × 2 m shrub layer quadrates and five 1 × 1 m herbaceous layer quadrates. Thus, these plots include 100 herbaceous layer quadrates and 100 shrub layers in the typical hilly center areas in August, 2011 (Tables 1 to 4).

Traditional indices

This is the most widely used indices, Simpson's index (Simpson, 1949). Simpson's index of diversity is generally calculated as the complement of D:

$$1 - D = 1 - \sum_i (p_i)^2$$

Where, p_i is the proportion of the sample belonging to the ith species.

RESULTS AND DISCUSSION

Dynamics in the dominant/companion tree species along disturbance gradient at pulsed field gradients (PFGs) levels

The results show that disturbance is an important environmental factor affecting weed species diversity in anthropogenic activities in the typical hilly areas center / the typical hilly boundary areas / the ridge of the typical hilly areas. The relationship between values of weighting disturbance and the weed species diversity investigated vary significantly in the herbaceous layer along the environmental gradient on varying disturbance scales along the different disturbance intensity gradient in the typical hilly areas (Figures 2 to 4).

The results indicate that herb diversity (e.g. Simpson indicator) is expected to increase when weighting values of disturbance is reduced along disturbance gradients in the typical hilly areas of Henan Province (Figures 2 to 4). Second, herbaceous diversity (e.g. Simpson indicator) is expected to decrease when weighting values of disturbance is increased along disturbance gradients in the typical hilly areas of Henan Province. Thus, our result implies that weak disturbance should lead to higher herbaceous species diversity in the typical hilly areas of Henan Province.

These results suggest that herb species diversity (e.g. Simpson indicator) is expected to increase when weighting values of disturbance is reduced along disturbance gradients in the farmland boundary areas and the ridge in the typical hilly areas (Figures 1 to 5). In addition, herb species diversity is expected to decrease when weighting values of disturbance is increased along disturbance gradients in the typical hilly areas boundary areas and the ridge in the typical hilly areas (Figures 3 to 4). Therefore, this result implies that weak disturbance should lead to higher herbaceous species diversity in the typical hilly areas of Henan Province. To do this, the

Table 2. The physical geographic conditions of *Fu-Niu* Mountain nature reserve.

Location	Climatic								Elevation (m) †	Area (hm²)	Vegetation
	Precipitation (mm)	Mean temperature (ºC)									
		Annual Mean		Maximum		Minimum					
		South slope	North slope	South slope	North slope	South slope	North slope				
Latitude (º): 32.75 - 34.00 Longitude(º): 110.50 - 113.01	800 - 1100	14.1- 15.1	12.1-12.7	26.5-28.5	26.5-28.5	1-2	-1.5- 2		351 - 1920	56000	Straddling mixed vegetation zones of the subtropical and warm-temperate zones of East China, the *Fu-Niu* Mountain National Reserve is representative of north-south climatic transition zones.

†Above sea level.

Table 3. The selection of the weighting values of disturbance.

Parameter	Differential ecosystem type
Values of weighting of disturbance	
Times of Disturbance	Typical hilly areas center
+ Frequency of Disturbance	/Ridge of the typical hilly areas
+100/ The width of road in the different types	/ Typical hilly areas boundary

Table 4. Investigation Index along the elevation gradient variable.

Investigation	Layer	Community	Species	Height	Crow	Diameter
Community Investigation	Tree/shrub /herbaceous	Coverage/community's age structure	Species/ individual number	Layer's Height	Crow height/ width	Basal diameter

Table 5. The negative correlation between Simpson and weighting values of disturbance.

Differential ecosystem type	Simpson
Typical hilly areas center	-0.962**
Ridge of the typical hilly areas	-0.981**
Typical hilly areas boundary	-0.917**

**P<0.01

correlation between species biodiversity and the weighting values of disturbance was then analyzed (Table 5). This results suggest that there is a linear relationship between weighting values of disturbance and herb species diversity (Simpson),

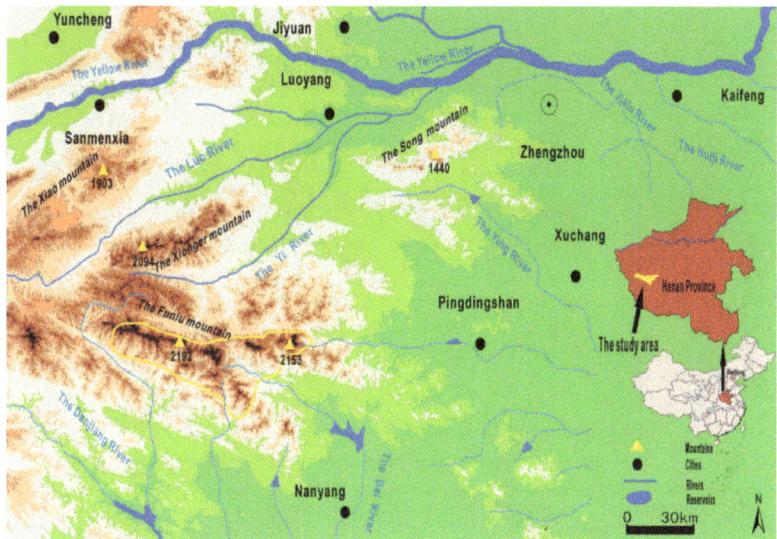

Figure 1. A digital cadastre map in the typical hilly areas of Henan Province.

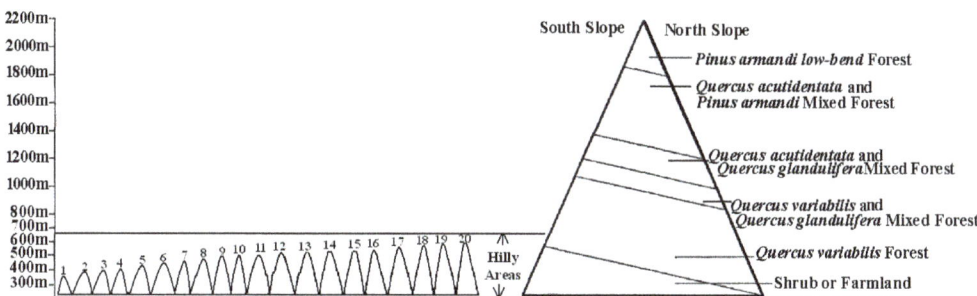

Figure 2. The Southern slope of *Fu-Niu* Mountain vegetations.

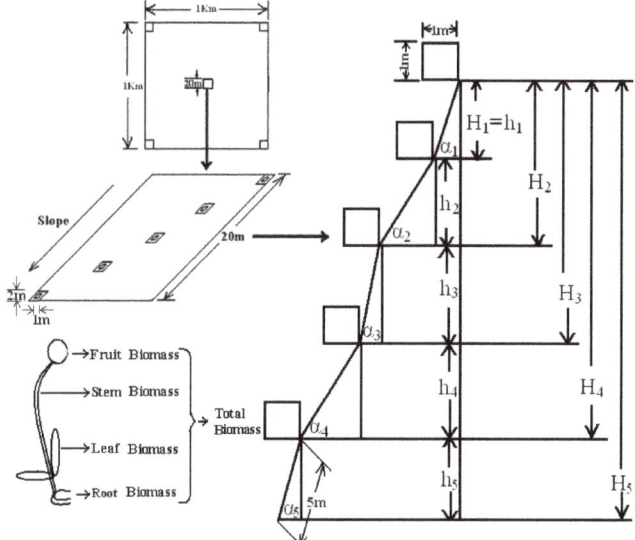

Figure 3. Quadrate settings and research methods. $H_1=h_1=5m \times Sin\alpha_1$; $H_2=h_1+h_2=5 m \times (Sin\alpha_1+ Sin\alpha_2)$; $H_3=h_1+h_2+h_3=5 m \times (Sin\alpha_1+ Sin\alpha_2+ Sin\alpha_3)$; $H_4=h_1+h_2+h_3+h_4=5 m \times (Sin\alpha_1+ Sin\alpha_2+ Sin\alpha_3+ Sin\alpha_4)$; $H_5=h_1+h_2+h_3+h_4+h_5=5 m \times (Sin\alpha_1+ Sin\alpha_2+ Sin\alpha_3+ Sin\alpha_4+Sin\alpha_5)$.

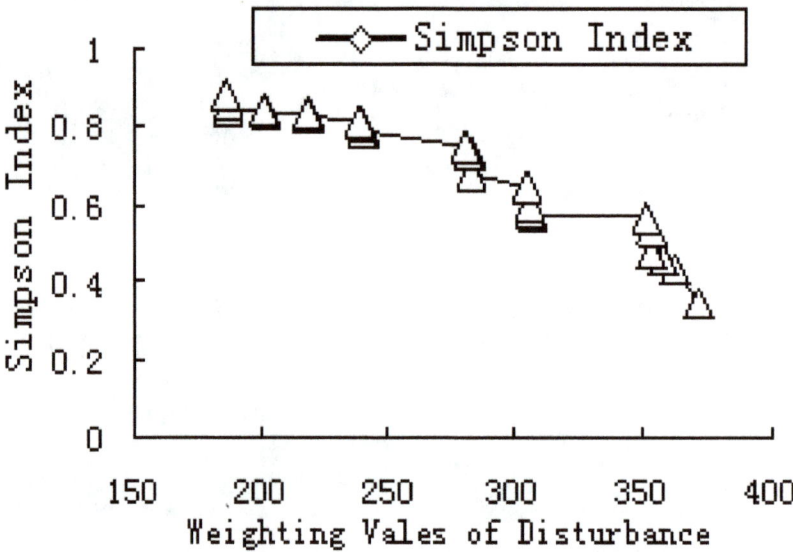

Figure 4. The relationship between Simpson and disturbance in the typical hilly areas boundary areas.

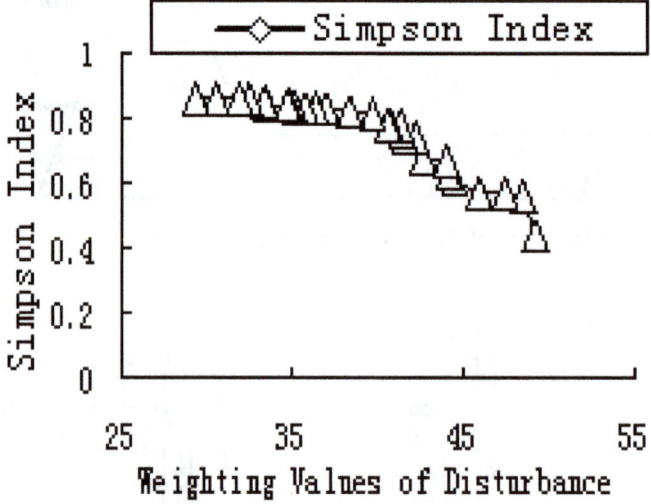

Figure 5. The relationship between Simpson and disturbance in the typical hilly areas center areas.

which were significantly negative correlated (P<0.01). The results indicate that disturbance, largely determined by weighting values of disturbance, are the most important environmental factors affecting the distribution of herb species diversity (Table 5).

The results indicate that elevation and disturbance are important environmental factors affecting the distribution of plant functional groups on southern slope of the *Fu-Niu* Mountain National Nature along elevation gradient, while distributions of individual species differed significantly along different elevation gradient (Figures 1 to 6, Table

6).

A general model of the links between herbaceous species diversity and disturbance gradient from IDH perspective (magnetohydrodynamics (MHD))

The IDH is used as a framework for investigating the linkages between disturbance and species diversity. Unfortunately, the various IDH makes it difficult to compare data from different studies in order to draw general conclusions (Connell, 1978;Aronson et al., 1995). It is therefore important to develop MHD which is

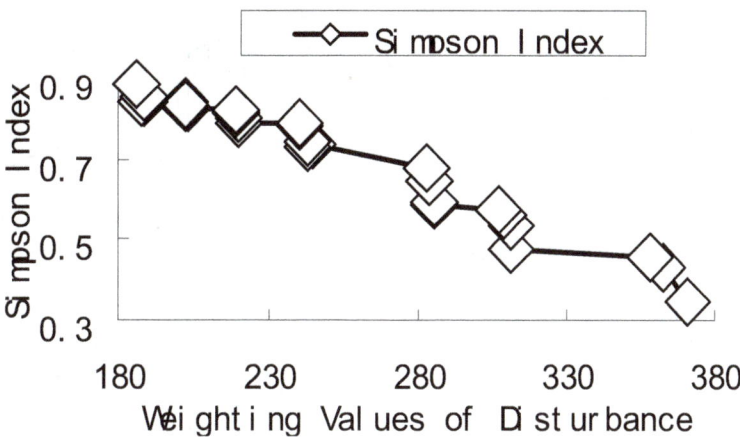

Figure 6. The relationship between Simpson and disturbance in the typical hilly areas.

Table 6. Dynamics of vegetation dominate species on the typical hill areas of the southern slope of the *Fu-Niu* Mountain National Nature at PFGs level along elevation gradient.

Number of site location	Elevation (m) †	South Slope
1	351	*Quercus variabills* and *Castanea mollissima* Mixed Forest
2	370	*Conyza canadensis* Herbs Vegetation
3	419	*Q. variabills* and *Lindera angustifolia* Mixed Forest
4	449	*Q. variabills* and *L. angustifolia* Mixed Forest
5	498	*Q. variabills* and *Cotinus coggygria* Mixed Forest
6	525	*Q.* and *C. mollissima* Mixed Forest
7	537	*Q. variabills* and *L. angustifolia* and *Melia azedarach* Mixed Forest
8	560	*C. canadensis* Herbs Vegetation
9	564	*Q. variabills* and *L. angustifolia* and *M. azedarach* Mixed Forest
10	568	*Populus L.* Forest
11	576	*Q. variabills* and *L. angustifolia* and *M. azedarach* Mixed Forest
12	577	*Phyllostachys glauca McClure* Forest
13	582	*Q. variabills* and *C. coggygria* and *L. angustifolia* Mixed Forest
14	584	*Populus L.* Forest
15	590	*Q. variabills* and *M. azedarach* Mixed Forest
16	602	*Cornus officinalis Sieb. et Zucc.* Forest
17	608	*Q. variabills* and *C. mollissima* and *L. angustifolia* Mixed Forest
18	620	*Q. variabills* and *L. angustifolia* Mixed Forest
19	636	*Pinus tabulaeformis Carr.* Forest
20	667	*Q. variabills* Forest

universally applicable across a range of ecosystems so that the relationship between herbaceous species diversity and disturbance gradient can be more accurately quantified with quantitative measurements of disturbance intensity. Here we proposed the use of a MHD framework that incorporates model (Figures 2 to 5) operating on varying spatio-temporal and disturbance scales for in-depth studies of the relationship between herbaceous species diversity and disturbance gradient

(e.g. anthropogenic disturbance and natural disturbance). Such a MHD will facilitate ecosystem studies that apply the links between herbaceous species diversity and disturbance gradient from IDH perspective (Figures 1 to 7, Tables 1 to 6).

The above MHD can help identify the relationship between herbaceous species diversity and disturbance intensity most relevant to tolerating environmental fluctuations or recovering from IDH perspective (Figure

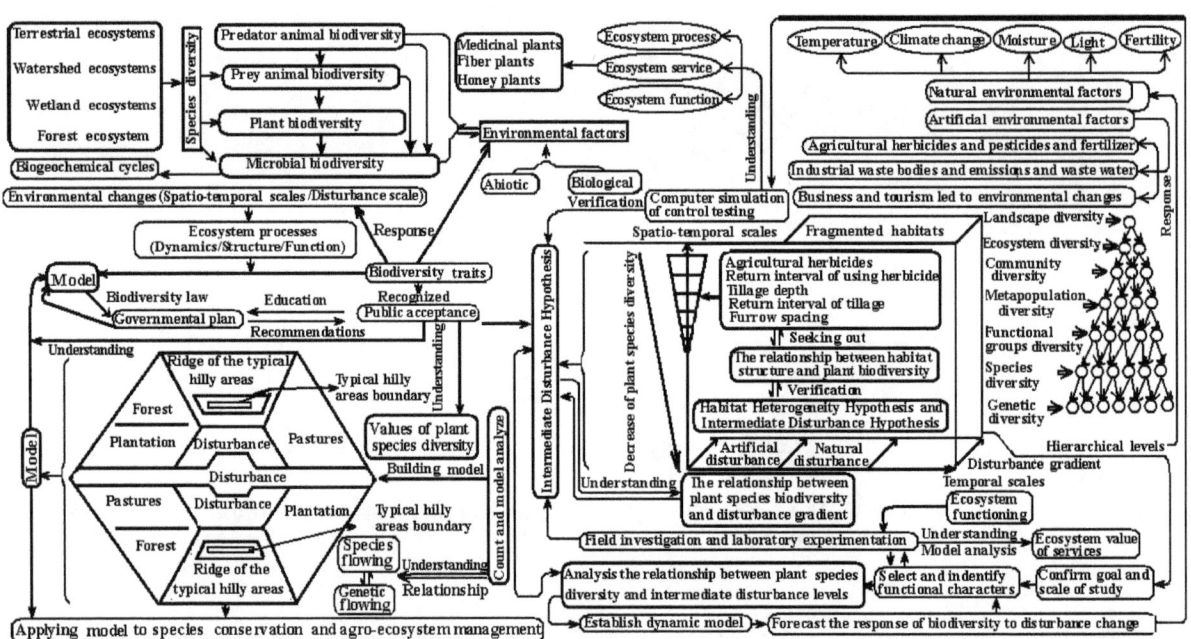

Figure 7. The proposed the framework of relationship between herbaceous species diversity and disturbance gradient that incorporates herbaceous species diversity dynamic from the intermediate disturbance level, and operating on varying spatio-temporal and disturbance scales in the forest ecosystem. The model was based on the study (Liao et al., 2010; 2011a, 2011b).

7). Based on the fact the MHD is useful to achieving the goal of ecosystem, sustainable development in the context of biodiversity conservation is needed. For example, Bartels et al. (2010) explained that whether resource quantity or resource heterogeneity is the determinant of understory plant diversity in individual studies was dependent on stand successional stage(s), presence or absence of intermediate disturbance, and forest biome within which the studies were conducted. Carvalheiro et al. (2011) suggest that presence of weeds allowed pollinators to persist within sunflower fields, maximizing the benefits of the remaining patches of natural habitat to productivity of this large-scale crop.

Applying general model to species conservation and forest ecosystem management

The above MHD can help identify theoretical ecology (e.g. diversity dynamics, species composition, food web structure, diversifying plant-microbial system, ecosystem stability/functioning, organic agriculture, biodiversity law, watershed ecosystem management, assessment of ecosystems and biodiversity, scale dependence) (Liao et al., 2010, 2011a, 2011b; Stutter et al., 2010; Shennan, 2008; Menalled et al., 2001, Bartels et al., 2010; Crowder et al., 2010; Bai et al., 2004; Davis et al., 2007; Bernez et al., 2002; Marris, 2010; Crawley et al., 2001). For instance, Menalled et al. (2001) suggest that ecosystem management systems can have both immediate and

long-term effects on weed species density, abundance, and diversity. Shennan (2008) proposed that important elements for understanding biotic interactions include consideration of the effects of diversity, species composition and food web structure on ecosystem processes; the impacts of timing, frequency and intensity of disturbance; and the importance of multitrophic interactions in ecosystem. Stutter et al. (2010) proposed a mechanism whereby the establishment of vegetated buffer strips between cropland and watercourses on previous agricultural land causes a diversifying plant-microbial system. Bartels et al. (2010) explained that whether resource quantity or resource heterogeneity is the determinant of understory plant diversity in individual studies was dependent on stand successional stage(s), presence or absence of intermediate disturbance, and forest biome within which the studies were conducted. Therefore, it will nonetheless be a substantial challenge to apply this MHD to specific real world policy problems (Gilbert 2010; James et al., 2010; Marris, 2010).

Conclusion

Progress in three key areas will substantially further efforts to gain a rigorous understanding of how the links between herbaceous species diversity and disturbance intensity, and their interactions, influence the response of ecosystem properties to changing biodiversity. First, better understanding of the relationship between herbaceous species diversity and disturbance intensity is correlated, particularly with respect to the predominant forces of human activities. Second, MHD is based on both field investigations of community ecology techniques of 100 herbaceous layer quadrates and scientific analysis, which can use more and more theoretical ecology. Finally, MHD can be useful to understand the plant species diversity and disturbance intensity from IDH perspective. Understanding MHD is essential for further research of local ecosystem functions and the goal of ecosystem sustainable development in the context of herbaceous biodiversity conservation in natural or artificial disturbance ecosystem. Therefore, MHD may help to ameliorate this situation about losses of plant species diversity (e.g. medicinal plants, honey plants, fiber plants, etc).

ACKNOWLEDGEMENT

This work was supported by the Forestry Department of Henan Province Science Foundation of China (No.201004044), the National Natural Science Foundation of China (No. 41071118) and by the many ideas of some researchers of "1st Biotechnology World Congress".

REFERENCES

Aronson RB, Precht WF (1995). Landscape patterns of reef coral diversity: a test of the intermediate disturbance hypothesis. J. Experimental Marine Biol. Ecol. 192(1): 1-14.

Bai YF, Han XG, Wu JG, Chuo ZZ, Li LH (2004). Ecosystem stability and compensatory effects in the Inner Mongolia grassland. Nature. 431: 181-184.

Bartels SF, Chen HY (2010). Is understory plant species diversity driven by resource quantity or resource heterogeneity? Ecology. 91(7):1931-1938.

Bernez I, Haury J, Ferreira MT (2002). Downstream effects of a hydroelectric reservoir on aquatic plant assemblages. Sci. World J. 2:740-750.

Biswas SR, Mallik AU (2010). Disturbance effects on species diversity and functional diversity in riparian and upland plant communities. Ecology. 91(1):28-35.

Boutin C (2008). Evaluating the relationship between floristic quality and measures of plant biodiversity along stream bank habitats. Ecological Indicators. 8(5): 466-475.

Carvalheiro LG, Veldtman R, Shenkute AG, Tesfay GB, Pirk CW, Donaldson JS, Nicolson SW (2011). Natural and within-farmland biodiversity enhances crop productivity. Ecology letters. 14(3):251-259.

Connell JH (1978). Diversity in tropical rain forests and coral reefs. Science. 199(4335): 1302-1310.

Crawley MJ, Harral JE (2001). Scale Dependence in Plant Biodiversity. Sci. 291: 864-868.

Crowder DW, Northfield TD, Strand MR, Snyder WE (2010). Organic agriculture promotes evenness and natural pest control. Nat. 466: 109-112.

Davis SR, Brown AG, Dinnin MH (2007). Floodplain connectivity, disturbance and change: a palaeoentomologi -cal investigation of floodplain ecology from south-west England. The J. Animal Ecol.

76(2): 276-288.

Gilbert N (2010). Biodiveristy law could stymie research. Nat. 463: 598.

James AN, Vorhies F (2010). Green development credits to forest global biodiversity. Nature. 465: 869.

Leis SA, Engle DM, Leslie DM Jr, Fehmi JS (2005). Effects of short- and long-term disturbance resulting from military maneuvers on vegetation and soils in a mixed prairie area. Environmental Management. 36(6): 849-861.

Liao BH, Wang XH (2010). Plant functional group classifications and a generalized hierarchical framework of plant functional traits. Afr. J. Biotechnol. 9: 9208-9213.

Liao BH, Ding SY, Hu Nan et al (2011a). Dynamics of environmental gradients on plant functional groups composition on the northern slope of the *Fu-Niu* Mountain Nature Reserve. Afr. J. Biotechnol.10:18939-18947.

Liao BH, Ding SY, Liang GF et al (2011b). Dynamics of plant functional groups composition along environmental gradients in the typical area of *Yi-Luo* River watershed. Afr. J. Biotechnol.10:14485-14492.

Marris E (2010). UN body will assess ecosystems and biodiversity. Nat. 465: 859.

Menalled FD, Gross KL, Mark H (2001). Weed aboveground and seedbank community responses to agricultural management systems. Ecological Applications. 11:1586-1601.

Shennan C (2008). Biotic interactions, ecological knowledge and agriculture. Philosophical transactions of the Rpyal Society of London. Series B, Biological sciences. 363(1492):717-739.

Simpson EH (1949). Measurement of diversity. Nature. Pp 163: 688.

Stutter MI, Langan SJ, Lumsdon DG (2009). Vegetated buffer strips can lead to increased release of phosphorus to waters: a biogeochemical assessment of the mechanisms. Environmental science & technology. 15;43(6): 1858-1863.

Phytosociology of some weeds of wheat communities around Kotli fields, Western Himalaya

Zahid Hussain Malik , Muhammad Shoaib Amjad, Sidra Rafique and Nafeesa Zahid Malik

Department of Botany, University of Azad Jammu and Kashmir Muzaffarabad, Pakistan.

There were ten plant communities of weeds in wheat fields of Kotli. The communities were *Euphorbia-Desmostachya-Coronopus, Parthenium-Galium-Taraxacum, Zanthium-Bidens-Bothriochloa, Silybum – Amaranthus-Avena, Ranunculus-Silybum-Imperata, Oxalis-Cannabis–Vicia, Calendula-Fimbristylis-Desmodium, Taraxacum-Geranium-Poa, Phalaris-Geranium-Cynoglossum* and *Themeda-Cardus-Urtica*. The soil texture differs from loam to clay loam, loam and sandy loam with basic pH. Organic matter was high in all the communities; saturation varied from 30 to 51%, Nitrogen differed from 0.53 to 0.87%, Phosphorus from 9 to 16 ppm, Potassium from 256 to 768 ppm; electrical conductivity varied from 2.3 to 5.6.

Key words: Kotli fields, plant communities, weeds.

INTRODUCTION

Weeds are those plants species which damage the yielding potential of the land surface or water volume managed by man (Anonymous, 1994). Weeds cause direct losses depriving crops of water, light, space, mineral nutrients (Reddy and Reddi, 2001), exhibit allelopathy (Tafera, 2002; Singh et al., 2003), harbour insects, pests, and diseases (Majid et al., 1998; Hussain et al., 1988). It exists internationally and both in Azad Jammu and Kashmir, Pakistan (Mortimer, 1990; McClly et al., 1991; Frick and Thomas, 1992; Smith and Smith, 1998; Majid et al., 1998; Anonymous, 2000; Reddy and Reddi, 2001; Nasir and Sultan, 2002; Rozsireni et al., 2003; Lososova et al., 2003; Kaar and Freyer, 2003; Ahmad and Shaikh, 2003; Oad et al., 2003; Bukun, 2004; Hussain et al., 2004; Lososova, 2004; Malik, 2004; Bukun, 2004; Bukun and Guler, 2005; Nasir and Sultan, 2006; Lososova et al., 2006; Kazi et al., 2007; Oad et al., 2007; Akhtar and Hussain, 2007; Ige et al., 2008; Abbas et al., 2009) (Table 1).

MATERIALS AND METHODS

Kalah, Tenda, Chowki, Thaler colony, Malahar, Dheri, Sarsawa, Panjeera, Dakhari, Chak Mir all localities lies 141 km from Muzaffarabad, were surveyed during May, 2009. Density, frequency and cover of each species were determined using 20, 0.5mx 0.5 m quadrats laid randomly in each community and importance values were determined (Malik et al., 2005). Plant communities were recognized on the basis of highest importance values of species and were named after the leading dominants following the Curtis and Macintosh, 1950 method. Nomenclature followed here is that of Stewart (1972). Soil was collected from each community up to the depth of 15 cm and analyzed physically and chemically in the soil and water research centre Kotli (Table 2).

RESULTS AND DISCUSSION

Euphorbia-Desmostachya-Coronopus community

This community was recorded from Thaler colony at an altitude of 400 m. The dominants were *Euphorbia*

Table 1. Metrological data recorded from District Kotli during (2001-2008).

Month	Rain fall (mm) Mean total	Temperature (°C) Maximum	Minimum	Humidity (%) At 8.00AM	At 5:00AM	Wind direction At 8:00AM	At 5:00AM	Wind speed (Knots) At 8:00 AM	At 5:00 AM
January	80.63	17.65	4.42	79.37	53.75	N-W	N-W	1.7	0.57
February	116.73	20.47	7.81	76	49.87	N-E	N-E	2.2	2.96
March	114.51	26.52	12.65	65.25	41.37	N-E	N-E	2.8	1.42
April	55.23	31.97	17.47	56	35.12	N-W	N-W	2.71	1.42
May	42.28	36.52	6.77	49	31.12	N-E	N-E	2.78	1.47
June	109.9	37.18	23.33	59.37	40.75	N-E	N-E	2.07	1.17
July	236.38	34.36	23.55	79.25	60.87	S-W	S-W	1.27	0.75
August	218.82	33.38	23.15	84	67.87	S-W	S-W	1.21	0.77
September	80.25	32.35	20.57	76.87	58.12	S-W	S-W	2.28	0.82
October	35.97	30.27	15.88	71.62	47.5	N-E	N-E	2.656	0.91
November	13.68	24.76	10.13	73.75	47.62	N-W & N-E	N-W & N-E	1.72	0.23
December	41.7	19.67	6.27	78.37	54.12	N-E & N-E	N-W & N-E	1.2	0.4
Average	95.50 mm	28.75°C	12.40°C	70.73%	49%	N-E	N-E	2.04	0.90

Source: Pakistan Metrological Department, Lahore.

Table 2. Physical and chemical characteristics of soil from 10 communities recorded from Kotli during May, 2009.

Community	Texture	pH	Saturation (%)	O.M	N.%	P (ppm)	K (ppm)	E.C
1	Clay loam	7.6	51.86	1.75	0.87	13	370	3.2
2	Loamy	7.6	34.76	1.75	0.87	14	398	2.5
3	Loamy	7.5	34.03	1.44	0.72	11	342	2.7
4	Loamy	6.9	37.04	1.75	0.87	16	382	2.3
5	Loamy	7.6	40.82	1.06	0.53	10	276	2.5
6	Loamy	7.7	33.14	1.75	0.87	13	272	2.7
7	Clay loam	7.4	47.76	1.44	0.72	15	390	5.6
8	Loam	7.3	36.45	1.06	0.53	09	256	4.5
9	Clay loam	7.1	49.48	1.44	0.72	11	768	5.0
10	Sandy loam	7.3	30.38	1.06	0.53	09	256	4.2

pH = Power of hydrogen ion concentration, O.M = organic matter, N = nitrogen, P = Phosphorus, K = potassium, E.C = electrical conductivity, ppm = parts per million.

helioscopia, Desmostachya bipinnata and *Coronopus didymus,* having I.V of 35.16, 30.36 and 27.53 respectively (Table 3). *Chenopodium album* and *Oxalis corniculata* having I.V of 27.22 and 25.90 were the co-dominant. *Rumex, Zanthium* and *Fimbristylis* were the associated species. The remaining two species namely: *Poa* and *Cirsium* were rare. The soil in this community was clayey loam having basic pH. Nitrogen 0.87%, phosphorus 13 ppm. Organic matter was very high (1.75), E.C 3.2 (Table 2).

Parthenium-Galium-Taraxacum community

Dakhari fields at an altitude of 420 m were dominated by *Parthenium, Galium Taraxacum* having I.V of 39.3, 37.75 and 34.73 respectively. *Oenothera rosea,* and *Phalaris minor* having I.V of 32.5 and 31.92 were the co-dominant. *Cynodon, Melilotus* and *Medicago* were the associated species. The remaining two species including *Sauromatum* and *Avena* were rare. The soil in this community was loamy having basic pH. Nitrogen 0.87%, Phosphorus 14 ppm, Organic matter was very high (1.75), E.C 2.5 (Table 2).

Zanthium-Bidens-Bothriochloa community

At an altitude of 460 m *Zanthium-Bidens-Bothriochloa* community was present. Importance values were 27.42, 24.37 and 23.62 respectively. *Chrysanthemum indicum*

Table 3. Weeds in the wheat fields of Kotli recorded during May, 2009.

Name of the species	I.V of 10 weed communities									
	E-D-C	P-G-T	Z-B-B	S-A-A	R-S-I	O-C-V	C-F-D	T-G-P	P-G-C	T-C-U
Amaranthus viridis L.	20.92	-	-	23.19	-	-	-	-	-	-
Scandix pectveneris L.	-	-	-	-	21.9	-	-	-	-	17.35
Sauromatum venosum (Ait.) Prodr	-	20.74	-	-	17.62	-	-	14.45	-	-
Calatropis procera (Ait.) Ait.f.	-	-	-	22.52	-	-	-	-	-	-
Artemisia scoparia Waldst. & kit.	-	-	-	22.44	-	-	-	-	-	-
Bidens bipinnata L.	-	-	24.37	-	-	-	22.97	-	-	-
Calendula officinale L.	-	-	14.00	-	-	-	21.06	-	13.80	25.93
Cardus edelberghii L.	-	-	-	17.68	-	-	-	-	19.14	30.27
Chrysanthemum indicum L.	-	-	21.86	-	-	13.31	-	-	-	-
Cirsium arvense Miller	10.60	-	-	-	-	-	-	-	-	-
Conyza canadensis L.	-	-	18.57	-	-	11.64	-	-	-	-
Parthenium parviflorum L.	-	39.30	-	-	-	-	23.78	-	12.81	-
Silybum murianum L.	-	-	-	25.83	24.26	-	-	-	-	-
Sonchus asper (L.) Pers	-	-	-	-	14.56	14.43	-	-	-	-
Taraxacum officinale Weber	-	34.73	-	-	-	-	22.11	44.92	-	-
Zanthium strumarium L.	23.78	-	27.42	-	-	-	28.94	-	-	-
Cynoglossum lanceolatum Forssk	-	-	-	-	-	-	-	-	25.14	-
Cannabis sativa L.	-	-	-	16.61	-	23.77	-	-	-	-
Chenopodium album L.	27.22	-	-	20.11	-	-	-	-	-	-
Convolvulus glomeratus Boiss	-	-	13.57	-	-	-	-	-	-	-
Evolvulus alsinoides (L.) Boiss	-	-	12.95	-	-	-	-	-	-	25.46
Coronopus didymus (L.) Smith	27.53	-	-	-	-	-	-	-	-	-
Lepidium capitatum Hook.f.&Thoms	-	-	-	-	22.49	17.75	-	35.00	-	-
Fimbristylis miliacea L.	21.91	-	-	-	-	-	29.81	-	-	-
Cyperus stoloniferus Retz	17.12	-	10.34	-	-	-	-	-	-	-
Euphorbia helioscopia L.	35.16	-	-	-	-	-	-	-	-	-
Euphorbia hirta L.	-	-	-	-	-	14.96	-	-	-	-
Geranium occellatum Camb	-	-	15.49	-	-	-	-	39.39	25.40	-
Malvestrum coromandelianum (L.) Carcke	-	-	-	-	-	9.17	-	-	25.06	-
Mollugo pentaphylla L.	-	-	-	16.31	17.29	-	-	-	21.51	-
Oxalis corniculata L.	25.9	-	-	-	-	25.40	-	-	20.25	-
Desmodium gangeticum (L.) Dc.	-	-	-	-	-	-	29.65	-	-	15.75
Lathyrus aphaca L.	-	-	-	-	23.62	-	-	-	-	-
Medicago laciniata (L.) Mill	-	21.53	19.20	-	-	12.24	-	-	22.72	-
Melilotus alba Desr	-	27.04	-	-	-	-	22.17	-	-	-

Table 3. Contd.

Species	E-D-C	P-G-T	Z-B-B	S-A-A	R-S-I	O-C-V	C-F-D	P-G-C	T-G-P	T-C-U
Trifolium dubium Smith	-	-	-	-	-	-	-	-	-	-
Vicia sativa L.	19.82	-	21.64	18.99	20.87	-	-	-	27.65	25.87
Agrostis viridis L.	21.25	21.61	-	-	-	-	-	-	-	-
Avena fatua L.	4.46	-	23.15	-	13.60	24.7	-	-	-	-
Catabrosa aquatica (L.) P.Beauv	-	-	18.85	-	-	27.90	-	14.59	-	26.27
Cynodon dactylon (L.) Pers	28.33	-	-	-	-	-	-	-	-	20.45
Desmostachya bipinnata (L.) Stapf	30.36	-	-	-	-	-	-	-	-	-
Dichanthium annulatum (L.) Stapf	-	21.85	-	-	-	-	-	-	-	-
Elymus repens (L.) Gould	-	-	7.57	-	-	-	-	-	-	-
Imperata cylindrica L.	-	-	-	23.91	16.12	-	-	10.86	-	-
Phalaris arundinacea L.	-	-	-	-	15.64	-	30.42	-	-	-
Phalaris arvensis L.	-	-	-	20.41	-	-	-	-	-	-
Phalaris minor Retz	31.92	-	-	-	-	-	-	31.75	-	25.74
Poa annua L.	14.52	-	-	-	-	-	-	24.88	-	-
Poa sinaica Steud	-	18.43	-	19.44	-	-	35.41	-	-	-
Saccharum officinarum L.	-	-	-	-	18.04	-	-	-	-	-
Themeda anathera (Nees ex Steud) Hack	-	17.08	-	-	16.33	-	-	-	-	34.83
Vetiveria zizanioides (L.) Nash Vilfa Adans	-	-	15.54	21.21	-	-	-	-	-	-
Polygonum molliaeforme Boiss	-	-	-	-	18.92	-	24.73	-	-	-
Rumex dentatus L.	24.52	-	-	-	-	-	-	-	-	-
Rumex hastatus D.Don.	-	19.43	-	23.55	-	-	24.72	-	-	32.92
Anagallis arvensis L.	-	-	15.17	-	-	-	-	-	-	-
Ranunculus muricatus L.	-	-	-	19.95	30.52	-	23.12	-	-	-
Oenothera rosea L.	32.5	-	-	-	-	-	-	15.70	-	-
Galium elegans Wall	37.75	-	-	-	-	-	-	16.18	-	-
Urtica dioica L.	-	-	17.43	-	-	15.36	-	-	-	-
Verbena officinalis L.	-	-	-	-	17.07	-	-	-	-	27.93
Viola canescens Wall ex Roxb	-	-	15.75	-	21.29	-	-	-	-	-

E-D-C = *Euphorbia-Desmostachya-Coronopus* community, P-G-T = *Parthenium-Galium-Taraxacum* community, Z-B-B = *Zanthium-Bidens-Bothriochloa* community, S-A-A = *Silybum-Amaranthus-Avena* community, R-S-I = *Ranunculus-Silybum-Imperata* community, O-C-V = *Oxalis-cannabis-Vicia* community, C-F-D = *Calendula-Fimbristylis-Desmodium* community, T-C-U = *Themeda-Cardus-Urtica* community, P-G-C = *Phalaris-Geranium-Cynoglossum* community, T-G-P = *Taraxacum-Geranium-Poa* community.

and *Dichanthium annulatum* having I.V of 21.86 and 21.85 were the co-dominant, *Agrostis, Rumex* and *Lathyrus* were the associated species. The remaining two species including *Evolvulus* and

Cyperus were rare. The soil in this community was loamy having basic pH. Nitrogen 0.72%, Phosphorus 11 ppm, Organic matter was very high (1.44), E.C was 2.7 (Table 2).

Silybum-Amaranthus-Avena community

Silybum-Amaranthus-Avena community was present in Malahar at an altitude of 490 m having I.V of 25.83, 23.19 and 23.15 respectively. *Calatropis procera and Artemisia scoparia* having I.V of 22.52 and 22.44 were the co-dominant. *Trifolium, Chenopodium* and *Catabrosa* were the associated species. The remaining two species viz *Vetiveria* and *Anagallis* were rare. The soil in this community was loamy having acidic pH. Nitrogen 0.87%, Phosphorus 16 ppm, Organic matter was 1.75, E.C 2.3 (Table 2).

Ranunculus-Silybum-Imperata community

From Dheri fields at an altitude of 500 m *Ranunculus-Silybum-Imperata* community was recorded having I.V of 30.52, 24.26 and 23.91 respectively. *Lathyrus aphaca* and *Rumex hastatus* having I.V of 23.62 and 23.55 were the co-dominant. *Lepidium, Vetiveria* and *Scandix* were the associated species. The remaining two species *Mollugo* and *Sonchus* were rare. The soil in this community was loamy having basic pH. Nitrogen 0.53%, Phosphorus 10 ppm, Organic matter 1.06, E.C 25 (Table 2).

Oxalis-Cannabis-Vicia community

From Chowki at an altitude of 510 m *Oxalis-Cannabis-Vicia* community was recorded having I.V of 25.4, 23.77 and 20.87 respectively. *Ranunculus muricatus* and *Polygonum molliaeforme* having I.V of 19.95 and 18.92 were the codominant. *Saccharum, Lepidium* and *Verbena* were the associated species. The remaining two species viz *Conyza* and *Malvestrum* were rare. The soil in this community was loamy having basic pH. Nitrogen 0.87%, Phosphorus 13 ppm, Organic matter was 1.75, E.C 2.7 (Table 2).

Calendula-Fimbristylis-Desmodium community

At an altitude of 520 m from the fields of Chak Mir *Calendula-Fimbristylis-Desmodium* community was present having I.V of 31.06, 29.81 and 29.65 res-pectively. *Zanthium strumarium* and *Catabrosa aquatica* having I.V of 28.94 and 27.9 were the co-dominant. *Avena, Parthenium* and *Bidens* were the associated species. The remaining two species viz *Viola* and *Urtica* were rare. The soil in this community was clayey loam having basic pH. Nitrogen 0.72%, Phosphorus 15 ppm, Organic matter was 1.44, E.C 5.6 (Table 2).

Taraxacm-Geranium-Poa community

From Sarsawa fields at an altitude of 580 m *Taraxacum-Geranium-Poa* community was present having I.V of 44.92, 39.39 and 35.41 respectively. *Lepidium Capitatum* and *Phalaris arundinacea* having I.V of 35 and 30.42 were the co-dominant. *Trifolium, polygonum* and *Rumex*

were the associated species. The remaining two species viz, *Ranunculus* and *Sauromatum* were rare. The soil in this community was loamy having basic pH. Nitrogen 0.53%, Phosphorus 9 ppm, Organic matter 1.06, E.C 4.5 (Table 2).

Phalaris-Geranium-Cynoglossum community

From Kalah fields *Phalaris-Geranium-Cynoglossum* community was recorded having I.V of 31.75, 25.4 and 25.14 respectively. *Malvestrum coromandelianum* and *Poa annua* having I.V of 25.06 and 24.88 were the co-dominant. *Medicago, Mollugo* and *Oxalis* were the associated species. The remaining two species viz *Parthenium* and *Imperata* were rare. The soil in this community was clayey loam having basic pH. Nitrogen 0.72%, Phosphorus 11 ppm, Organic matter 1.44, E.C 5.0 (Table 2).

Themeda-Cardus-Urtica community

From Panjeera at an altitude of 600 m *Themeda-Cardus-Urtica* community was recorded having I.V of 34.83, 30.27 and 27.93 respectively. *Catabrosa aquatic* and *Calendula officinale* having I.V of 26.27 and 25.93 were the co-dominant. *Vicia, Phalaris* and *Evolvulus* were the associated species. The remaining two species viz *Scandix* and *Desmodium* were rare. The soil in this community was sandy loam having basic pH. Nitrogen 0.53%, Phosphorus 9 ppm, Organic matter was high (1.06), E.C 4.2 (Table 2).

Ten weed plant communities were recorded from Kalah and its outskirts of District Kotli. The investigated area differs from 400 to 600 m. Every community has different dominants due to altitude and climatic conditions. Most of the dominant communities were Euphorbia-Desmostachya-Coronopus, Parthenium-Galium-Taraxacum, Zanthium-Bidens-Bothriochloa, Silybam–Amaranthus-Avena, Ranunculus-Silybum-Imperata, Oxalis-Cannabis–Vicia, Calendula-Fimbristylis-Desmodium, Taraxacum-Geranium-Poa, Phalaris-Geranium-Cynoglossu, Themeda-Cardus-Urtica. In Kalah the most common weed species were *Oxalis, Phalaris, Oenothera, Cynoglossum, Medicago, Galium, Malvestrum, Parthenium. Catabrosa. Chenopodium, Amaranthus, Urtica, Cynodon, Euphorbia, Rumex, Cirsium, Anagallis, Zanthium* and *Medicago* were the most common weeds in all other localities.

At the base (400 m) *Euphorbia-Desmostachya-Coronopus* was dominant with clayey loam soil having basic pH and high organic matter. The dominant weeds are annuals that can easily be eradicated before flowering and fruiting. From 420 to 500 m *Parthenium, Galium, Taraxacum, Zanthium, Bidens, Bothriochloa, Silibum, Amaranthus, Avena, Ranunculus, Imperata, Oxalis* and *Vicia* were dominant species with loamy soil and high organic matter. Most of the species in these localities were annual except *Bothriochloa* and *Imperata*.

Avena fatua, Cynodon dactylon, Calendula officinale, Ranunculus muricatus, Cannabis sativa and *Imperata cylindrica* are *allelopathic* plants which can suppress the growth of other plants and are susceptible (Hussain et al., 1987; Zebun Nisa, 1984; Hussain and Khan, 1987; Inam et al., 1989; Allien, 1979; Qurashi et al., 1987). From 510 to 600 m species such as *Calendula, Fimbristylis, Themeda, Cardus* and *Urtica* were dominant weeds with clay loam and sandy loam with high organic matter. *Poa annua* is a cosmopolitan plant which exists everywhere, *Phalaris* is a notorious weed. *Themeda* is also an allelopathic weed that suppresses the other species found in its vicinity. It is a fine fodder grass (Malik et al., 2005).

Conclusion

Annual weeds were dominant in the era that can be controlled by eradicating then before flowering and fruiting. Weeds reduce the crop yield production. Several control mechanisms could be employed to control weeds.

REFERENCES

Abbas SH, Saleem M, Maqsood M, Mujahid MY, Ul-Hassaan M, Saleem R (2009). Weed density and grain yield of wheat under rainfed conditions of Pothohar. Pak. J. Agri. Sci. 46(4):242-247

Ahmad R, Shaikh AS (2003). Common weeds of wheat and their control. Pak. J. water Res. 1(1):71-73.

Akhtar N, Hussain F (2007). Weeds of wheat fields of village Qambar, Distt. Swat, Pl. J. Sci.13(1):33-37. (was not cited in the article)

Bukun B (2004). The weed flora of winter wheat from Thurkey. Pak. J. Bio. Sci. 7(9):1530-1534.

Bukun B, Guler BH (2005). Densities and importance values of weeds in lentil production. Int. J. Bot. 1(1):15-18.

Hussain F, Malik ZH (1986). The distribution of some weeds in maize (*Zea may* L) Fields of Kotli (Azad Jammu and Kashmir) Sar. J. Agri 2(3):561-569

Hussain F, Chaghtai SR, Marwat Q (1987). Distribution of some weeds in apple orchards of Quetta. Pak. J. Agric. Res. 8 (3): 260-265.

Hussain F, Khan TW (1987). Allelopathic effects of *Cynodon dactylon*. Pak. J. Weed Sci. Res. 1: 8-18.

Hussain F, Murad A, Durrani MJ (2004). Weed communities in the wheat fields of Mastuj, Distt. Chitral. Pak. J. Weed Sci. Res.10(3-4):101-108.

Ige OE, Olumekun VO, Olagbadegun JO (2008). A Phytosociological study of weed flora in three abandoned farmland of Nigeria. Pak. J. Weed Sci. Res. 14(1-2):81-89.

Kaar B, Freyer B (2003). Weed diversity and cover-abundance in organic and conventional winter cereal fields. Div. Org. Farming; Uni. Natural Res. and App. Life Sci. (BOKU), Gregor-Mendelstr. 33,A-1180 Vienna.

Kazi BR, Buriro AH, Kubar RA, Jagirani AW (2007). Weed spectrum frequency and density in wheat under Tandojam conditions. Pak. J. Weed Sci. Res. 13(3-4):241-246.

Lososova Z, Danihelka J, Chytry M (2003). Seasonal dynamics and diversity of weed vegetation in tilled and mulched vineyards. Bio. Bratis, 58/1:49-57.

Lososova Z, Chytry M, Cimalova S, Otypkova Z, Pysek P, Tichy L (2006). Classification of weed vegetation of arable land in the Chzech Republic and Slovakia. Folia Geobotanica 41:259-273.

Lososova Z (2003). Estimating past distribution of vanishing weed vegetation in South Moravia. Preslia, Praha, 75:71-79. (Was not cited in the article)

Lososova Z (2004). Weed vegetation in Southern Moravia (Chzech Republic): a formalized Phytosociological classification. Preslia, Praha, 76:65-85.

Malik MA (2004). Distribution of Chenopodium album in some irrigated and rieverian wheat fields of distt. Lahore. Pak. J. Weed Sci. Res. 10(3-4):133-138.

Malik RS, Yadav A, Malik RK, Singh S., (2005). Performance of weed control treatments in mungbean under different sowing methods. Indian J. Weed Sci., 37:273-274

Nasir ZA, Sultan S (2002). Floristic, Biological and leaf size spectra of weeds in Gram, Lentil, Mustard and Wheat fields of Distt. Chakwal. Pak. J. Bio. Sci. 5(7):758-762.

Nasir ZA, Sultan S (2006). Noxious weeds of winter crops in Distt. Chakwal. Int. J. Agri. Res.1(5):480-487.

Oad FC, Siddiqui MH, Buriro UA (2007). Growth and yield losses in wheat due to different weed densities. Asian. J. Pl. Sci. 6(1):173-176.

Oad FC, Agha SK, Jamro GH, Solangi GS (2003). Weed spectrum frequency and density in wheat under Tandojam conditions. Pak. J. App. Sci. 3(3):170-172.

13

Study of flora of Miandasht Wildlife Refuge in Northern Khorassan Province, Iran (a)

Rahimi A.[1] and Atri M.[2]

[1]Department of Biology, Bojnourd Branch, Islamic Azad University, Bojnourd, Iran.
[2]Department of Biology, Bu Ali-Sina University, Hamedan, Iran.

A wide area of Iran is covered by arid and semiarid regions. In this survey, flora of an area of the Miandasht Wildlife Refuge, out of the safe part, was studied. This region covers 84435 Ha, situated in the west of Khorassan province in Iran. The climate of the area according to de Martone system is semiarid. The mean annual precipitation is 275 mm and the altitude varies from 931 to 1021 m above sea level. Plants were collected from 2008 to 2011. A total of 256 taxa belonging to 152 genera and 35 families from Angiospermae and Gymnospermae were found. Asteraceae, Chenopodiaceae, Brassicaceae and Fabaceae were the greatest families, respectively. Geraniaceae, Ixioliriaceae, Orobanchaceae, Plantaginaceae, Primulaceae, Resedaceae and Rosaceae, each included one species. Based on Raunkiaer life form classification system, majority of the species (55.86%) were therophytes. Other life forms in descending order were hemicryptophytes (15.62%), chamaephytes (10.16%), phanerophytes (8.6%) and geophytes (9.38%). Chorologicaly, most of the species were Irano-Turanian. Flora of Miandasht Wildlife Refuge include 20 low risk species and 29 (11.6%) endemic of Iran species. 67 pasture species and 38 medicinal species were distinguished. Most of the species were invasive plants. They are established in this area because of overgrazing and natural disturbance.

Key words: Flora, Miandasht Wildlife Refuge, Iran endemic, life form, chorotype.

INTRODUCTION

Iran in terms of topography, climate, vegetation and geographical features is one of the most important and unique countries in the Middle East. According to a recent study (Mozaffarian, 2007), flora of Iran includes 8000 species belonging to 1450 genera and 150 families. These families include 124 dicotyledons, 22 monocotyledons and 4 gymnosperms.

Some resources related to vegetation of Iran are as follows: Flora Orientalis (Boissier, 1936), Flora Keredjensis (Bornmuller and Gauba, 1935-1940), Flore de l'Iran (Parsa, 1948-1952), Flora Iranica (Rechinger, 1963-2005) Flora of Iran, Tracheophyta (Mobayen, 1975-1995), Colored Flora of Iran (Ghahreman, 1977-2007) and Flora of Iran (Assadi et al., 1988-2011).

One of the most extensive areas for speciation in holarctic kingdom is located in Iran (Akhani, 2006). Also, some studies in the field of semi deserts and deserts of Iran have been conducted. For instance, studies on the autumn plants of Kavir, Iran (Assadi, 1984), plants of the Kavir Protected Area, Iran (Rechinger and Wendelbo, 1976), Plants of the Touran Protected Area, Iran (Rechinger, 1977), Notes on the distribution, climate and flora of the sand deserts of Iran and Afghanistan (Freitag, 1986).

Some case studies have been performed in deserts of Iran, for example: A contribution to the vegetation and flora of Kavire Meyghan, Iran (Akhani, 1989), floristic and cartographic study of protected area of Ghamishloo

(Yousofinajafabadi, 1996), study of the flora of the Kabar dam of Ghom (Tavakkoli and Mozaffarian, 2005) and flora of halophytes in Iran (Asri, 2007).

A fundamental role of government conservation agencies is to set priorities for the conservation and management of biodiversity (David and Kenneth, 2001). To evaluate the status of biodiversity and to determine how current conservation efforts can be improved, biodiversity monitoring is crucial (Kerstin et al., 2013). The nature and quality of vegetation cover is an important factor for soil conservation through its role in reducing the erosive impact of precipitation degraded areas in semi-arid regions (Turan and Filiz, 2011).

The objective of this study was to provide urgently needed scientific support for programs of biodiversity conservation. The Miandasht Wildlife Refuge (MWR) covers 84435 Ha, and is located in North Khorassan province (56°, 26′ to 56°, 57′ longitude and 47°, 30′, 36′ to 37°, 30′ latitude) Figure 1. The mean annual precipitation is 250 mm and the altitude varies from 912 to 1085 m above sea level. The climate is semiarid and chorologicaly located in the Irano-Turanian region. A wide area of MWR is rangeland. Early and uncontrolled grazing in these rangelands led to the decrease of the production by pastured plants, imbalances in the ecosystem, disruption in water and food cycles, increasing the unfavorable species and decreasing the quantity of vegetation and soil erosion. Also, a wide part of this region is covered by sand and saline soils. Totally, approximately 40 to 50% of critical points of erosion around of Jajarm city is located in this area, and the plants are permanently exposed to environmental stresses. MWR is habitat for important animal species, such as: the Asiatic cheetah which is rare worldwide and its survival is one of the most important goals of the Environmental Protection Agency, so the study of various environmental aspects of this area is essential for the balance and stability of the ecosystem, particularly vegetation which is the first loop of the animal food chains.

Aims of this study were to introduce the flora of MWR, to detect endemism, vulnerability and chorology of species, and to distinguish medicinal and pasture plants in 12000 Ha out of the safe part of the MWR. This study is intended as a useful tool for policy markers and scientists to advocate for modifications in national legislation and policy aimed at conservation and combating desertification. Analyzing species richness, extinction level and distribution drivers are important preliminary steps to set conservation priorities and to test environmental policies (Giuseppe, 2013).

In order to determine the influence of protection from grazing on diversity of plant species, flora of the safe part was studied which will be written in another article.

MATERIALS AND METHODS

Basic information on MWR was obtained using geological and topographical maps 1/50000 and aerial photos of this area. Then,

by scrolling in the area, its boundaries were determined using handheld Garmin's GPS map76CS. All plants were photographed by means of a digital camera (Nikon D70S). Plants were identified using stereomicroscope, keys and descriptions in available scientific resources, specially flora Iranica (Rechinger, 1963-2005), Flora of Iran (Assadi et al., 1988-2011) and flora of the adjacent countries, namely flora of Turkey and the East Aegean Island (Davis, 1965-1985). Flora of U.S.S.R (Komarov, 1934-1957), Flora of West Pakistan (Nasir and Qaser, 1970-2001) and Flora of Iraq (Townsend and Guest, 1966-1986).

Finally, names of all plants were confirmed by taxonomists in the research institute of forests and rangelands of Iran. The life forms of species were distinguished according to the life form classification (Raunkiaer, 1934). The geographical distribution of each species was assessed from reviews, monographs and distributional data in the floras, particularly Flora Iranica (Rechinger, 1963-2005). The terminology and delimitation of the main phytogeographic areas, that is, Euro-Siberian (ES), Irano-Turanian (IT), Mediterranean (M) and within Euro-Siberian region relate to standard works of reference, particularly those of Zohary (1973), Takhtajan(1986) and flora Iranica (Rechinger, 1963-2005). The IUCN Red List Categories (Jalili and Jamzad, 1999) are used to designate the threat categories.

Pasture plants were detected by asking the villagers by direct observations, and by using the Codes of Pasture Plants (Publication Committee and Propaganda of the Research Institute of Forestsand Rangelands of Iran 1982).

Medicinal plants were determined using the available scientific resources, for example: Amin (1991), Zargary (1999) and Javidtalesh (2001).

RESULTS AND DISCUSSION

Totally, 256 taxa belonging to 152 genera and 35 families of Angiospermae and Gymnospermae were determined in the study area (Table 1, Figure 5-12). These families consist of 1 Gymnospermae, 29 Dicotyledons and 5 Monocotyledons.

According to Table 1, the following families had the highest number of species: 43 (16.8%) Asteraceae, 41 (16.02%) Chenopodiaceae, 32 (12.5%) Brassicaceae and 22 (8.59%) Fabaceae (Figure 2). These results are consistent with the results of most studies in similar areas (Rechinger and Wendelbo, 1977; Asri, 2003; Yousofinajafabadi, 1996).

Presumably uncontrolled grazing caused the maximum number of Asteraceae species (Tavakkoli and Mozaffarian, 2005). Since the studied area is at risk due to early and excessive grazing, the maximum number of Asteraceae species is justifiable. In the study of halophytes of Iran (Asri, 2007), the numbers of Chenopodiaceae species were large. The large number of this taxon species can be explained by saline soils which are widely spread in this region.

The large number of Chenopodiaceae, Asteraceae and Brassicaceae are indicator of desert conditions (Saberamoli et al., 2001). Table 1 illustrates that the genera of *Astragalus, Salsola, Atriplex* and *Valerianella* have the highest number of species with 17, 9, 6 and 6 species, respectively. Irano-Turanian region is the major origin of *Astragalus*, and 91% of *Astragalus* species Iran, grow in

Table 1. List of plants in the Miandasht Wildlife Refuge (out of the safe part) and some of their features.

Scientific name of taxon	L.F	Chor	Pa	Me	En	Vu
Apiaceae						
Cuminum setifolium L.	Th	IT		*		
Dorema aitchisonii Korov. ex M. Pimen.	Hem	IT				
Ducrosia anetifolia (DC.) Boiss.	Hem	IT-SS	*	*		
Eryngium bungei Boiss.	Hem	IT	*			
Psammogeton canescens (DC.) V.	Th	IT				LR
Schumannia karelinii (Bunge) Korov.	G.t	IT-SS				
Asteraceae						
Acantholepis orientalis Less.	Th	IT	*			
Acroptilon repens (L.) DC.	Hem	IT		*		
Amberboa turanica LIjin	Th	IT				
Amberboa nana (Boiss.) LIjin.	Th	IT				
Anthemis austero-iranica Rech. f., Aell. & Esfand.	Th	IT-SS			*	LR
Anthemis rhodocentra Iranshahr	Th	IT-SS				
Artemisia scoparia Waldst. & Kit.	Ch	IT-ES	*	*		
Artemisia sieberi Besser.	Ch	IT	*	*		
Carthamus oxyacantha M. B.	Th	IT-M-SS				
Centaurea bruguieriana (DC.)Hand.	Th	IT				
Centaurea pulchella Ledeb.	Th	IT				
Cousinia lasiandra Bunge.	Hem	IT			*	LR
Cousinia neurocentra Bunge.	Hem	IT			*	LR
Cousinia piptocephala Bunge.	Hem	IT			*	LR
Cousinia prolifera Jaub. & Spach.	Th	IT				
Cousinia turkmenorum Bornm	Th	IT				
Crepis sancta (L.) Babcock.	Th	IT-M				
Dipterocome pusilla Fisch & C. A. Mey.	Th	IT				
Echinops leucographus Bunge	Hem	IT				
Echinops pungens Trautv.	Hem	IT				
Epilasia acrolasia (Bunge.) C. B. Clarke.	Th	IT				
Epilasia hemilasia (Bunge.) C. B. Clarke.	Th	IT				
Filago arenaria L.	Th	IT				
Gymnarrhena micrantha Desf.	Th	IT-SS				
Heteroderis pusilla (Bunge) Boiss.	Th	IT				
Koelpinia linearis Pall.	Th	IT-SS				
Koelpinia tenuissima Pavl. & Lipsch	Th	IT				
Lactuca serriola L.	Hem	IT	*	*		
Lasiopogon muscoides (Desf.) DC.	Th	IT-SS				
Launaea acanthodes (Boiss.)	Hem	IT			*	DD
Mausolea eriocarpa (Bge.) Poljak. ex Podl.						
Microcephala lamellata (Bunge) Pobed.	Th	IT				
Oligochaeta minima (DC.) C. koch.	Th	IT				
Pulicaria gnaphalodes (Vent.) Boiss.	Hem	IT-SS				
Senecio glaucus L.	Th	IT-M-SS				
Scariola orientalis (Boiss.) Sojak.	Hem	IT				
Scorzonera lituisinowa Fisch. & C.A.Mey.	G.t	IT				
Scorzonera paradoxa Fisch. & C.A. Mey.	G.t	IT				
Scorzonera pusilla Pall.	G. t	IT				
Scorzonera raddeana C. Winkl.	G. t	IT				
Scorzonera rigida Auch.	G. t	IT				
Thevenotia persica DC.	Th	IT				

Table 1. Contd.

Xanthium stromarium L.	Th	IT		*	
Boraginaceae					
Arnebia decumbens (Vent.) Coss. & Kral.	Th	IT-SS	*		
Arnebia linearifolia DC.	Th	IT-SS			
Gastrocotyle hispida (Forssk.) C. B. Clarke	Th	IT-SS			
Heliotropium aucheri DC.	Hem	IT			
Heliotropium europaeum L.	Th	IT -ES	*	*	
Heliotropium dasycarpum Ledeb.	Hem	IT			
Heterocaryum subsessile Vatke, Zeitschr. Gesammt.	Th	IT-SS			
Lappula ceratophora (M. Pop.) M. Pop.	Th	IT			
Lappula semiglabra (Ledeb.) Gurke	Th	IT			
Lappula sesiliflora (Boiss.) Gurke	Th	IT			
Lappula spinocarpus (Forssk.) Ascherson & O. Kuntze	Th	IT-SS			
Microparacary bungei (Boiss.) Khatamsaz, comb .Nov.	Th	IT-SS			
Nonnea caspica (Willd.) G. Don	Th	IT			
Brassicaceae					
Aethionema carneum (Banks & Soland.) B. Fedtsch.	Th	IT			
Alyssum dasycarpum Steph. ex Willd.	Th	IT	*	*	
Alyssum linifolium Steph. ex Willd.	Th	IT-ES-M	*	*	
Alyssum marginatum Steud. ex Boiss.	Th	IT		*	
Arabidopsis pumila (Steph.) N. Bosch.	Th	IT			
Cardaria draba (L.) Desv.	G. r	IT	*	*	
Cryptospora falcata Kar. & Kir.	Th	IT			
Descurainia sophia Webb. & Berth.	Th	IT-M-ES	*	*	
Erysimum crassicaule (Boiss.) Boiss.	Hem	IT	*		* LR
Euclidium syriacum (L.) R. Br.	Th	IT	*		
Goldbachia laevigata DC.	Th	IT	*	*	
Goldbachia verrucosa DC.	Th	IT			
Isatis buschiana Schischk.	Hem	IT			
Isatis emarginata Kar. & Kir.	Th	IT			
Isatis minima Bge.	Th	IT			
Lepidium perfoliatum L.	Th	IT-M-ES	*		
Lepidium vesicarium L.	Th	IT			
Leptaleum filifolium (Willd.) DC.	Th	IT	*		
Malcolmia africana (L.) R. Br.	Th	IT-M-SS	*		
Malcolmia turkestanica Litw.	Th	IT			
Matthiola chenopodifolia Fisch. & C. A. Mey	Th	IT			
Matthiola dumulosa Boiss. & Buhse.	Ch	IT		*	LR
Octoceras lehmannianum Bunge.	Th	IT			
Sameraria armena (L.) Desv.	Th	IT			
Sameraria elegans Boiss.	Th	IT		*	DD
Sinapis arvensis L.	Th	ES-IT-M-SS	*	*	
Sisymbrium septolatum DC.	Th	IT-SS			
Sterigmostemum acanthocarpum Fish. & C. A. Mey.	Th	IT			
Sterigmostemum rhodanthum Rech. f.	Th	IT		*	DD
Tetracme recurvata Bge.	Th	IT			
Thlaspi perfoliatum L.	Th	IT			
Torularia torulosa (Desf.) O. E. Schulz.	Th	IT-SS			

Table 1. Contd.

Capparidaceae						
Buhsea trinervia (DC.) Stapf.	G.r	IT				
Capparis spinosa L.	Ch	IT-M-SS	*	*		
Caryophyllaceae						
Acanthophyllum acerosum Sosn.	Ch	IT				
Acanthophyllum crassifolium Boiss.	Ch	IT				
Acanthophyllum glandulosum Bunge ex. Boiss	Ch	IT				
Gypsophila linearifolia (Fisch. & C. A. Mey.) Boiss.	Th	IT				
Holosteum umbellatum L.	Th	IT-SS				
Chenopodiaceae						
Anabasis annua Bge.	Th	IT	*			
Anabasis setifera Moq.	Ch	IT-SS	*	*		
Atriplex dimorphostegia var. *dimorphostegia* Kar & Kir.	Th	IT-SS	*			
Atriplex dimorphostegia var. *sagitiformis* Allen.	Th	IT	*			
Atriplex leucoclada (Boiss.) Allen.	Hem	IT-SS	*			
Atriplex moneta Bge.	Th	IT				
Atriplex tatarica L.	Th	IT-M				
Atriplex verrucifera M. B.	Hem	IT	*			
Ceratocarpus arenarius L.	Th	IT	*			
Gamanthus gamocarpus (Moq.) Bge.	Th	IT	*			
Girgensohnia oppositiflora (Pall.) Fenzel in Ledb.	Th	IT				
Halimocnemis mamamensis (Bge.) Assadi, comb. nov.	Th	IT-ES			*	DD
Halimocnemis pilifera Moq.	Th	IT				
Halocharis sulphurea Moq.	Th	IT	*			
Halocharis violaceus Bge.	Th	IT				
Halocnemum strobilaceum (Pall.) M. B.	Ch	IT-SS-M	*			
Halostachys belangeriana (Moq.) Botsch.	Ph	IT	*			
Halothamnus glaucus subsp. *cinerascens* (Moq.) Assadi.	Ch	IT			*	LR
Halothamnus glaucus subsp. *vestitus* (Allen.) Assadi, comb. nov.	Ch	IT			*	LR
Halothamnus subaphyllus Botsch.	Ph	IT				
Haloxylon ammodendron (C. A. Mey.) Bge.	Ph	IT	*			
Haloxylon persicum Bge. & Boiss. et Buhse.	Ph	IT-SS	*			
Horaninowia anomala (C. A. Mey.) Moq.	Th	IT				
Kalidium caspicum (L.) Ung-Sterb.	Ch	IT	*			
Kalidium foliatum (Pall.) Moq.	Ch	IT				
Kochia stellaris Moq.	Th	IT	*			
Petrosimonia glauca (Pall.) Bge.	Th	IT				
Salsola arbuscula Pall.	Ph	IT				
Salsola arbusculiformis Drob.	Ph	IT				
Salsola crasaa M. B.	Th	IT	*			
Salsola dendroides Pall.	Ch	IT	*			
Salsola kali L.	Th	IT	*	*		
Salsola kerneri (Wol.) Botsch.	Ch	IT				
Salsola orientalis S. Gmelin.	Ch	IT	*			
Salsola tomentosa (Moq.) Spach.	Ch	IT				
Salsola turcomanica litw.	Th	IT-SS	*			
Seidlitzia florida (M. B.) Boiss.	Th	IT	*			
Seidlitzia rosmarinus (Ehrh.) Bge.	Ph	IT-SS	*			

Table 1. Contd.

Suaeda acuminata (C. A. Mey.) Moq.	Th	IT				
Suaeda microphylla Pall.	Ph	IT-ES				
Suaeda microsperma (C. A. Mey.) Fenzel in Ledeb.	Th	IT				
Convolvulaceae						
Convolvulus eremophilus Boiss. et Buhse.	Ch	IT-SS			*	DD
Convolvulus pilosellaefolius Desr.	Hem	IT-SS	*			
Cressa cretica L.	Hem	IT-M-SS		*		
Cyperaceae						
Carex physodes Bieb.	G.r	IT				
Scirpus maritimus L.	G.r	IT				
Dipsaceae						
Scabiosa olivieri Coult.	Th	IT-SS-ES				
Scabiosa rotata M. B.	Th	IT				
Ephedraceae						
Ephedra sarcocarpa Aitch. et Hemsl.	Ph	IT		*		
Ephedra strobilacea Bge. ex Lehm.	Ph	IT		*		
Euphorbiaceae						
Chrozophora tinctoria (L.) Juss.	Th	IT-M		*		
Euphorbia cheirolepioides Rech. f.	Th	IT			*	DD
Euphorbia densa Schrenk.	Th	IT				
Euphorbia heteradenia Iaub. & Spach.	Hem	IT				
Euphorbia sororia Schrenk	Th	IT				
Euphorbia turcomanica Boiss.	Th	IT				
Fumariaceae						
Fumaria parviflora Lam.	Th	IT-SS		*		
Geraniaceae						
Erodium oxyrrhynchum M. B.	Th	IT-SS-ES				
Iridaceae						
Iris kopetdaghensis (Vved.) Mathew & Wendelbo.	G. b	IT-ES				
Iris songarica Schrenk	G. r	IT-ES				
Ixioliriaceae						
Ixiolirion tataricum Fisch. ex Herb.	G. b	IT-SS-ES	*			
Lamiaceae						
Eremostachys hyoscyamoides Boiss & Buhse.	Hem	IT			*	LR
Lallemantia royleana (Benth. In Walt.) Benth.	Hem	IT	*	*		
Salvia reuterana Boiss.	Hem	IT		*	*	DD
Stachys trinervis Aitch &Hemsl.	Ch	IT				
Thuspeinanta persica (Boiss.) Briq.	Th	IT				
Ziziphora tenuir L.	Th	IT	*	*		
Liliaceae						
Allium borszczowii Regel.	G. b	IT				

Table 1. Contd.

Allium capsicum (Pall.) M. B.	G. b	IT			
Eremorus inderiensis Stev.)Boiss.	G. t	IT			
Gagea reticulata (Pall.) Schultes	G. b	IT			
Tulipa biflora Pall.	G. b	IT			
Tulipa montana Lindl. var.*montana*	G. b	IT	*	*	DD
Orobanchaceae					
Orobanch picridis FW. Schultz.	Par	IT			
Papaveraceae					
Glaucium elegans Fisch. & C. A. Mey.	Th	IT			
Hypecoum pendulum L.	Th	IT-M-SS	*		
Romeria hybrida (L.) DC.	Th	IT-SS			
Papilionaceae					
Alhagi persarum Boiss. & Buhse.	Hem	IT	*	*	
Astragalus angustatus Boiss.	Hem	IT	*		* LR
Astragalus argyroides G. Beck.	Hem	IT	*		
Astragalus arpilobus (Boiss.) Podl.	Th	IT-SS			
Astragalus bakaliensis Bunge.	Th	IT			
Astragalus campylorrhyncus F. & M.	Th	IT			
Astragalus commixtus Bunge	Th	IT			
Astragaluscorronilla Gazer & Podl.	Th	IT			
Astragalus crenatus Schultes	Th	IT-SS			
Astragalus dactylocarpus Emend. Ott.	Ch	IT			
Astragalus eremophilus subsp. *eremophilus* Emend. Podlech	Th	IT-SS			
Astragalus eremophilus subsp. *makranicus* Podlech	Th	IT-SS			
Astragalus kahiricus DC.	Hem	IT-SS		*	
Astragalus macrobotrys Bunge	Ch	IT			
Astragalus nigricans Barneby	P	IT			
Astragalus oxyglottis Bieb.	Th	IT			
Astragalus pellitus Bunge	Hem	IT			
Astragalus podolobus Boiss. & Hohen.	Ch	IT			
Astragalus tribuloides Delile	Th	IT-SS			
Ophiocarpus sp.	Th	IT			
Sophora pachycarpa C. A. Mey.	Hem	IT			
Trigonella calliceras Fisch.	Th	IT	*		
Plantaginaceae					
Plantago evacina Boiss.	Th	IT		*	
Plumbaginaceae					
Acantholimon acmostegium Boiss. & Buhse.	Ch	IT		*	LR
Acantholimon scorpius (Jaub. & Spach)	Ch	IT		*	LR
Acantholimon acerosum (Wild.) Boiss.	Ch	IT			
Poaceae					
Aeluropus littoralis (Gouan) Parl.	G. r	IT-M-SS	*		
Boissiera squarrosa Hochst. ex Steud.	Th	IT	*		
Bromus danthoniae Trin.	Th	IT	*		
Bromus sericeus Drobv.	Th	IT			

Table 1. Contd.

Bromus tectorum L.	Th	Cosm	*	*		
Eremopyrum bonaepartis (Speng.) Nevski	Th	IT	*			
Eremopyrum distans (C. Koch.) Nevski	Th	IT				
Hordeum glaucum Steud.	Th	IT-M	*			
Nardurus subulatus (Banks & Soland.) Bor.	Th	IT				
Phalaris minor Retz.	Th	IT-M	*			
Poa bulbosa L.	G. b	IT-M-ES	*			
Schismus arabicus Nees.	Th	IT-M-SS				
Stipa caucasica Schmalh.	Hem	IT-ES				
Stipa lessingiana Trin. & Rupr.	Hem	IT-Es	*			
Stipagrostis pennata (Trin.) De. Winter	G. r	IT	*			
Stipagrostis plumosa (L.) Munro. ex. T. Anders	Hem	IT-SS	*			
Polygonaceae						
Atraphaxis spinosa L.	Ph	IT-M		*		
Calligonum denticulatum Bge. ex Boiss.	Ph	IT			*	
Calligonum junceum (Fisch. &C.A.Mey.) Litw.	Ph	IT				
Polygonum hyrcanicum Rech. f.	Th	IT-ES			*	LR
Polygonum olivascens Rech. f. & Schiman- Czeika	Th	IT-SS			*	
Polygonum patulum M. B.	Th	IT-M				
Pteropyrum aucheri Jaub. & Spach.	Ph	IT	*		*	
Primulaceae						
Anaghalis arvensis L.	Th	IT-M-ES	*	*		
Ranunculaceae						
Ceratocephalus falcatus (L.) Pers.	Th	IT-M				
Consolida rugulosa (Boiss.) Schrod.	Th	IT				
Resedaceae						
Reseda buhseana Mull-Arg. var. *buhseana* Mull- Arg.	Hem	IT				
Rosaceae						
Rosa persica Michx.ex Juss.	Ph	IT				
Rubiaceae						
Callipeltis cucullaria Stev.	Th	IT				
Leptunis trichodes (J. Gay) Schischk	Th	IT				
Gaillonia brungieri A. Rich.	G. r	IT-SS			*	LR
Rutaceae						
Haplophyllum glaberrimum Bge. ex Boiss.	Hem	IT			*	LR
Haplophyllum sp.	Hem	IT				
Solanaceae						
Hyoscyamus pusillus L.	Th	IT				
Lycium ruthenicum Murr.	Ph	IT				
Tamaricaceae						
Reaumuria cistoides Adam.	Ch	IT-ES				
Reaumuria oxiana (Ledeb.) Boiss var. *persica* (Boiss.) Assadi	Ch	IT			*	LR

Table 1. Contd.

Tamarix gallica L.	Ph	IT-SS	*	
Tamarix macrocarpa (Ehrenberg.) Bge.	Ph	IT-SS		
Valerianaceae				
Valerianella cymbicarpa C. A. Mey.	Th	IT		
Valerianella dufresnia Bge. ex Boiss.	Th	IT		
Valerianella oxyrrhynca Fisch & C. A. Mey.	Th	IT		
Valerianella szwitsiana Fisch. & C. A. Mey.	Th	IT		
Valerianella triplaris Boiss & Buhse.	Th	IT		
Valerianella turkestanica Regel & Schmalh. ex Regel.	Th	IT		LR
Zygophyllaceae				
Nitraria schoberi L.	Ph	IT	*	
Peganum harmala L. var *harmala*	Hem	IT-M-SS	*	
Tribulus macropterus Boiss.	Th	IT-SS		
Tribulus terrestris L. var *terrestris*	Th	IT-SS-ES	*	
Zygophyllum atriplicoides Fisch. & C. A. Mey.	Ph	IT-SS	*	
Zygophyllum eurypterum Boiss. & Buhse.	Ph	IT-SS-M		
Zygophyllum miniatum Cham. & Schlechtend.	Hem	IT	*	

L.F: Life form, Chor: chorotype, Pa: pasture, Me: medicinal, En: endemic, Vu: vulnerability, Th: therophyte; Hem: hemicryptophyte, Ch:Chamaephyte, Ph:phanerophyte, G.t geophyte tuberous, G.b: geophytes bulbous; G.r: geophytes rhizomous, IT: Irano-Turanian, M :Mediterranian, SS: Sahara-Sindian, ES: Europa-Siberian, DD: data deficiency, LR: lowrisk, Par: parasite.

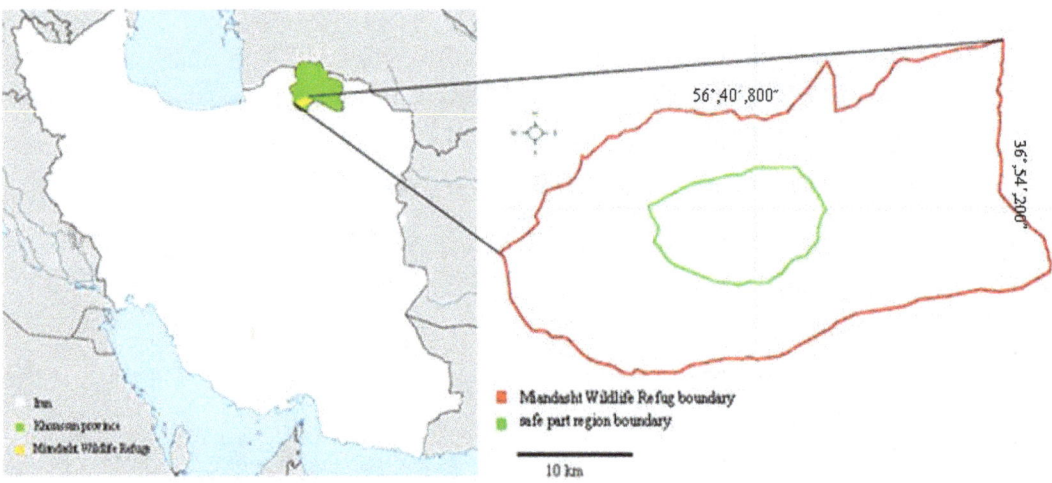

Figure 1. Location of Miandasht Wildlife Refuge on Iran map (drawn using Adobe Photoshop CS4).

this region (Maassoumi, 1986-2005). Hence, the maximum number of *Astragalus* species was expected in Miandasht because it is situated in the Irano-Turanian region.

The floristic composition of the vegetation expressed the climatical and edaphical conditions of this region. In this composition, relatively high presence of certain groups of plants, that each of them grows *in* specific environmental conditions, could be observed. These groups include the desert plants (*Artemisia sieberi, Cousinia neurocentra, Heliotropium aucheri, Acantholimon crassifolium*), psammophytes (*Carex physodes, Calligonum junceum, Schomannia karelinii, Psammogeton canescens*), desert halophytes (*Anabasis annua, Atriplex moneta, atriplex dimorphostegia, Halimocnemis pilifera, Halocharis sulphurea, Seidlitzia rosmarinus, Salsola turcomanica*), marsh halophytes (*Kalidium capsicum, Halocnemum strobilaceum,*

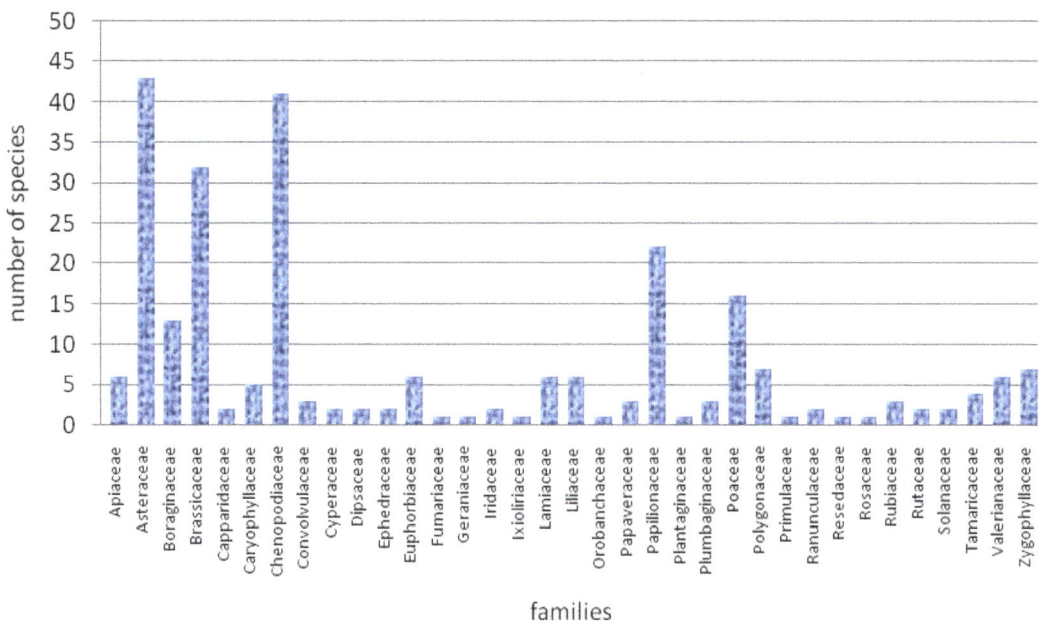

Figure 2. Chart of the number of species belonging to families in flora of Miandasht Wildlife Refuge (out of the safe part).

Aeloropus littoralis, Phragmites australis) and a large number of ruderals, for example: *Buhsea trinervia, Cardaria draba, Launaea acanthodes,Peganum harmala, Sophora Pachycarpa, Acantholepis orientalis*. Thelarge number of annual and perennial anthropogenicand ruderal plants in this area indicate excessive degradation, especially as a result of overgrazing and undermining by livestock. If rangelands are allowed to regenerate properly and grazing controlled in selected area, conservation can be achieved and the long term stability of the pastural life-style can be enhanced (Shahina and Ghazanfar, 1998).

Figure 3 illustrates that the therophytes with 143 species (55.86%) were the maximum number of life form on the flora of this area, then there were 42 (15.62) hemicryptophytes, 26 (10.16%) chamephytes, 23 (8.6%) phanerophytes and totally 20 (9.38%) geophytes.

The *Orobanch picridis* is a parasitic species that grows in this area. Among the geophytes, 8 (3.125%) rhizomous species, 8 (3.125%) tuberous species and 8 (3.125%) bulbous species were found. Therophytes are reproduced by the seeds. Compatibility of this procedure for reproduction in arid areas is more than other ways, because seeds are reproduced in small size and large number. So, they are distributed very easily. Seeds usually survived on unfavorable conditions. Also, genetic diversity of seeds which are produced through sexual reproduction leads to genetic flexible populations (Neishabouri, 1995). Most of the therophytes that grow in MWR are adapted with short duration of precipitation and high temperature, and they complete the life cycle and produce the seeds. Saberamoli (2001) said that when the

numbers of therophytes exceed the other life forms it shows desert conditions.

According to Table 1, 29 (11.33%) Iran endemic species grow in MWR. The presence of endemic species is among the fundamental criteria for characterizing biodiversity of a territory (Giuseppe, 2013). On the other hand, there are no obvious correlation between modern climate and endemism (Linder, 2001). Iran is one of the main centers of endemism in the world (Saberamoli, 2001). Totally 2000 endemic species grow in Iran and Irano-Turanian region contain 85% of endemic plant species of Iran (Dehshiri, 2005). The results of this research show that in the MWR which is located in Irano-Turanian region grow 1.45% of all endemic species of Iran. Since extent of this area is much less than the Irano-Turanian region in Iran, the above result was justifiable. Chamephytes, for example *Artemisia* and some species of *salsola*, are relatively big shrubs and their production is more than that of therophytes. Also, they are significant component of vegetation in a long period of the year. By increasing drought trend from spring to late summer, also by decreasing temperature until autumn, dominance of the chenopodiaceae in the plant formation of the study area was clearly evident. They are resistant to salinity and drought. Therefore, they grow very well in spite of intense environmental conditions (Saberamoli, 2001). Chamephytes and Phanerophytes are resistant to drought and they are morphologically adapted instead of adaptation in life cycle (Saberamoli, 2001). Although these two groups were not large in the life form spectrum of MWR, they were the main component of vegetation. Also, most of them are

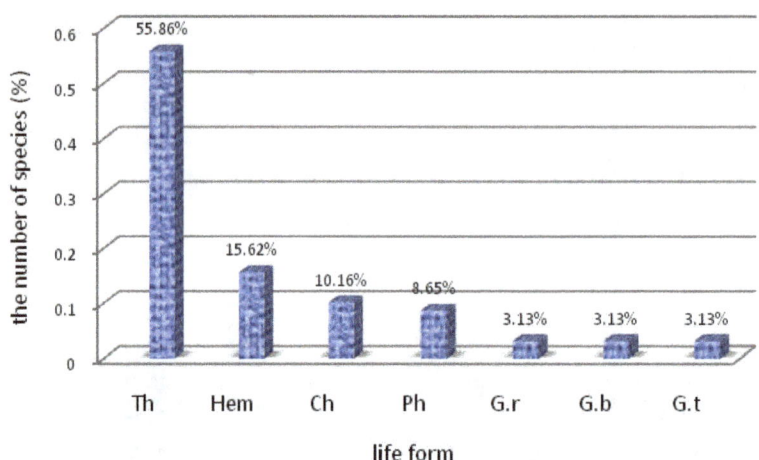

Figure 3. Life form spectrum of flora of Miandasht Wildlife Refuge (out of the safe part). Th: therophyte, Hem: hemicryptophyte, Ch: chamephyte, Ph: phanerophyte, G.r: geophyte with rhizome, G.b: geophyte with bulb, G.t: geophyte with tuber.

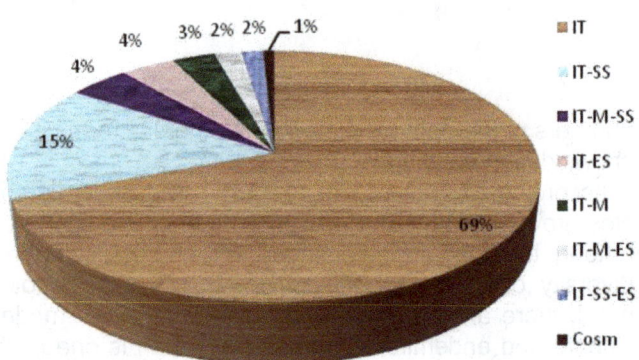

Figure 4. Weight of chorotype on flora of Miandasht Wildlife Refuge (out of the safe part).

Figure 5. *Salsola turcomanica.*Litw.

effective in preventing erosion. In addition to these, majority of them are pasture plants and provide food for wildlife. Therefore, these facts should be considered in management of this area. Because geophytes have underground organs like tubers, bulbs and rhizomes, they need suitable soil depth. Relatively high presence of geophytes showed that this area had capability of pedogenes.

Figure 4 show that the majority of species were Irano-Turanian elements. The remaining plants were common elements between Irano-Turanian and other regions. Ascending order were Irano-Turanian and Sahara-Sindian, Irano-Turanian and Euro-Siberian, Irano-Turanian and Mediterranian and Sahara–Sindian, Irano-Turanian and Euro-Siberian, Irano-Turanian and Mediterranian, Irano-Turanian and Mediterranian and Euro-Siberian, and Cosmopolitan and Irano-Turanian and Sahara-Sindianand Euro-Siberian.

Table 1 shows that from the 256 species, 64 were pasture plants and 38 were medicinal. Shrubs like *Artemisia*, especially different shrub and herbal *Chenopodiaceae* species have high forage value. They were widely distributed in this area, and they had high production. Therefore, MWR is a good pasture.

As the number of rare and threatened species has multiplied, it has become increasingly important to select species for conservation management and to provide information on the causes of decline (Perrine et al., 2013). Fortunately, this study showed that any of the species existing in MWR were not at risk or vulnerable. The current picture of extinction risk is still incomplete because many species in flora of MWR were never assessed.

We concluded on the plant diversity of MWR, in spite of environmental stresses such as drought, salinity, erosion

Figure 6. *Halocnemum strobilaceum* (Pall)M. B.

Figure 7. *Atriplex moneta* Bge.

Figure 8. *Tamarix macrocarpa* (Ehrenberg.) Bge.

Figure 9. *Nitraria schoberi* L.

Figure 10. *Lycium ruthenicum* Murr.

Figure 11. *Acantholimon acmostegium* Boiss. & Buhse.

and uncontrolled grazing by livestock is high. Moreover, this diversity is manifested not only in number of species but also in terms of the chorotype and life-form presents. In this region, the diversity of microclimates and the

Figure 12. *Anabasis setifera* Moq.

physiographic units of dry and saline areas as hillsides plains, sandy deserts, seasonal streamlets and Kavir areas have made many different local edaphic conditions.

Among the plants, Chenopodiaceae species are very important for providing the food for wildlife. Specially, the shrubs which have large amounts of proteins and minerals are important. These plants are a valuable source of food for wildlife. Overall, the flora consists of plants that have adapted to climatic and edaphic conditions of the region in life-form or morphology or life cycle. Also, many plants are invasive. These plants are distributed in this area due to the natural erosion and degradation resulting from human interventions. So, proper management and protection of the existing sources can increase species richness.

ACKNOWLEDGEMENTS

This study was funded by Islamic Azad University, Bojnourd Branch. We are thankful to the Northern Khorassan Provincial Directorate of Environment Protection and Jajarm Department of Environmental Protection for providing necessary facilities during the field surveys. We would like to thank Dr. Maassoumi and Dr. Mozaffarian for helping us in identifying the plants.

REFERENCES

Akhani H (1989). A contribution to the vegetation and flora of kavire Meyghan (NE.Arak), Iran. J. Sci.Univ. Tehran 18:75-84.

Akhani H (2006). Flora Iranica, Facts and Figures and a List of Publications by K.H. Rechinger on Iran and adjacent areas. Rostaniha 7 (suppl.2).

Amin Gh (1991). Medicinal plant of Iran. Ministry of Health, Therapy and Medical teaching press, Tehran.

Asri Y (2003). Plant diversity in Kavir biosphere reserve. Research Institute of Forests and Rangelands press,Tehran.

Asri Y (2007). Flora of halophytes of Iran. The proceeding of 1st national Plant Taxonomy Conference of Iran. Research Institute of Forests and Rangelands,Tehran.

Assadi M (1984). Studies on the autumn plants of kavir, Iran. Iran. J. Bot. 2 (2): 125 – 148.

Assadi M et al (1988-2011). Flora of Iran. Vol: 1-71 Research Institute of Forests and Rangelands press,Tehran.

Boissier E (1936). Flora Orientalis. Genevae & Basileae, Lugduni.

Bornmuller J, Gauba E (1935 -1940). Flora keredjensis Fundamenta. Cambridge University Press, Cambridge.

David JC, Kenneth AA (2001).Prioritysetting and the conservation of Western Australia's diverse and highly endemic flora. Biol. Conserv. 97(2): 251-263.

Davis PH (1965 - 1985). Flora of Turkey and the East Aegean Islands. Vols 1-9 Edinburg Univ. Press,Edinburgh.

Dehshiri MM (2005). Study of plant associations of Dizin, Gagereh and Velayatrood regions.Dessertation, Science and Research branch of Islamic Azad University of Tehran.

Freitag H (1986). Notes on the distribution, climate and flora of the sand deserts of Iran and Afghanistan. Proceeding of the Royal Society of Edinburg. Edinburg.

Ghahreman A (1977-2007). Colored flora of Iran.Vols 1-26, Research Institute of Forests & Rangelands press,Tehran.

Giuseppe B (2013). Adaptive management as a tool to improve the conservation of endemic floras:the case of Sicily, Malta and their satellite islands. Biodivers. Conserv. 22(6-7): 1317-1354.

Javidtalesh I (2001). Medicinal plant of Fars province. Res. Med. Aromat. Plants 11: 103-148.

Komarov VL (1934-1957). Flora of U. S. S. R. Vols 1-30, The Botanical Institute of science of the U.S.S.R., Leningrad.

Linder HP (2001). Plant diversity and endemism in sub-saharian tropical Africa. J. Biogeogr. 28(2): 169-182.

Maassoumi AA (1986-2005). The genus Astragalus in Iran. Vols 1-5, Research Institute of Forests & Rangelands,Tehran.

Mobayen S (1975-1995). Flora of Iran, Teracheophyta Flora. Vols 1-4, Tehran University.Tehran.

Mozaffarian V (2007). Plant distribution In Iran and endemism in Iran. The proceedings of 1st national Plant Taxonomy conference of Iran. Tehran.

Nasir E, Qaiser M (1970-2001). Flora of West Pakistan.Vols 1-202, B.C.C. and T Press,University of Karachi.

Neishabouri A (1995). Biogeography. Study and Collection of human science books organization. Samt, Tehran.

Parsa A (1948-1952). Flore de l'Iran. Offset Press Ink.

Perrine G, Yoann, Olivier J, John DT (2013).Quantifying habitate vulnerability to assess species priorities for conservation management. Biol. Conserv.158:321-325.

Raunkaier C (1934). Plant Life Forms. Clarendon Press, Oxford.

Rechinger KH, Wendelbo P (1976). Plants of the Kavir Protected Area, Iran. Iran. J. Bot. 1: 23-56.

Rechinger KH (1963 - 2005). Flora Iranica. Nos 1-176, Akademische Druck velsanstalt, Graz – Aust.

Rechinger KH (1977). Plants of the Touran Protected Area, Iran. Iran. J. Bot. 1(2):155 – 180.

Saberamoli S (2001). Floristic study and preparation of vegetation map of Mahroieh wildlife refuge of Kerman. Dissertation, Tarbiat Moallem Univ.press , Tehran.

Shahina A, Ghazanfar (1998).Status of the flora and plant conservation in the Sultanate of Oman. Biol. conserv.85: 287-295.

Takhtajan A (1986). Floristic Regions of the World.University of California Press, California.

Tavakkoli Z, Mozaffarian V (2005). Study of the flora of Kabar dam of Ghom. Pajouhesh and Sazandegi J. 17 (1): 22-29.

Townsend CC, Guest E (1966-1986). Flora of Iraq.Vols 1-9 Ministry of Agriculture and Agrarian Reform, Baghdad.

Turan Y, Filiz Y (2011). The effects of restoration on soil properties in degraded land in the semi-arid region of turkey. Catena 84(1-2) 47-53.

Yousofinajafabadi M (1996). The study of flora and preparation of plant vegetation of Ghomeshlou preserved region. Dissertation, Shahid Beheshti University,Tehran.

Zargary A (1999). Medicinal plants. Vol 1-5, Tehran University Press, Tehran.

Zohary M (1973), Geobotanical Foundation of the Middle East. 2 Vols, Gustav Fischer Verlag, Stuttgart.

14

A preliminary simulation model of individual and synergistic impacts of elephants and fire on the structure of semi-arid miombo woodlands in northwestern Zimbabwe

Isaac Mapaure

Department of Biological Sciences, University of Namibia, P. Bag 13301, Windhoek, Namibia.

Sustainable management of plant-herbivore systems requires an understanding of their long-term dynamics through modeling approaches. A preliminary simulation model was developed using STELLA to predict the impacts of elephants and fire on the structure of semi-arid miombo woodlands of north-western Zimbabwe. Elephants alone at a density of 0.27 km^{-2} will convert the woodland into coppice in 120 years due to resulting massive declines of large trees. The same result is achieved in only 10 years if elephant density is at 2 km^{-2}. The pattern remains similar with simultaneous application of fire once every 4.7 years with elephants at 0.27 km^{-2}. When elephants are culled at 30% whenever their densities reach 1 km^{-2}, the woodland does not degenerate into coppice despite fire occurrence once every 4.7 years. Therefore, elephants alone can degrade and maintain semi-arid miombo woodland into coppice, largely due to their damaging impacts on mature canopy trees and fire acts to speed up the process by suppression of an already low recruitment. Fire alone has a lesser influence on woodland structure than elephants because of low fuel loads due to heavy grazing and low grass production as a result of low rainfall and inherently poor soils in the area. A maximum elephant density of 0.1 km^{-2} is recom-mended to achieve equilibrium in the area.

Key words: Zimbabwe, Sengwa, STELLA, elephants, fire, miombo woodlands, simulation model.

INTRODUCTION

Spatial and temporal dynamics of plant communities are often strongly influenced by one or more types of disturbance events (Richards et al., 1999) of which fire and herbivory are two important determinants of semi-arid savannas. The impacts of these two determinants on vegetation largely depend on their frequency and timing of occurrence, as well as on the severity and the length of period of influence. The literature is awash with reports on the impacts of fire and elephant (*Loxodonta africana*

Blumenbach) herbivory on woodlands in general (Bond, 1993; Ribeiro, 2007; Valeix et al., 2011). The impacts of elephants and fire on semi-arid miombo woodlands were highlighted by, among others, Mapaure and Moe (2009) and Joseph (2012). Their influence in other savanna eco-systems such as Kalahari sand woodlands have also been reported (Holdo, 2007; Gambiza, 2001). The ability to predict the long-term impacts of fire and herbivory on the structure and composition of ecosystems and how ecosystems may respond to given levels of impacts should be an important consideration in ecosystem management. In semi-arid ecosystems, complex interactions among species, disturbance events, and unpredictable and low rainfall, make it difficult to assess probabilities and time scales of vegetation changes (Wiegand and Milton, 1996), making it a challenge to manage these ecosystems on a sustainable basis. In such situations, modelling becomes an essential approach which gives insights and understanding of future dynamics of ecosystems. In addition to exploring the long-term effects of disturbance factors on vegetation change over time, models also help in directing management strategies and decision-making with respect to fire and grazing/browsing management. Models come in various forms, including deterministic (in which relationships are fixed) and stochastic (in which parameters can vary), the latter being more appropriate for dynamic systems (Baxter and Getz, 2005; Taylor and Karlin, 1998).

In the last few decades, modelling has become a common approach to predict disturbance-induced vegetation change. Comparatively, more work has been done on modelling dynamics of moist forests and related ecosystems than on arid and semi-arid ecosystems. Rangeland dynamics are generally described with respect to the Clementsian concepts of single equilibrium (Clements, 1916) or state-and-transition model based on non-equilibrium dynamics (Westoby et al., 1989). The state-and-transition model holds that the transition of vegetation from one state to another is triggered by external shocks, and management regime is thought to alter the sensitivity of the rangeland to these shocks (Westoby et al., 1989; Perrings and Walker, 1997). Most current range dynamics models are constructed on the basis of the state-and-transition model. This approach allows inclusion of spatial variations and dynamics, event-driven changes, lag effects and thresholds of events (Walker, 1993). This approach is better in improving our understanding of dynamic ecosystems where functional processes may be quite complex.

In tropical savannas, a number of recent studies attempted to model the effects of grazing and/or fires on ecosystem dynamics. Pivello and Coutinho (1996) developed a predictive qualitative model on successional trends in Brazilian savannas under the influence of fire, grazing, wood cutting, drought, frost and weed invasion. Pivello and Norton (1996) developed an expert system with an ability to deal with qualitative information and recommended prescribed fire procedures accordingly for Brazilian savannas. Caughley (1976) developed a model on dynamics of *Colophospermum mopane* woodlands under the impacts of elephants in the Luangwa Valley, Zambia, in which he predicted a stable-limits cycle with a periodicity of 200 years. Caughley (1976) model was, however, recently doubted as unlikely (Mapaure and Mhlanga, 1998; Duffy et al., 1999). Predictions of miombo

woodland dynamics under the influence of various disturbance factors have largely been presented in conceptual models in the form of flow diagrams. These conceptual models were mainly based on observations in long-term experiments within the southern African region, some of which dealt with influences of fire (Trapnell, 1959) and regeneration after clearing (Robertson, 1984).

Frost (1996) indicated that shifts in the state of miombo from a woodland state to shrubland or grassland were driven by wood-clearing, elephants and fire. Such a situation was becoming evident in some areas of Sengwa Wildlife Research Area (SWRA) (Mapaure, 2001), the study site for this research. Starfield et al. (1993) developed a frame-based model to investigate interactions among rainfall, elephants and fire, which they applied to miombo woodlands of western Zimbabwe. Further to this, Desanker and Prentice (1994) and Desanker (1996) developed a computer model called MIOMBO on miombo woodland dynamics, with particular application to Malawian woodlands. Their model was based on individual-tree gap models and included a drought routine. Light conditions were modelled as the only restriction for a species to establish. This is an over simplification of ecosystem processes. Much more recently, Baxter and Getz (2005) used a grid-based model to evaluate elephant impacts on savanna dynamics in Kruger National Park, South Africa. Their model constitutes what could be considered a much more realistic representation of ecosystem dynamics under the influence of elephants and fire. However, most of the models discussed earlier largely dealt with structural dynamics rather than species composition, the latter being more challenging to model.

The need to model miombo woodlands dynamics has long been recognised by the Miombo Network. The Miombo network comprises an international network of researchers which aims to understand how land use practices affect land cover and ecosystem processes in miombo ecosystems of central, eastern and southern Africa. They also aim to predict the effects of global change on land use dynamics, structure and function of miombo ecosystems. One of their objectives is 'to develop a predictive understanding of miombo woodland structure and functioning' (Desanker et al., 1997).

The objective of this study was, therefore, to build a simple preliminary simulation model which best describes changes in the structure of semi-arid miombo woodland under the influence of elephant herbivory and fire. Modelling was done for structure because Baxter and Getz (2005) noted that even in cases where elephant impacts on woody plants may not significantly affect species composition, the structure is usually considerably altered. Also, the STELLA modelling platform used was developed to deal with changes in stocks. It is hoped that this will help decision-makers and resource managers to improve their elephant and fire management programmes in the miombo ecoregion in East and southern Africa.

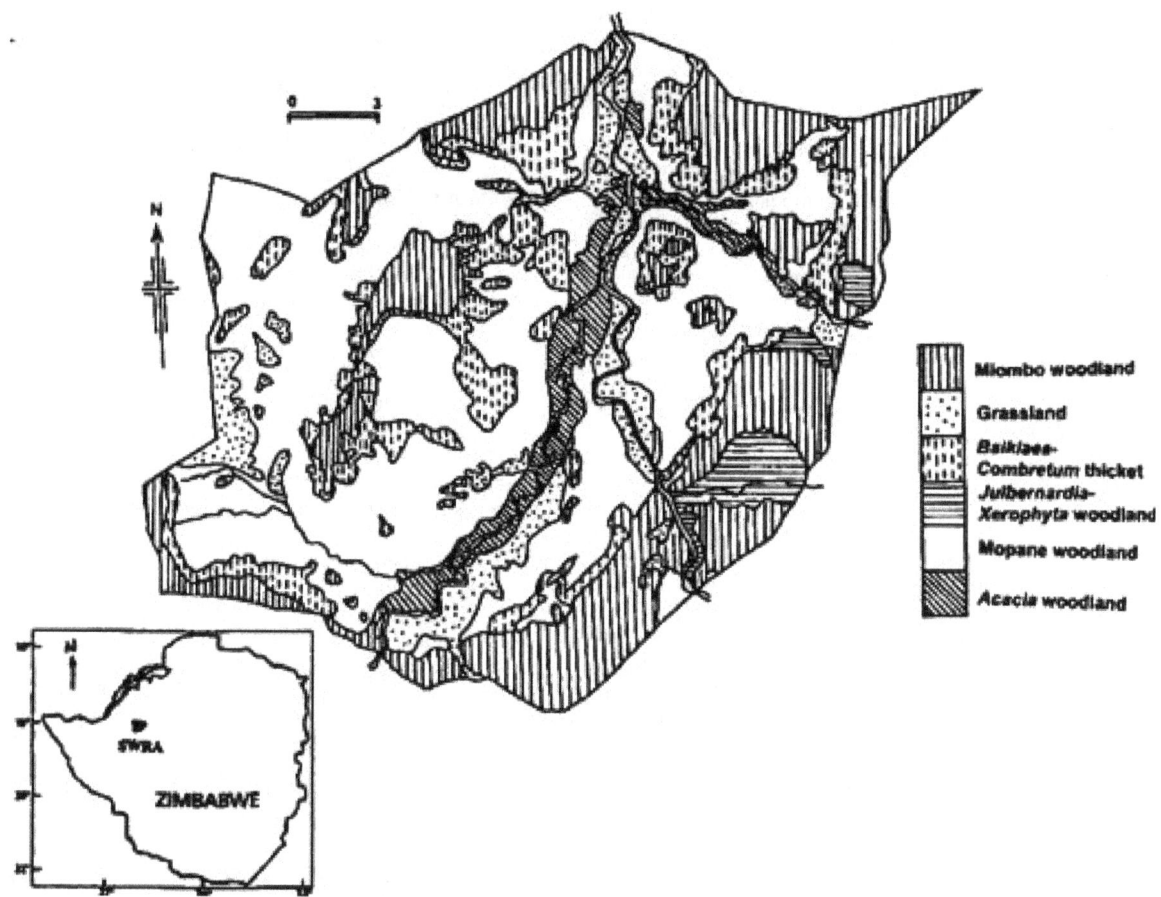

Figure 1. A map showing the location of Sengwa Wildlife Research Area (SWRA) in Zimbabwe and a detailed map of SWRA showing the major vegetation types in the area.

MATERIALS AND METHODS

Study area

This model was calibrated and based on data collected from various studies done by the author and other researchers (Anderson and Walker, 1974; Coulson, 1996; Craig, 1983; Guy, 1989; Mapaure, 2001; Mapaure and Campbell, 2002; Mapaure and Moe, 2009; Mapaure et al., 2009) in SWRA, north-western Zimbabwe (Figure 1). SWRA lies between 28° 03' and 28° 20' E and 18° 01' and 18° 13' S, covering an area of 373 km². It is bounded by communal lands on all but the northern side, where it shares a border with Chirisa Safari Area, a state protected hunting area. SWRA experiences three climatic seasons: a hot wet period from November to April, a cool dry period from May to July and a hot dry period from August to October. Mean annual rainfall is 642 mm while mean annual temperature is 24°C. October is the hottest month and July is the coldest. Altitude varies from 808 to 1043 m. The area is drained by three major rivers, the Sengwa, Manyoni and Lutope. Two main soil types occur, one formed on sandstones of the Escarpment Grits and another formed on mudstones (Selibas, 1974; Bennett et al., 1983). The vegetation is generally deciduous *Brachystegia–Julbernardia* (miombo) woodland on sandy soils and dry early deciduous woodland dominated by *Colophospermum mopane* on the lower heavier soils.

Other vegetation types are riverine *Acacia* woodlands and mixed

Combretum thickets on sands. These habitats are home to a diverse large mammal community of seven species of large carnivores and eighteen species of large herbivores (Cumming, 1983).

Model description

The model was developed using STELLA (High Performance Systems, 1996). STELLA is an icon-based program which allows both graphical and tabular outputs (Blankenship et al., 1995). It is an object-oriented programming language designed specifically for modelling dynamic systems (Costanza and Gottlieb, 1998; Costanza et al., 1998). Once a structural diagram has been completed using the icons, STELLA can write the equations internally in the form of first-order difference equations and provides a list of rate variables and auxiliary variables necessary for mathematical formulation (Pan and Raynal, 1995). STELLA is recommended as an excellent modelling tool and provides a potential to break new ground for simulating biologically complex systems (Costanza, 1987; Hannon and Ruth, 1997). The model simulates the effects of fire and elephants on woody plants. Fire regime is regulated through grass biomass production; itself is a function of rainfall patterns. By varying the rainfall seasonal patterns and grass biomass production, the fire intensity and frequency would change accordingly. These factors, in turn influence recruitment of shrubs, saplings and seedlings into small and large tree classes. The possibility of the woodland shifting into a coppice woodland state was also built into the model. Sets of rules were defined to control these dynamic possibilities. The model was calibrated for SWRA using data from the present study as well as other literature sources from similar or comparable ecosystems, particularly Campbell (1996), Kundlande et al. (2000), Mapaure and Mhlanga (1998), Campbell et al. (1996), Gambiza et al. (2008), Mapaure et al. (2009), Mapaure (2001) and Mapaure and Moe (2009). The model can be modified and adapted for other semi-arid systems such as teak woodlands on poor sandy soils by altering relevant parameters on rainfall, grass biomass production, recruitment rates, maximum tree densities and probabilities of fire occurence.

The model consists of five interactive sub-models covering rainfall, grass, trees, fire and elephants. Parameters used, assumptions and sources of data are given under each sub-model. Diagrammatic representations of the different sub-models and the details of the parameters used (including justifications for the values used in the calculations and their sources) are given in the Appendix.

Rainfall sub-model

This sub-model generates the annual rainfall, which influences grass production. Rainfall was generated using the long-term mean annual rainfall for SWRA, the standard deviation of the mean and the sine wave function to simulate decade-scale fluctuations in rainfall around the long term average. The long-term period was set at 20 years as this broadly corresponds to the observed pattern of rainfall in southern Africa where there is an 18 to 20 year cycle comprising approximately 9 to 10 years above average rainfall alternating with a corresponding period of below average rainfall (Tyson, 1986). Recent changes in rainfall patterns due to climatic change were not included in the model due to their stochastic nature and inapplicability of the Global Circulation Models on local-specific events (IPCC, 2007). However, it may be possible to build this in the model by inclusion of carefully-thought out randomization climatic change generator.

Grass sub-model

This sub-model simulates grass biomass production; hence, grass fuel load. Grass production was calculated using regression equations relating production to rainfall in savannas (Dye and Spear, 1982) and shading by the tree canopy (Robertson, 1984; Frost, 1996). One millimetre of rain was assumed to produce 2 kg dry matter (DM) ha^{-1} yr^{-1} on cleared areas on sandy soils (Dye and Spear, 1982) while 1 mm of rain produces 1 kg DM ha^{-1} yr^{-1} under a high woody vegetation canopy cover (Frost, 1996). Fuel load was simulated as a function of grass growth rate, grass production, grazing rate, proportions of grass burnt off and grass decay. Successive years of above-average rainfall may result in the accumulation of grass fuel loads. Processes reducing grass biomass are decomposition, burning and grazing. Grass decay was estimated using a decomposition constant of 0.88 yr^{-1} for grass litter. It was estimated that the proportion of grass removed by herbivores was about 15% per annum, since the stocking rates of multi-species systems of wild herbivores were tricky to estimate compared to livestock production systems (Noy-Meir, 1981).

Fire sub-model

The occurrence of fires is simulated as a function of time-since-the-last fire and the probability of a fire spreading (a function of grass biomass). Trollope (1993) found out that fire only spreads in savannas when grass fuel loads were greater than 1000 kg DM ha^{-1}, though Rushworth (1975) earlier reported that fires could spread when the fuel load was as low as 800 kg DM ha^{-1} but it is noted that such fires did not cause top-kill of shrubs (Gambiza et al., 2000). The fire spread constant was therefore set at a grass biomass of 800 kg DM ha^{-2}. Natural fire was calculated as a random event determined by the probability of ignition. The probability of ignition was calculated as a function of the ignition constant and time-since-last fire; hence, the ignition constant was set at 0.3.

Elephant sub-model

This sub-model simulates densities of elephants in the area. The numbers of elephants at any one time is a function of additions through immigration and births, and subtractions due to culling, emigration and natural mortality. Immigration and births constitute natural increase. A natural rate of increase of 5% per annum in most elephant populations was reported by Cumming et al. (1997), a figure similar to that reported from Kenya (Armbruster and Lande, 1993). Between 1965 and 1979, elephant populations in SWRA increased at a rate of 4.19% per annum (Gibson, 1983). Therefore, a natural rate of increase (r) of 0.05 (that is, 5%) was used in the model and the growth rate was described by the logistic equation whose upper limit was set at 1500 elephants, limited by vegetation availability. Natural mortality of elephants in the Sebungwe was reportedly very low (Craig, 1996). Natural mortality was, therefore, set at 2% per annum, a mean of several values for SWRA given by Department of National Parks and Wildlife Management (1996). Poaching and animals killed through problem animal control were included in the culling variable, and this can be set by management. Both population increase and mortality are functions of the current population numbers, the base of which was set at 100, since the maximum recommended number for SWRA was 250 (Guy, 1989).

To simplify issues of differential use of habitats by elephants (with respect to availability of favoured species, extent of habitat relative to other habitats, proximity to water sources by elephants, spatial heterogeneity of food sources and variations in seasonality of elephant occupancy (including local migrations due to hunting disturbances during the hunting season), it was assumed that miombo woodland was the most available habitat for elephants at all times.

Tree sub-model

Woody vegetation was divided into four size classes representing shrubs (<3 m in height), small trees (3 to 5 m in height), large trees (>5 m in height), and coppice (converted trees and other elephant-and/or fire-suppressed trees). Transitions among size classes were set as a function of mortality, the proportion of individuals escaping a fire and growth rates of the respective strata. The current relative proportions of shrubs, small trees and shrubs in SWRA were used as base values for the model. Recruitment of individuals into the shrub layer was from seedlings whose establishment success is a function of grass biomass (incorporating competitive interactions between the two components). Elephants were assumed not to have any effect on shrubs but on small, large and coppice trees. Tree death was modelled as either natural or elephant induced while that of shrubs was either fire-caused or natural. The woodland was set to convert to coppice only when elephant density exceeded 2 km^{-2}. Reversions of small trees due to fire augmented numbers of shrubs but reversions of large trees due to elephants resulted in coppice woodland. Some of these dynamics are informed by Frost (1996).

Simulations

All simulations were run for a period of 200 years. This period was chosen because Caughley (1976) had indicated that equilibrium conditions between elephants and woodlands may be reached

RESULTS

Interactive effects of elephants and fire

Simulations clearly demonstrated changes in the structure of miombo woodland when subjected to elephants and fire. Without elephants and fire, the woodland assumed a structure characterized by well-defined proportions of the three woody strata: shrubs, small trees and large trees (Figure 2a). Since initial conditions were set at current (hence disturbed) proportions of these three strata, the figure indicates that if the two disturbance factors were removed from the system, it would take up to 120 years before stabilizing again. An initial elephant density of 0.27 km⁻² (100 elephants) without culling and without fire, will push the system completely into a predominantly coppice woodland in about 120 years, characterised by a massive decline in large trees (Figure 2b).

The length of time before the woodland degenerates into coppice depends upon the initial density of elephants, where an elephant density of 2 km⁻² without culling would result in conversion to coppice in about 10 years. Applying fire once in

elephants and fire. Without culling did not significantly change the pattern (Figure 2c). When elephants are culled at a rate of 30% when densities reach about 1 km⁻² (starting with an initial density of 0.27 km⁻²), the woodland does not degenerate into coppice despite the occurrence of fire once every 4.7 years (Figure 2d). The woodland would quickly establish some relatively constant proportions of the three strata.

If fire is applied alone without elephants, the woodland does not degenerate into coppice but there seems to be If fire is applied alone without elephants, the woodland does not degenerate into coppice but there seems to be an initial marked decline in the densities of small trees, apparently due to lack of shrub recruitment into small trees (Figure 2e) but trends largely remain the same as in the base run (Figure 2a). Simulations indicate that the current density of elephants estimated

4.7 years to the woodland, with an initial 100 Elephants without culling did not significantly change

after 200 years, from his case study of mopane woodlands in Zambia. For any given scenario, five runs were performed. Simulations were run with randomly generated rainfall. They were run with varying combinations of elephant densities and fire frequencies. Effects of elephants and fire, applied individually or interactively, on woodland structure were monitored. For most simulations, the model was run with an initial number of 100 elephants (that is, 0.27 km⁻²), with or without culling. In some runs, fire occurrence was set at zero and where present, it was run with a return-time of 4.7 years (Mapaure et al., 2009). The effects of fire on woody plant community structure were therefore investigated by varying the frequency of fire between annual occurrence and once in several years. Elephant densities were also varied between 0 and 2 km⁻² to compare sensitivity of woodlands to the impacts of elephants.

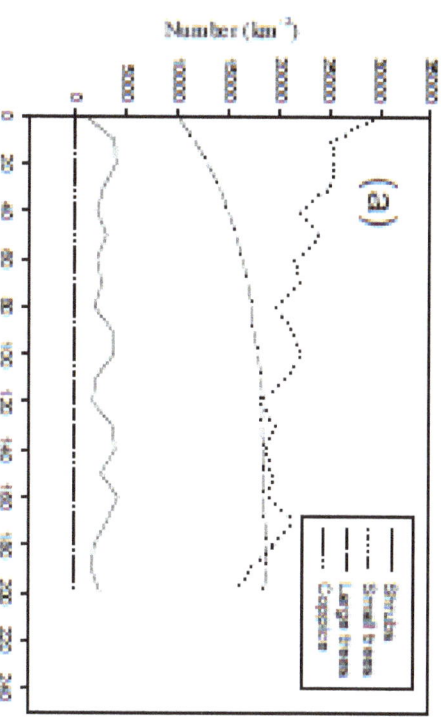

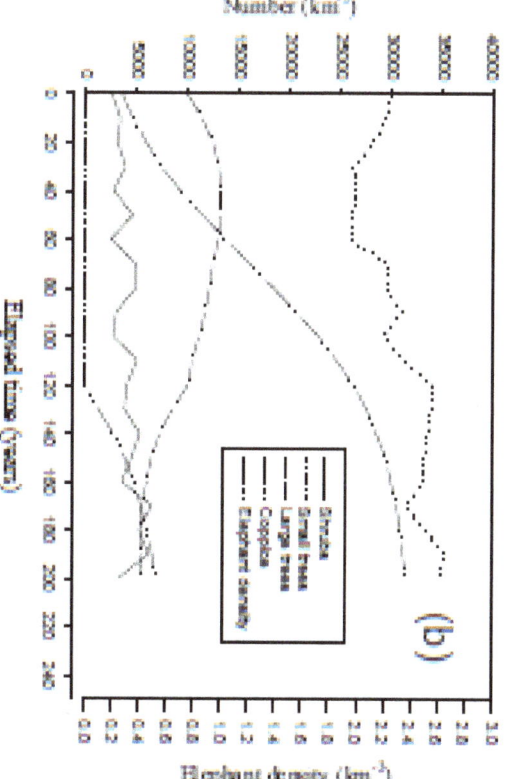

Figure 2. Initial structure (base run) of miombo woodland structure at current proportions of trees and shrubs with neither fire nor elephants (a) and changes that take place with elephant density of 0.27 km⁻² (without culling) and no fire (b).

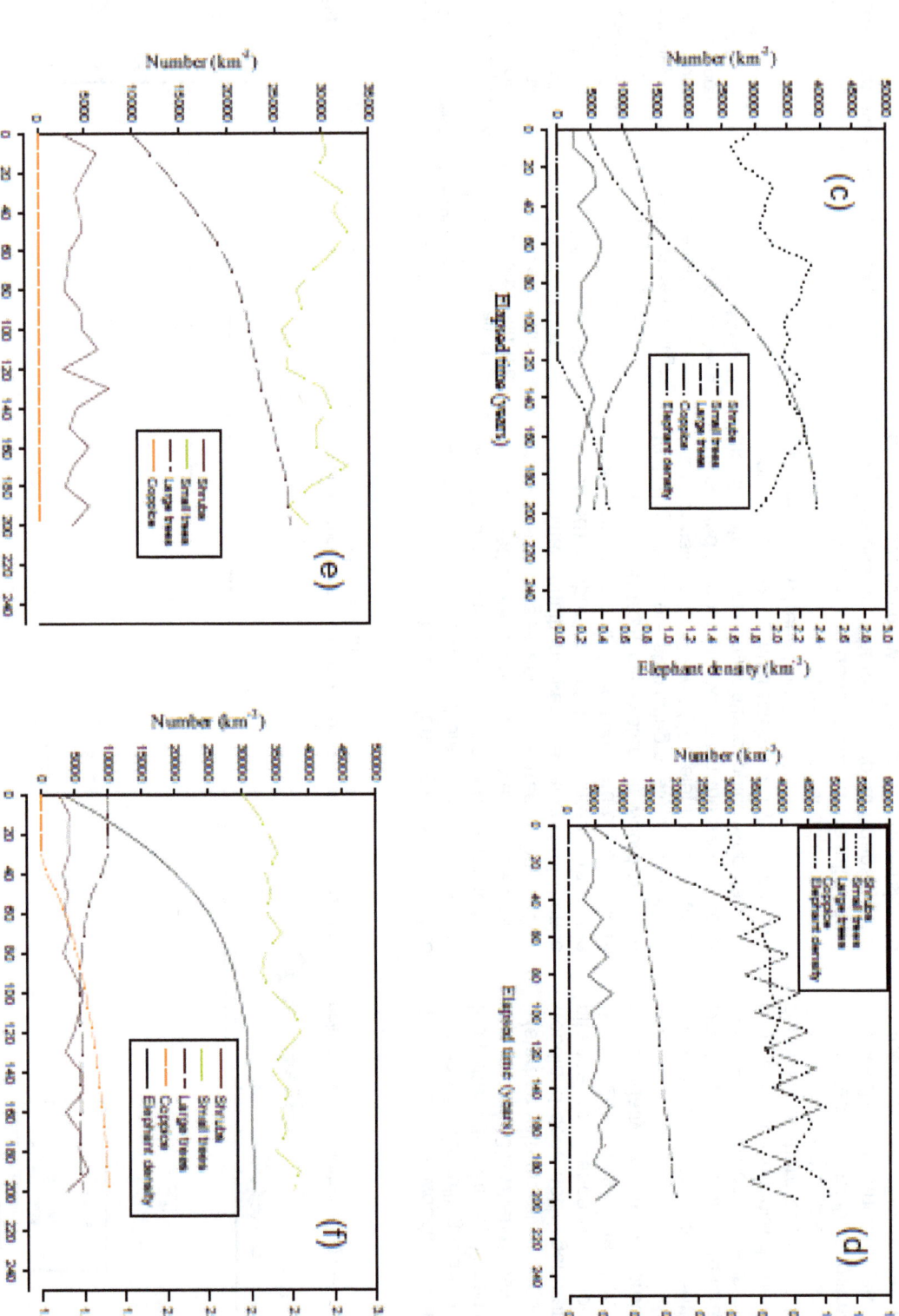

Figure 2 Contd. Initial structure (base run) of miombo woodland structure at current proportions of trees and shrubs with neither fire nor elephants. The structure in (c) shows changes when fire is applied once every 4.7 years, keeping elephant density the same (without culling). Culling elephants at 30% when their density reaches 1 km⁻² while applying fire once every 4.7 years results in the structure shown in (d) while, (e) shows the structure of the woodland when fire is applied once every 4.7 years without elephants in the woodland. Applying an elephant density of 1.5 km⁻² without fire results in woodland structure is shown in (f).

at 1.5 km^{-2}, if not culled, will push the woodland into coppice dominated by small trees (reversions) in 30 to 40 years (even without fire) and further deplete the coppice possibly into shrubland (Figure 2f).

Effects of varying elephant densities and fire return period

Increasing elephant densities from 0 to 2 km^{-2} resulted in small variations in the abundance of small trees at the start of the simulations but abundances increased with increase in elephant densities after about 40 years of simulations (Figure 3a). Shrubs were less affected by variations in elephant densities but showed an increase during the first 10 years (Figure 3b), which is not surprising since elephants do not have a marked direct effect on shrubs. There were sharp declines in the abundance of large trees with increases in elephant density (Figure 3c). At an elephant density of 2 km^{-2}, large trees dropped to their lowest abundance after about 60 years. This resulted in the formation of coppice woodland (Figure 3d). No coppice was formed without elephants while coppice was formed almost immediately at an initial elephant density of 2 km^{-2} without culling. An elephant density of 0.5 km^{-2} without culling will take the woodland about 100 years before degenerating into predominantly coppice. The effects of fire on trees were not evident. When fire return period was varied between non-occurrences and once in 20 years, differences in the abundance of shrubs were minimal (Figure 4a). However, the abundance of shrubs increased during the first 10 years, with increases in abundance at high fire return periods during that period.

Grazing limited the accumulation of grass fuel but this was ultimately determined by long-term rainfall patterns as evidenced by the stabilisation of the two curves in Figure 4b.

DISCUSSION

Impacts of elephants and fire

A preliminary simulation model presented in this paper has demonstrated that the impacts of elephants on miombo woodland structure were more marked than that of fire, but fire has been shown to act as a catalyst that speeds up the process through its suppression effects on woody plant recruitment into the tree layer. This situation seems to be unique to this area because of its semi-arid nature and limited grass biomass production compared to what has been reported elsewhere such as East African rangelands (Dublin et al., 1990) and various savanna woodlands of Botswana (Ben-Shahar, 1996).

However, Starfield et al. (1993) illustrated that the probability of woodland being trapped in a shrubland state increased with increasing probability of ignition. Since the major influence of fire in savanna woodlands is mainly on shrubs, it is therefore not surprising that its simulated effect on the tree layer in this model was minimal. Changes in vegetation state alluded to by Dublin et al. (1990) which can take place even in the absence of fire, confirms findings reported by Mapaure (2001) where some patches of miombo woodland had been converted to woodland-thickets in SWRA. Norton-Griffiths (1979) reported different extents to which fire may influence the process of vegetation change where some areas of the Serengeti, fires were significantly fuelled by increases in grass production due to the opening up of the canopy by elephants but in the central woodlands of the same area, fire merely tipped the balance so that the overall trend was towards a decrease in tree density.

A re-examination of hypotheses proposed by Dublin et al. (1990) implies that elephants alone are capable of degrading miombo woodlands into coppice by negatively impacting heavily on trees. Elephants are also capable of maintaining the vegetation in the new state, while fire only speeds up the process of change. It is therefore hypothesised that elephants alone can degrade semi-arid miombo woodland into coppice and can maintain it in that state, while fire only serves to accelerate the process by suppression of woody plant recruitment. Similar trends have been shown in other savanna ecosystems elsewhere (Staver and Levin, 2012).

In the model, various fire frequencies produced no major differences in woodland structure, probably due to low grass biomass production because of low rainfall, impoverished soils, as well as removal of grass fuel by wild herbivores. Gambiza et al. (2008) indicated that grazing can be very effective in reducing fire impacts on woodlands, and should be considered as a management option where reduction of fire intensity is desired. The model requires the amount of grass biomass to be above a specific value for fire to spread, but given the low rainfall regime, conditions fulfilling this requirement are often not met. Gambiza et al. (2000) indicated that removal of trees and reduction of livestock grazing caused increases in grass fuel loads resulting in a corresponding increase in fire frequency. By implication, therefore, grazing would result in low fire occurrence because of reduced fuel loads through consumption.

The occurrence of natural fires in SWRA was shown to be lower than expected (Mapaure et al., 2009), an observation supported by the model. The model also demonstrated that to maintain the current vegetation structure or to arrest its further decline into coppice, more emphasis should be put towards the management of elephant densities. Elephant densities of about 1.5 km^{-2} (possibly higher) (Mapaure and Campbell, 2002) in the area are clearly detrimental to the woodland.

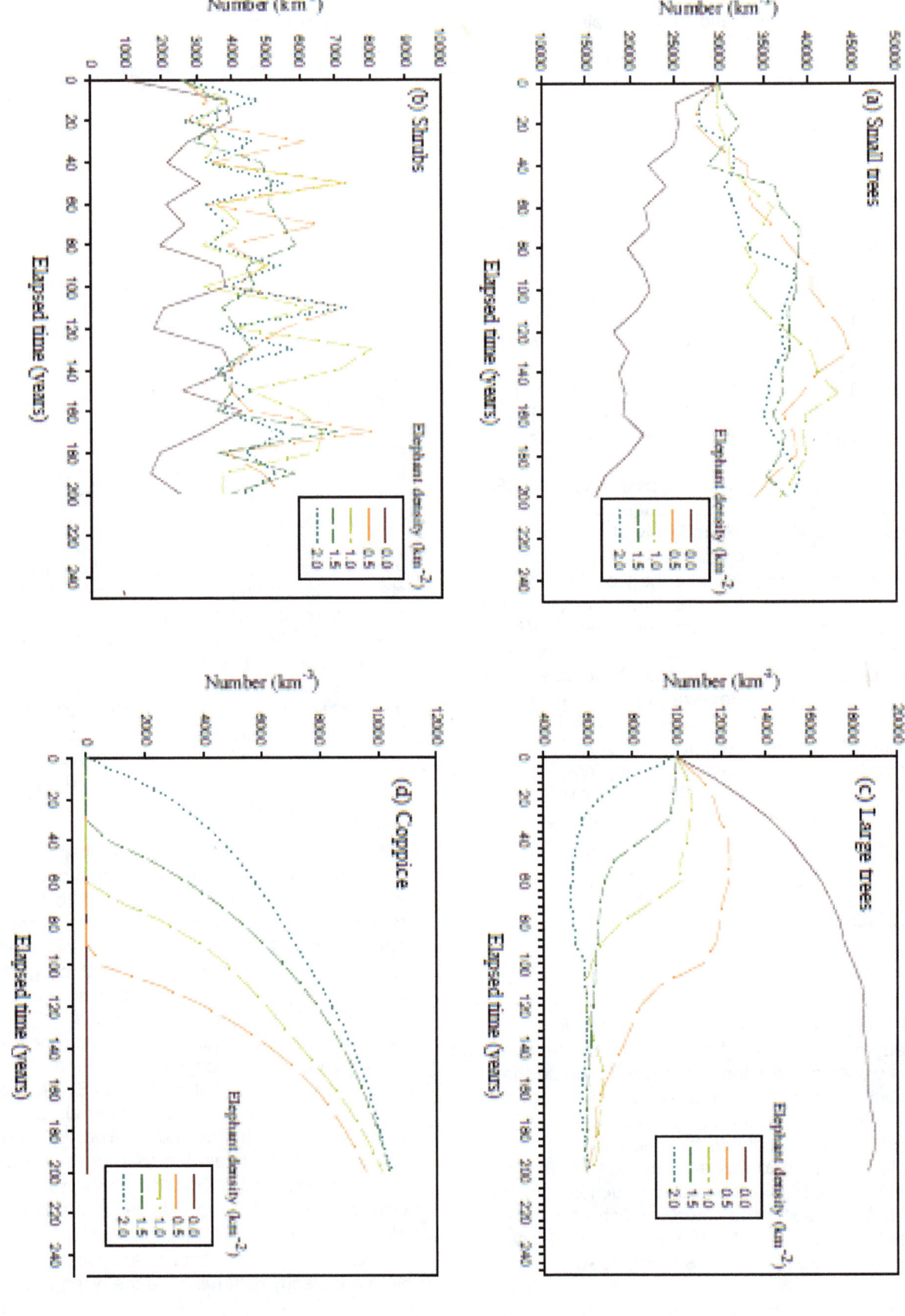

Figure 3. Changes in the abundance of small trees (a), shrubs (b), large trees (c) and coppice (d) subjected to various densities of elephants without culling and no fire occurrence.

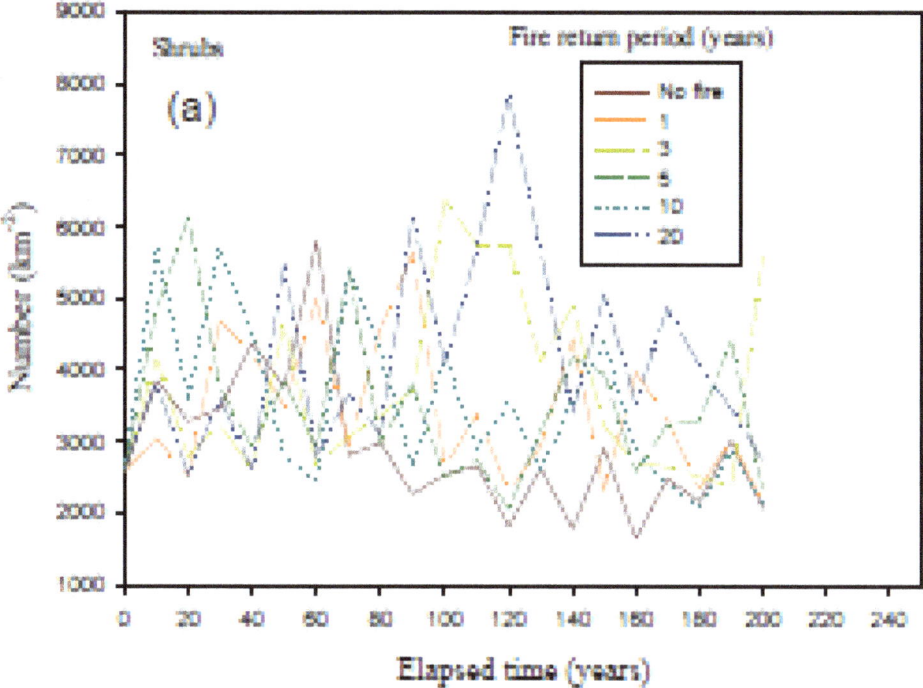

Figure 4a. Changes in the structure of the shrub layer when fire return period is varied between non-occurrence and once every 20 years, with no elephants in the woodland.

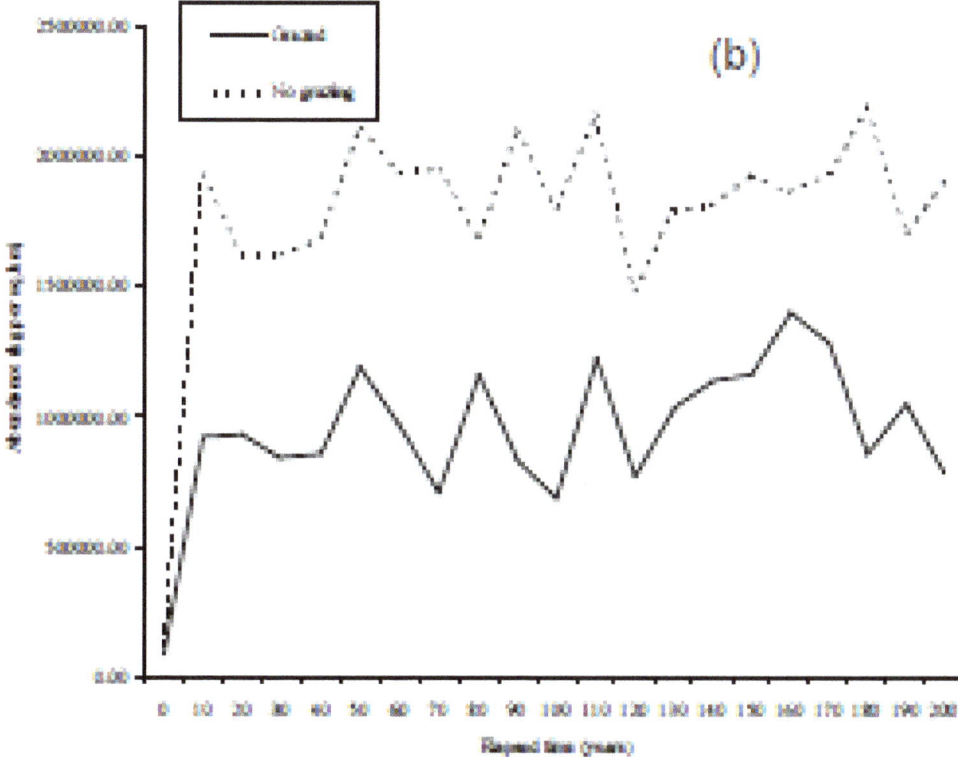

Figure 4b. Changes in the structure of the shrub layer when fire return period is varied between non-occurrence and once every 20 years, with no variations in grass fuel production in grazed and non-grazed woodland.

This observation is contrary to Baxter and Getz (2005) results which showed that at an elephant density of 1.0 km^{-2}, woody plants persisted in a Kruger National Park savanna for over a century. Therefore, caution should be exercised in applying these models as some of them may be limited to local application.

In SWRA, if elephant densities are reduced to well below 0.5 (to about 0.1) km^{-2}, the woodland may be maintained in a desirable (less degraded) state as long as the current fire management practices are maintained. Moreover, the recent transformation of the SWRA into a hunting area will have important considerations on the population structure and dynamics of the elephant populations themselves, since mostly bulls are targeted. If annual hunting quotas are known, these would have to be included in the culling factor in the simulation model. However, the changes in elephant dispersal patterns (as a result of hunting pressure) relative to the spatial distribution of the habitats would be more challenging to predict.

Future directions

Since this model deals with changes in the structure of miombo woodland based on abundances of its woody components, more insights would be derived if an element of vertical structure was built into the model. This would clearly capture the responses of the shrub layer to fire. Fire should be modeled with clear differences between early and late dry season occurrences since differences in seasonal effects of fire are well known in African savannas (Gambiza et al., 2008). Such an undertaking would pave the way to include fire intensities in the model.

An additional condition for conversion from coppice or shrub land into grassland should be built into the model. This, however, can only be possible if the model incorporates an element of floristic composition in addition to structure, which STELLA modeling may not be able to deal with since it models stocks. The overall applicability of the model beyond the study area would require some caution. Savannas are very floristically and structurally heterogenous and dynamic; hence, one would have to be cautious in generalizing this across different ecosystems.

Conclusions

Elephants alone can degrade and maintain semi-arid miombo woodland into coppice, largely due to their impacts on mature canopy trees, but fire acts to speed up the process by suppression of an already low recruitment. Fire has a lesser influence on the structure of the woodlands than elephants and does not result in degradation of miombo woodland in the area due to the low fuel loads available as a result of heavy grazing and low grass production (due to relatively low rainfall) and inhe-

rently poor soils. Fire in semi-arid miombo woodland does not necessarily lead to an increase in densities of shrubs but would certainly lead to height reversions. Current elephant densities in SWRA may degrade the ecosystem into predominantly coppice woodland in less than 40 years and should, therefore, be reduced to about 0.1 km^{-2} [which seems to be within the carrying capacity of (at least) the miombo ecosystem]. Recent introduction of hunting in the area may help to reduce the elephant populations to manageable densities or to keep their population under control, depending on the annual quotas set for hunting.

ACKNOWLEDGEMENTS

I am grateful to the Norwegian Universities' Committee for Development Research and Education (NUFU) and the European Commission who provided funding for this research. I thank Professor David Cumming, Professor Bruce Campbell and Professor Johan du Toit for their helpful suggestions. Permission to carry out research in Sengwa Wildlife Research Area was granted by the Director of the then Department of National Parks and Wildlife Management of Zimbabwe. My thanks also go to the staff at Sengwa Wildlife Research Institute (SWRI) who assisted me in various ways. All local field assistants who rendered their services during data collection are gratefully acknowledged.

REFERENCES

Anderson GD, Walker BH (1974). Vegetation composition and elephant damage in the Sengwa Wildlife Research Area, Rhodesia. J. South Afr. Wildl. Manage. Assoc. 4:1-14.

Armbruster P, Lande R (1993). A population viability analysis for African elephants (*Loxodonta africana*) - How big should reserves be? Conserv. Biol. 7:602-610.

Baxter PWJ, Getz WM (2005). A model-framed evaluation of elephant effects on tree and fire dynamics in African savannas. Ecol. Appl. 15(4):1331-1341.

Bennett J, Hopkins M, Garikayi A (1983). The soils of Sengwa Wildlife Research Area. Report No. A502, Ref. No. CS/3/7/5, October 1983. Chemistry and Soil Research Institute, Department of Research and Specialist Services, Harare.

Ben-Shahar R (1996). Woodland dynamics under the influence of elephants and fire in northern Botswana. Vegetatio 123:153-163.

Blankenship V, Tomlinson J, Sims MA (1995). A STELLA II teaching simulation of the dynamics of action model. Behav. Res. Meth. Instrum. Comput. 27:244-250.

Bond WJ (1993). Keystone species. In: Schulze ED, Mooney HA (Eds.), Biodiversity and Ecosystem Function. Springer-Verlag, pp. 237-253.

Bond WJ, van Wilgen BW (1996). Fire and plants. Population and community Biology Series 14. Chapman and Hall, London.

Campbell B (1996). The miombo in transition: woodlands and welfare in Africa. Centre for International Forestry Research, Bogor. (ed.).

Campbell BM, Butler JR, Mapaure I, Vermeulen S, Mushove P (1996). Elephant damage and safari hunting in *Pterocarpus angolensis* woodland in north-west Matabeleland, Zimbabwe. Afr. J. Ecol. 34:380-388.

Caughley G (1976). The elephant problem - an alternative hypothesis. East Afr. Wildl. J. 14:265-284.

Clements FE (1916). Plant succession: an analysis of the development of vegetation. Carnegie Institute, Washington.

Costanza R (1987). Simulation modelling on the MacIntosh using STELLA. BioSci. 37: 129-132.

Costanza R, Duplisea D, Kautsky U (1998). Modelling ecological and economic systems with STELLA. Ecol. Modell. 110:1-4.

Costanza R, Gottlieb S (1998). Modelling ecological and economic systems with STELLA: Part II. Ecol. Modell 112:81-84.

Coulson IC (1996). Elephants and vegetation in the Sengwa Wildlife Research Area. In: Martin RB, Craig GC and Booth VR (eds), Elephant management in Zimbabwe, Third Edition, Department of National Parks and Wildlife Management, Harare, pp. 59-65.

Craig GC (1983). Vegetation survey of Sengwa. Bothalia 14:759-763.

Craig GC (1996). Population dynamics of elephants. In: Martin RB, Craig GC and Booth VR (eds). Elephant management in Zimbabwe, Third Edition. Department of National Parks and Wildlife Management, Harare, pp.75-80.

Cumming DHM (1981). The management of elephant and other large mammals in Zimbabwe. In: Jewel PA and Holt S (eds). Problems in managing locally abundant wild animals, Academic Press, New York, pp. 91-118.

Cumming DHM (1983). The Sengwa Wildlife Research Area and Institute. Zim. Sci. News 17: 32-37.

Cumming DHM, Fenton MB, Rautenbach IL, Taylor RD, Cumming GS, Cumming MS, Dunlop JM, Ford AG, Havorka MD, Johnston DS, Kalcounis M, Mahlangu Z, Portfors CVR (1997). Elephants, woodlands and biodiversity in southern Africa. S. Afr. J. Sci. 93:231-236.

Department of National Parks and Wildlife Management (1996). An overview of elephant populations status in Zimbabwe: 1980-1995. Unpublished report prepared for Zimbabwe CITES Technical Committee, 22 February 1996, Department of National Parks and Wildlife Management, Harare.

Desanker PV (1996). Development of a miombo woodland dynamics model in Zambezian Africa using Malawi as a case study. Clim. Change 34:279-288.

Desanker PV, Frost PGH, Justice CO, Scholes RJ (eds). (1997). The Miombo Network: Framework for a terrestrial transect study of land-use and land-cover change in the miombo ecosystems of central Africa. IGBP Report 41. The International Geosphere-Biosphere Programme (IGBP), Stockholm, Sweden.

Desanker PV, Prentice IC (1994). MIOMBO - a vegetation dynamics model for the miombo woodlands of Zambezian Africa. For. Ecol. Manage. 69:87-95.

Dublin HT, Sinclair ARE, McGlade J (1990). Elephants and fire as causes of multiple stable states in the Serengeti-Mara woodlands. J. Anim. Ecol. 59: 1147-1164.

Duffy KJ, Page BR, Swart JH,Bajic VB (1999). Realistic parameter assessment for a well known elephant-tree ecosystem model reveals that limit cycles are unlikely. Ecol. Modell 121:115-125.

Dye PJ, Spear PT (1982). The effect of bush clearing and rainfall variability on grass yield and composition in southwest Zimbabwe. Zimb. J. Agric. Res. 20: 103-117.

Frost P (1996). The ecology of miombo woodlands. In: Campbell B (ed.), The miombo in transition: Woodlands and welfare in Africa. Centre for International Forestry Research (CIFOR), Bogor, pp.11-57.

Gambiza J (2001). The regeneration of Zambezi teak forests after logging: Influence of fire and herbivory. PhD Thesis, University of Zimbabwe, Harare.

Gambiza J, Bond W, Frost PGH,Higgins S (2000). A simulation model of miombo woodland dynamics under different management regimes. Ecol. Econ. 33:353-368.

Gambiza J, Campbell B, Moe SR,Mapaure I (2008). Season of grazing and stocking rate interactively affect fuel loads in a Baikiaea plurijuga woodland in north-western Zimbabwe. Afr. J. Ecol. 46: 637-645.

Gibson D (1983). Transect monitoring of large mammal populations in Sengwa Wildlife Research Area. Progress Report No. 9, Project N0. SWRA/B4/3a/3, Department of National Parks and Wildlife Management, Sengwa Wildlife Research Institute, Gokwe.

Guy PR (1989). The influence of elephants and fire on a Brachystegia-Julbernardia woodland in Zimbabwe. J. Trop. Ecol. 5:215-226.

Hannon B, Ruth M (1997). Modeling dynamic biological systems. Springer-Verlag, New York.

High Performance Systems (1996). Stella and Stella Research software. High performance systems Inc. Hanover, NH.

Holdo RM (2007). Elephants, fire and frost can determine community structure and composition in Kalahari woodlands. Eco. Appl. 17: 558-568.

IPCC (Intergovernmental Panel on Climate Change) (2007). Climate Change 2007: The Physical Science Basis. Contribution of Working Group I to the Fourth Assessment Report of the Intergovernmental Panel on Climate Change. Cambridge University Press, Cambridge, U. K., available at http://ipcc-wg1.ucar.edu/wg1/wg1-report.html.

Johnson EA , Gutsell SL (1994). Fire frequency models, methods and interpretations. Adv. Ecol. Res. 25: 239-287.

Joseph GS (2012). Understanding pattern-process relationships in a heterogenous landscape: effects of termitaria on diversity and disturbance regimes in miombo woodlands of northern Zimbabwe. PhD thesis, Percy FitzPatrick Institute, Department of Zoology, University of Cape Town, South Africa.

Kundhlande G, Adamowicz WL, Mapaure I (2000). Valuing ecological services in a savanna ecosystem: a case study from Zimbabwe. Ecol. Econ. 33(3):401-412.(Reference was not cited in the main text. Please cite.

Mapaure I (2001). Small-scale variations in species composition of miombo woodland in Sengwa, Zimbabwe: the influence of edaphic factors, fire and elephant herbivory. Syst. Geogr. Pl. 71: 935-947.

Mapaure I (2013). Short-term responses of shrub layer communities to dry season fires and tree thinning in semi-arid miombo woodlands of north-western Zimbabwe. Accepted for publication, Afr. J. Plant Sci.

Mapaure I, Campbell BM, Gambiza J (2009). Evaluation of the effectiveness of an early peripheral burning strategy in controlling wild fires in north-western Zimbabwe. Afr. J. Ecol. 47:518-527

Mapaure I, Mhlanga L (1998). Miombo woodlands and fire: The impact of elephant damage on Colophospermum mopane on Namembere Island, Lake Kariba, Zimbabwe. Zimb. Sci. News 32(1):15-19.

Mapaure I, Moe SR (2009). Changes in the structure and composition of miombo woodlands mediated by elephants and fire over a 26-year period in north-western Zimbabwe. Afr. J. Ecol. 47:175-183.

Mapaure I,Campbell BM (2002). Changes in miombo woodland cover in and around Sengwa Wildlife Research Area, Zimbabwe, in relation to elephants and fire. Afr. J. Ecol. 40:212-219.

Norton-Griffiths M (1979). The influence of grazing, browsing and fire on the vegetation dynamics of the Serengeti. In: Sinclair ARE and Norton-Griffiths M (eds), Serengeti: Dynamics of an ecosystem, University of Chicago Press, Chicago, pp. 310-352.

Noy-Meir I (1981). Responses of vegetation to the abundance of mammalian herbivores. In: Jewel PA and Holt S (eds), Problems in management of locally abundant wild animals, Academic Press, New York, pp. 233-246.

Pan Y, Raynal DJ (1995). Decomposing tree annual volume increments and constructing a system dynamic model of tree growth. Ecol. Modell. 82:299-312.

Perrings C, Walker B (1997). Biodiversity, resilience and the control of ecological-economic systems: the case of fire-driven rangelands. Ecol. Econ. 22:73-83.

Pivello VR, Coutinho LM (1996). A quantitative successional model to assist in the management of Brazilian cerrados. For. Ecol. Manage. 87:127-138.

Pivello VR, Norton GA (1996). FIRETOOL: an expert system for use of prescribed fires in Brazilian savannas. J. Appl. Ecol. 33:348-356.

Polakow DA, Dunne TT (1999). Modelling fire-return interval T: stochasticity and censoring in the two-parameter Weibull model. Ecol. Modell. 121:79-102.

Ribeiro NS (2007). Interaction between fires and elephants in relation to vegetation structure and composition of miombo woodlands in northern Mozambique. PhD Thesis, University of Virginia, Charlottwesville, Va, USA.

Richards SA, Possingham HP, Tizard J (1999). Optimal fire management for maintaining community diversity. Ecol. Appl. 9:880-892.

Robertson FE (1984). Regrowth of two African woodland types after shifting cultivation. Ph.D. thesis, University of Aberdeen, Aberdeen, Scotland.

Rushworth JE (1975). The floristic, physiognomic and biomass structure of Kalahari sand vegetation in relation to fire and frost in Wankie National Park, Rhodesia. M.Sc. Thesis, University of Rhodesia, Salisbury.

Scholes RJ, Kendall J, Justice CO (1996). The quantity of biomass burned in southern Africa. J. Geophys. Res. 101:23 667-23 676.

Selibas NJ (1974). Notes on the geology of the Sengwa Wildlife Research Area, and a preliminary basin analysis. B.Sc. (Special Honours) dissertation, University of Rhodesia, Salisbury.

Sharp GJ (1982). Seasonal variations in the diet and condition of the African elephant in Sengwa Wildlife Research Area, Zimbabwe. M.Sc. Thesis, University of Zimbabwe, Harare.

Starfield AM, Cumming DHM, Taylor RD, Quadling MS (1993). A frame-based paradigm for dynamic ecosystems models. Artif. Intel. Appl. 7:1-13.

Staver AC, Levin SA (2012). Integrating theoretical climate and fire effects on savanna and forest systems. Am. Nat. 180(2):211-224.

Stronach NRH, McNaughton SJ (1989). Grassland fire dynamics in the Serengeti ecosystem and a potential method for retrospectively estimating fire energy. J. Appl. Ecol. 26:1025-1033.

Taylor HM, Karlin S (1998). An introduction to stochastic modelling, Third Edition, Academic Press, London.

Trapnell CG (1959). Ecological results of woodland burning experiments in Northern Rhodesia. J. Ecol. 47:129-168.

Trollope WSW (1993). Fire regime of the Kruger National Park for the period 1980-92. Koedoe 36:45-52.

Tyson PD (1986). Climatic change and variability in southern Africa. Oxford University Press, London.

Valeix M, Fritz H, Sabatier R, Murindagomo F, Cumming D, Duncan P (2011). Elephant-induced structural changes in the vegetation and habitat selection by large herbivores in an African savanna. Biol. Conserv. 144:902-912.

Walker BH (1993). Rangeland ecology: understanding and managing change. Ambio 22:80-87.

Westoby M, Walker B, Noy-Meir I (1989). Opportunistic management for rangelands not at equilibrium. J. Range Manage. 42:266-274.

Wiegand T, Milton SJ (1996). Vegetation change in semiarid communities. Vegetation 125:169-183.

Appendix. The five interactive sub-models built in STELLA showing rainfall sub-model (a), grass sub-model (b), fire sub-model (c), elephant sub-model (d) and tree sub-model (e).

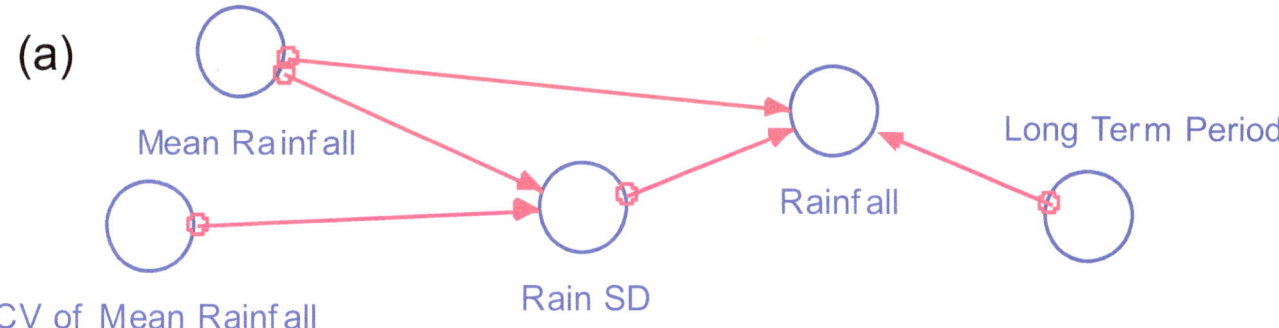

CV_of_Mean_Rainfall = 0.26.
The standard deviation of rainfall for SWRA for the period of 1965/1966 to 1996/1997 is 165.7 mm. This gives a coefficient of variation of 25.8%. This parameter is used here to adjust the standard deviation of the mean rainfall when mean rainfall is varied (auxiliary variable RAIN_SD.).

Long_Term_Period = 20.
Rainfall in the summer rainfall area of southern Africa shows quasi-periodicity with a cycle length of approximately 20 years, with about 10 years of above-average rainfall alternating with 10 years below-average rainfall (Tyson, 1986).

Mean_Rainfall = 642 mm p.a.
The mean annual rainfall for SWRA for the period of 1965/1966 to 1996/1997 is 641.9 mm (unpublished data at SWRI internal files). This rainfall amount sets the basic level of rainfall for Gokwe South, NW Zimbabwe, but it can be varied to simulate other rainfall conditions.

Rainfall = NORMAL(Mean_Rainfall, Rain_SD) + 0.1*Mean_Rainfall*SINWAVE(1,Long_Term_Period).
This is a rainfall generator that takes into account periodicity in rainfall (long-term period = 20 years). Annual rainfall is a random variable drawn from a normal distribution defined by the mean annual rainfall and standard deviation. The amplitude of the long-term periodicity is set at 10% of the mean annual rainfall. This is in turn scaled to 1 within the sine wave function.

Rain_SD = Mean_Rainfall*CV_of_Mean_Rainfall.
This variable is calculated as the product of Mean_Rainfall and CV_of_Mean_Rainfall. The standard deviation of mean rainfall therefore varies as a function of mean rainfall, thus allowing mean rainfall to be varied in a consistent manner.

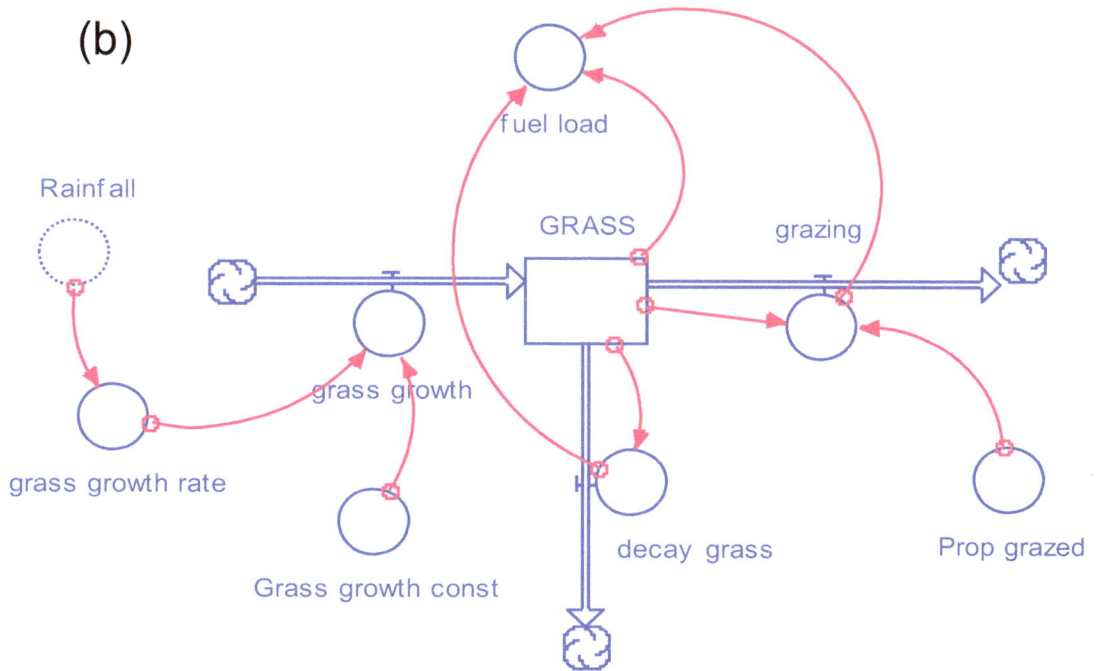

GRASS(t) = GRASS(t - dt) + (grass_growth - decay_grass - grazing) * dt.

INIT GRASS = 200000 kg.
Initial grass biomass. Maximum biomass at Samapakwa in SWRA was 186000 kg km^{-2}; hence, initial biomass set at 200000. Maximum set at 300000 for complete grassland.
grass_growth = grass_growth_rate*Grass_growth_const.

decay_grass = GRASS*EXP(-0.88).
This flow reduces grass biomass through decay. The grass decay rate (k = -0.88 yr^{-1}) is derived from Frost (1996:38), though note that this value comes from a litter bag experiment carried out at Marondera, Zimbabwe (MAP = 885 mm p.a.), not from measurements of the decay of standing grass in a drier environment. Equation relating the decay rate (K) to grass biomass (F) is: Ft = Fo*Exp(-Kt) (Scholes et al., 1996).

grazing = GRASS*Prop_grazed.
This flow reduces grass biomass as a function of grazing pressure, itself a product of grass biomass and grazing rate.

fuel_load = GRASS-(grazing+decay_grass).
Grass biomass less grazing and decay (= fuel load).

Grass_growth_const = 1000.
Constant adjusts for effects of shading, etc, that also limit grass growth. This is the minimum amount of grass per km^2 generated with the lowest rainfall received in the area.

grass_growth_rate = 2*Rainfall.
1 mm rain produces 2 kg ha^{-1} of grass on sandy soils. Dye and Spear (1982).(= 200 kg km^{-2}). 1 mm rain produces 1 kg ha^{-1} under a heavy canopy (Frost, 1996). Scholes et al. (1996) give a figure of 1.25 kg ha^{-1} for all infertile savannas.

Prop_grazed = 0.15.
This proportion has been set as a function of amount of graze available rather than the stocking rate since it is a bit difficult to calculate stocking rate in livestock units for a multispecies system.

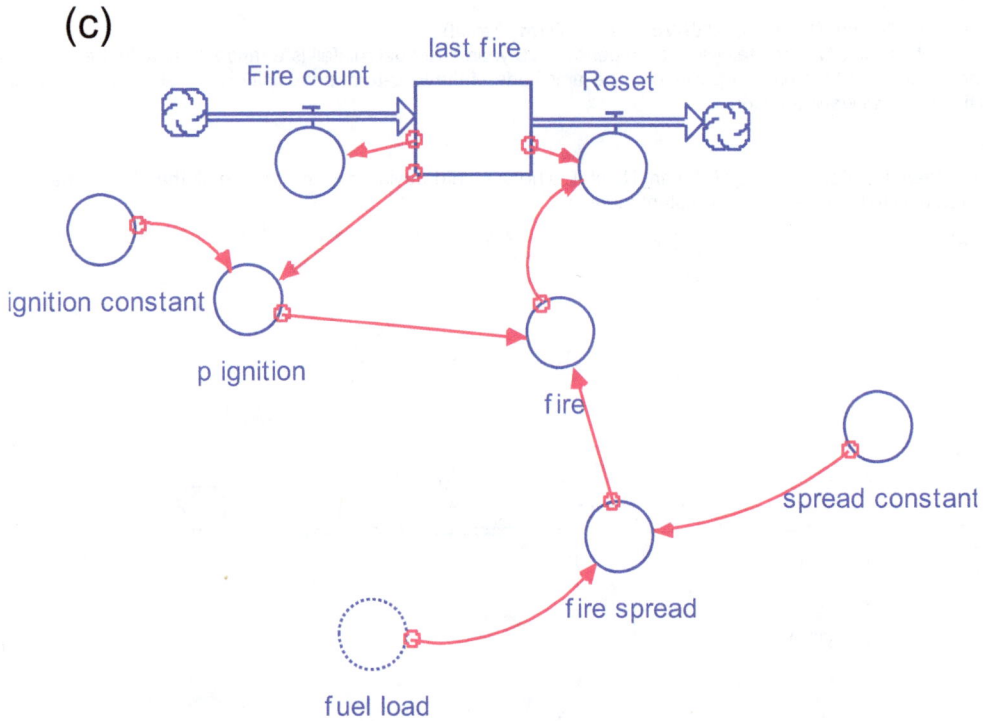

(c)

last_fire(t) = last_fire(t - dt) + (Fire_count - Reset) * dt.

INIT last_fire = 4.7.
Fire return period in the area (Mapaure et al., 2009).

Fire_count = 1-EXP(-(dt/last_fire)).
The distribution function of Fire return interval (T) is given by F(t) = 1-exp(-t/y) (Polakow and Dunne, 1999; Johnson and Gutsell, 1994).

Reset = if fire = 1 then (last_fire/dt) else 0.

fire = if(fire_spread > 0) then(fire_spread*p_ignition) else (0).

fire_spread = if fuel_load > spread_constant then 1 else 0.

ignition_constant = 0.3.

p_ignition = (1- (1/last_fire))*ignition_constant.
Relationship derived from the Weibull equation given by Polakow and Dunne (1999:87). Note - Probability of ignition without management fires in SWAR is 0.32 (Mapaure et al., 2009).

spread_constant = 800 kg ha^{-1}.

800 kg ha^{-1} minimum needed to get a fire spreading (Stronach and McNaughton, 1989; Scholes et al., 1996) (= 5000 kg km^{-2}).

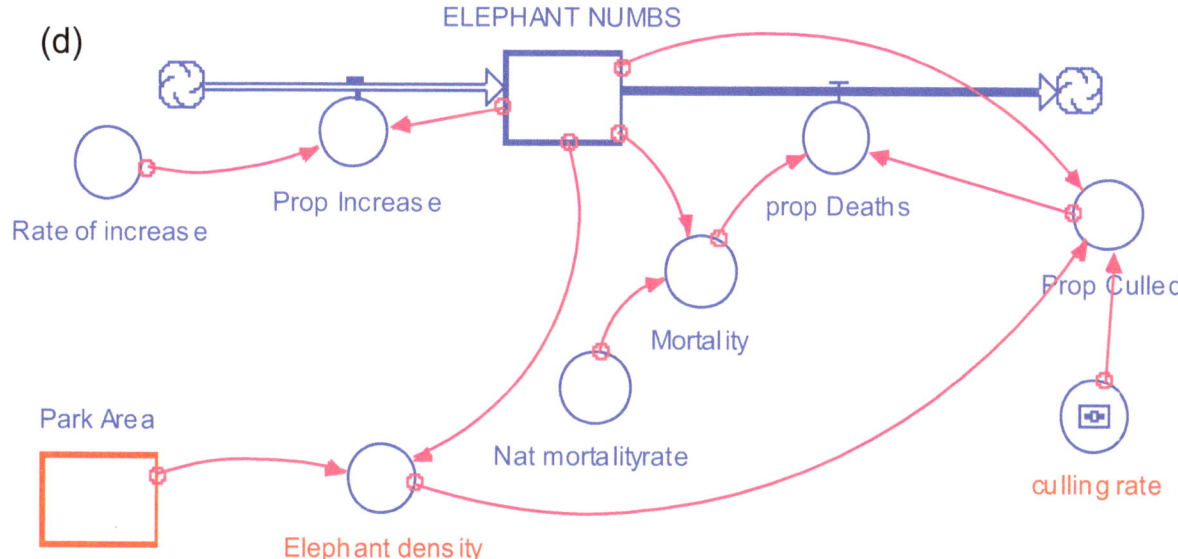

(d)

ELEPHANT_NUMBS(t) = ELEPHANT_NUMBS(t - dt) + (Prop_Increase - prop_Deaths) * dt.

INIT ELEPHANT_NUMBS = 100.
The maximum elephant density at which no adverse effects on vegetation are caused. Guy (1989) indicated an upper limit of 250 elephants in 373 sq. km.

Prop_Increase = Rate_of_increase*ELEPHANT_NUMBS*((1500-ELEPHANT_NUMBS)/1500).
This is the logistic growth equation indicating an upper limit (K, carrying capacity) of 1500 elephants for the whole of SWRA.

prop_Deaths = Prop_Culled+Mortality.
Sum of culling proportion and natural mortality scaled by density. This gives total elephant deaths.

Park_Area(t) = Park_Area(t - dt).

INIT Park_Area = 373.

culling_rate = 0.3 (when switched on).

Elephant_density = (ELEPHANT_NUMBS)/(Park_Area).

Mortality = ELEPHANT_NUMBS*Nat_mortalityrate.

Nat_mortalityrate = 0.02.
The propotion of elephants that die of causes other than culling. This has been set low at 0.02 (that is 2% per annum). This is the average derived from mortality of 1.2% in 1989, 2.4% in 1994 and 2.4% in 1995 (DNPWLM, 1996) (Note: figures were records of carcases; hence, may include poached animals since there was no distinction made).

Prop_Culled = IF(Elephant_density >1)THEN (ELEPHANT_NUMBS*culling_rate)ELSE(0).

Rate_of_increase = 0.05.

Intrinsic growth rate of elephant populations. Cumming et al. (1997) reported an increase of 5% per annum in Zimbabwe and Armbruster and Lande (1993) reported an increase of 5% per annum in Tsavo, Kenya. These figures are LESS (minus) mortality. Hence, actual additions before accounting for mortality are higher than these. Gibson (1983) reported increases of 4.19 and 3.79% per annum in SWRA; hence, the average 4.5% has been used.

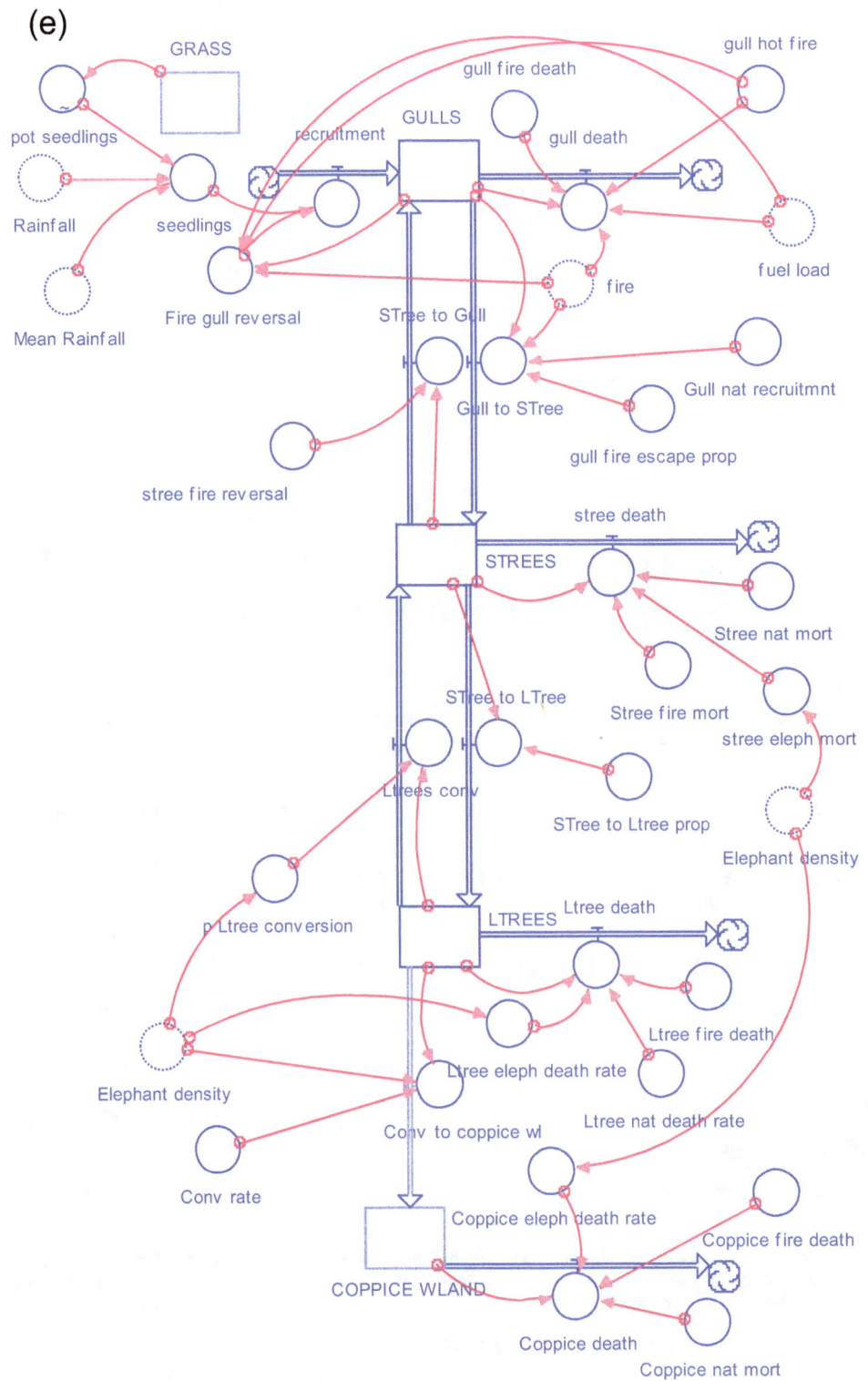

(e)

COPPICE_WLAND(t) = COPPICE_WLAND(t - dt) + (Conv_to_coppice_wl - Coppice_death) * dt.

INIT COPPICE_WLAND = If (Conv_to_coppice_wl>0)then(Conv_to_coppice_wl*LTREES)else 0.
Conv_to_coppice_wl = IF Elephant_density >2 THEN (Conv_rate*LTREES) ELSE(0).
Proportion of trees converted to shrubland. This change of state can only take place when elephant numbers are above 250 per km^{-2}; hence, can only take place if product of conversion rate and elephant numbers (that is, 0.11*250) is above this level, otherwise no conversion to different state takes place.

Coppice_death = COPPICE_WLAND*(Coppice_eleph_death_rate+Coppice_fire_death+Coppice_nat_mort).
GULLS(t) = GULLS(t - dt) + (recruitment + STree_to_Gull - Gull_to_STree - gull_death) * dt.
INIT GULLS = 1000.
recruitment = seedlings+Fire_gull_reversal.
This assumes that the number recruited into the gull stage is the sum of the number of seedlings (cf seedling rule) and the number of resprouts produced by elephant-felled trees from both the small and large trees (the proportions of resprouts have been set at a low rate).

STree_to_Gull = STREES*stree_fire_reversal.

Gull_to_STree = if(fire = 1) then (gull_fire_escape_prop*GULLS) else GULLS*Gull_nat_recruitmnt.
An estimated proportion of gullivers/shrubs that get above 2 m.

gull_death = if(fire = 1) and (fuel_load > gull_hot_fire) then (gull_fire_death*GULLS) else (0).
Total number of gullivers that die from various reasons.

LTREES(t) = LTREES(t - dt) + (STree_to_LTree - Ltree_death - Ltrees_conv - Conv_to_coppice_wl) * dt.

INIT LTREES = 10000.
Trees > 20 cm dbh. About 25% of trees are in this category (Mapaure and Moe, 2009). Taking an initial tree density of 400 trees/ha (40000/sq. km), then there are 10000 trees/sq. km in this category.

STree_to_LTree = STree_to_Ltree_prop*STREES.
Refers to the number of Strees that grow to Ltrees. This was assumed to take 15 years and was scaled by the maximum woody basal area. Calculated as the product of the proportion of Strees that grow into Ltrees and the reciprocal of the number of years taken to mature, scaled by basal area.

Ltree_death = LTREES*(Ltree_nat_death_rate+Ltree_eleph_death_rate+Ltree_fire_death).
Estimated tree longevity is 200 years. Fire assumed to have no detrimental effect on large trees.

Ltrees_conv = LTREES*p_Ltree_conversion.
Proportion of Ltrees converted to Strees by elephant herbivory.

Conv_to_coppice_wl = IF Elephant_density>2THEN(Conv_rate*LTREES) ELSE(0).
Proportion of trees converted to shrubs and/coppice. This change of state can only take place when elephant densities exceeded 2 km^{-2}); hence, can only take place if product of conversion rate and elephant numbers (that is, 0.11*250) is above this level, otherwise no conversion to different state takes place.

STREES(t) = STREES(t - dt) + (Gull_to_STree + Ltrees_conv - stree_death - STree_to_Gull - STree_to_LTree) * dt.

INIT STREES = 30000.
Trees 2 to 5 m tall. 75% of total trees is in this category, while the remainder is in the large tree category. Initial overall density is 40000 trees km^{-2} (present study).

Gull_to_STree = if (fire = 1) then (gull_fire_escape_prop*GULLS) else GULLS*Gull_nat_recruitmnt.
An estimated proportion of gullivers/shrubs that get above 2 m.

Ltrees_conv = LTREES*p_Ltree_conversion.
Proportion of Ltrees converted to Strees by elephant herbivory.

stree_death = STREES*(stree_eleph_mort+Stree_fire_mort+Stree_nat_mort).
Estimated about 3% die each year.

STree_to_Gull = STREES*stree_fire_reversal.

STree_to_LTree = STree_to_Ltree_prop*STREES.
Refers to the number of Strees that grow to Ltrees. This was assumed to take 15 years.

Conv_rate = 0.021.
There was a loss of 1.1% of vegetation cover (Mapaure and Campbell, 2002). This has been taken to translate into change in state but above a critical elephant density (conv to shrubland). Calculations from damage assessments (Mapaure and Moe, 2009) give conversion at 2.1% per year.

Coppice_eleph_death_rate = 0.00015*Elephant_density.

(conversion rate for L-trees).

Coppice_fire_death = 0.006.
Fire becomes important again here (as opposed to Ltrees) and kills similar proportions as gullivers/shrubs.

Coppice_nat_mort = 0.0038.
Coppice death in the absence of elephants. (Ltrees nat mortality).

Fire_gull_reversal = IF(fire = 1)AND(fuel_load>gull_hot_fire)THEN(0.30*GULLS)ELSE(0).
The proportion of shrubs reversed by fire (that is, 1-(fire escape+fire mort+shade mort). Ben Shahar (1996) recorded 70% fire reversal in new seedlings and 31% fire reversal in shrubs in *Baikiaea* woodland.

gull_fire_death = 0.015.
The proporiton of gullivers that die in a fire. Present study shows that 3.1% (= 0.031) died (Mapaure, in press) after two fires/2 years (and 1.1% had died after one fire). Hence, mortality was 1.5% per year (0.015).

gull_fire_escape_prop = 0.1.
Mapaure and Moe (2009) have shown an increase in small tree proportions equivalent to 2.5% per year, implying similar net recruitment rate. But this was inclusive of other influences. Hence, actual escape rates may even be twice this figure since fire is patchy.

gull_hot_fire = 1000.
This is a constant referring to the minimum fuel load required for a fire to kill 10% of the gullivers. It was set at 1000 kg DM per ha. [Trollope (1993) found out that fire can only spread in savannas at grass fuel >1000 kg DM/ha; fires as low as 800 kg DM/ha do not cause top-kill of gullivers (Rushworth, 1975)].

Gull_nat_recruitmnt = 0.3.

Ltree_eleph_death_rate = Elephant_density*0.00015.
Elephant- induced death rate of large trees. This is a function of the number of trees available and the elephant density. In SWRA, each elephant pushes down 1500 trees per year (Cumming, 1981) and I assumed that a large proportion of pushed down trees dies (which is not always true). The original number of trees has been incorporated into the calculation because availability of trees is likely to affect the number of trees pushed down per elephant per year. The proportion is therefore 1500/10000 (= 0.15). But Sharp (1982) reported that trees were being lost at 9% per year at about 1.6 elephants km^{-2} (that is, 1 elephant km^{-2} accounts for 5.63%, = 0.00015 elephant/tree year (Duffy et al., 1999) gave range 5.9 to 7.5×10^{-6}).

Ltree_fire_death = 0.
Negligible numbers of trees killed by fire.

Ltree_nat_death_rate = 0.01.
Large tree mortality in the absence of elephants (for example, caused by drought, old age). Bond and van Wilgen (1996) gave 0.38%. In SWRA, this could be much lower than this per year.

p_Ltree_conversion = Elephant_density*0.013.
Tree height reduction over 16 years was 2.1% per year, at an average elephant density of 1.6 km^{-2} (Mapaure and Moe, 2009). Hence, conversion factor is 1.3% per elephant per sq. km.

seedlings = if Rainfall>Mean_Rainfall*1.25 then pot_seedlings else 0.
This assumes that we only get seed production in above average rainfall years and that grass biomass influences how many seedlings establish (as gulls). Seedling production is therefore independent of the total number of trees producing seed in the stand - this is because it is assumed that seed production is non-limiting. Perhaps, we should be able to switch seed production off when the number of small trees and large trees gets below a critical level.

stree_eleph_mort = 0.0001*Elephant_density.
The proportion of small trees that are killed by elephants. About a 1.3% (that is, 0.01) per annum loss in small trees was recorded (Mapaure and Moe, 2009). Norton-Griffiths (1979) reported a 1% mortality in Serengeti.

Stree_fire_mort = 0.001.
The proportion of small trees that are killed by fire. As most trees survive the fire, this constant has been set at a very low rate. Note: 28% of trees in 2 to 3 m height class were burnt back but not necessarily killed (Norton-Griffiths, 1979). Bond and van Wilgen (1996) give 0.63% (0.006 prop) and 1.58% for early and dry season fires, respectively. Again this is much lower (than 0.0063) in SWRA because of patchy fires.

stree_fire_reversal = 0.028.
The proportion of small trees that are reversed by fire. Observations from Serengeti indicate that tree survival from fire was 8, 32, and 72% in three height classes of the trees affected and that fire does not affect trees larger than 3 m in height (Norton-Griffiths, 1979). In SWRA, however, the proportion is very low (100 minus 72). Ben Shahar (1996) reported fire reversal rate on trees of 28% in *Baikiaea* woodland. Since the fires are very patchy in SWRA, his figure could be more than 10 times that of SWRA.

Stree_nat_mort = 0.004.
It is assumed that natural mortality in small trees (2 to 5 m) is negligible (Norton-Griffiths, 1979). (0.38%) (Bond and van Wilgen, 1996).

STree_to_Ltree_prop = 0.01.
The proportion of small trees that recruit into large trees each year.

15

Assessment of biomass carbon stock in an *Ailanthus excelsa* Roxb. plantation Uttarakhand, India

Nishita Giri and Laxmi Rawat

Forest Ecology and Environment Division, FRI, Dehradun, India.

The article presents biomass carbon stock for an *Ailanthus excelsa* plantation in Dehradun Forest Division, Uttarakhand, India. Destructive sampling was used to calculate the biomass of *A. excelsa* and undergrowth vegetation (shrubs and herbs); volumetric equations were used for estimating the biomass of associated tree species. The total biomass of *A. excelsa* was calculated as 126.07 t ha^{-1} with above ground biomass (AGB) 102.96 t ha^{-1} and below ground biomass (BGB) 23.11 t ha^{-1}. The total biomass of the two associated tree species was estimated as 43.91 t ha^{-1} (AGB 34.01 and BGB 9.9 t ha^{-1}). The total biomass of shrub and herb species was calculated as 1.62 and 0.98 t ha^{-1}, respectively. Litter biomass was calculated as 0.98 t ha^{-1}. The estimated total biomass of the whole ecosystem (173.56 t ha^{-1}) was obtained as the sum of these component biomass values. Carbon content of the main tree species, associated tree species, and understory vegetation (shrubs+herbs), was estimated in AGB pool (63.76 Mg ha^{-1}) and BGB pool (14.84 Mg ha^{-1}), and added to the litter carbon (0.35 Mg ha^{-1}) and soil organic carbon (SOC) (46.27 Mg ha^{-1}) to estimate the carbon stock in the whole ecosystem (125.22 Mg ha^{-1}). The SOC to AGB ratio was 0.72.

Key words: Biomass, carbon stock, litter, *Ailanthus excelsa* Roxb. plantation ecosystem, above ground biomass and below ground biomass.

INTRODUCTION

Forest ecosystems are deemed to be an important factor in climate change because they can be both sources and sinks of atmospheric CO2. They can assimilate CO2 via photosynthesis and store carbon in biomass and in soil (Trexler and Haugen, 1994; Brown et al., 1996; Watson et al., 2000). Plantations or naturally regenerated trees can protect watersheds against droughts, flash floods or landslides thought to be more prevalent due to climate change. Sustainable forestry practices can increase the ability of forests to sequester atmospheric carbon, while simultaneously enhancing other ecosystem services, such as improved soil and water quality. Carbon sequestration is also a good indicator of the health and functioning of ecosystems. Forests may help local communities to cope with climate change in a numerous

ways (Robledo and Forner, 2005).

Ailanthus excelsa Roxb. commonly known as 'Ardu' or 'Mahanimb' is a fast growing tree and is extensively cultivated in many parts of India. Its wood is very light, soft and perishable. The timber is used for packing cases, fishing floats, boats, spear sheaths, sword handles, toys and drums. The bark is bitter, astringent, anthelmintic and it is used in diseases like dysentery, bronchitis, asthma, dyspepsia and ear ache. It is also used for environmental conservation as it is resistant to drought and soil conditions. It grows well on slopes. The pulp is obtained from debarked wood and is used in paper industry as a substitute for aspen, for printing papers, the leaves are rated as highly palatable and protein rich nutritious fodder for sheep and goats and are said to augment milk production

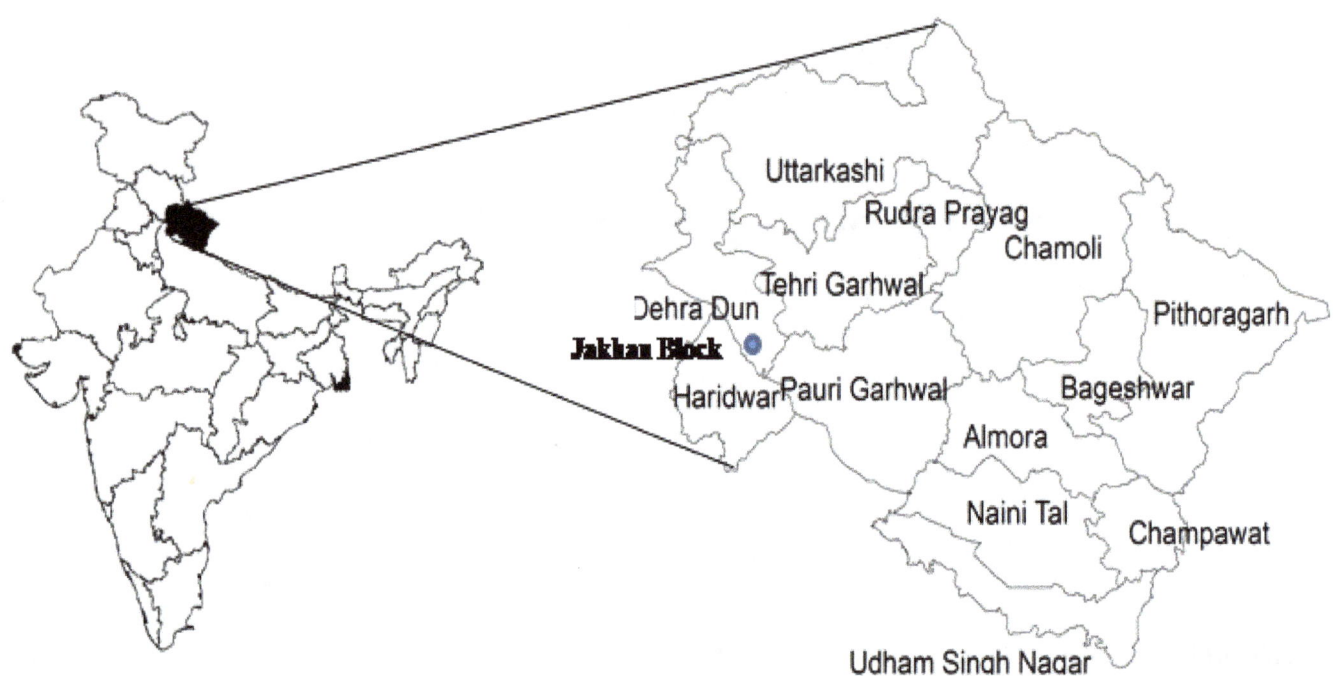

Map 1. Study site.

(Jat et al., 2011).

Above ground biomass (AGB) has been given the highest importance in carbon inventories and in most mitigation projects and is the most important pool for afforestation and reforestation CDM projects under the Kyoto Protocol. However, below ground biomass (BGB) has been shown to be an important carbon pool for many vegetation types and land-use systems and accounts for about 20% (Santantonio et al., 1997) to 26% (Cairns et al., 1997) of the total tree biomass. BGB accumulation is linked to the dynamics of AGB. The greatest proportion of root biomass occurs in the top 30 cm of the soil surface (Bohm, 1979; Jackson et al., 1996).

The carbon (C) sequestration potential of a forest eco-system depends on initial soil organic carbon (SOC) con-tent, stand growth rates, the biological carrying capacity of the stand and stand age. In particular, C sequestration and storage may be increased significantly, if forests are harvested and trees are converted into wood products (Skog and Nicholson, 1998). Some researchers suggest that sequestration of C in tree biomass and litter is a de-laying tactic that only buys time for finding more perma-nent solutions for C sequestration (IPCC, 2000). Making an effort to maximize the productivity of the restored forest is also worthwhile because forest C pools can vary five-fold within a local edaphic gradient as a function of site quality (Burger and Zipper, 2002).

This article presents complete stand level (ecosystem level) estimates of biomass by component. This is the first report of C stock / C pool estimation of AGB, BGB of all existing vegetation and litter, as well as soil organic carbon (SOC) at the ecosystem level in India.

MATERIALS AND METHODS

Study area

This study was conducted in a 39 year old *A. excelsa* plantation in the Jakhan block, Barkot Range of Dehradun Forest Division, Uttarakhand, India (Map 1), nearly 25 km east of Haridwar and 30 km south east side of Dehradun city. The area lies in a subtropical region at an altitude of 449 m msl at 30°04'37.2"N and 78°12'11.1"E. It has a very gentle slope with a south aspect. The maximum, minimum and mean temperatures of the area (1980 to 2010) were 28.11 13.52 and 20.32°C, respectively. The mean annual rainfall during this period was 1901.03 mm when averaging monthly and approximately 80% of the rainfall occurred during the south-west monsoon period (June to September) (Figure 1).

Soil analysis

Texture

It is the proportion of particle size distribution (soil texture) into classified grades expressed as percentage of sand, silt and clay. After air drying of samples, big stones were removed and the soil was passed through 2 mm sieve. Part of the soil samples having particle size less than 2 mm were subjected for texture analysis by Hydrometric method (Black, 1965) and percentage of different fractions namely: sand, silt and clay was estimated in each sample and textural class was determined using the Triangular diagram by U.S.D.A (Black, 1965).

Soil moisture

Soil moisture percentage (%) was measured by means of moisture

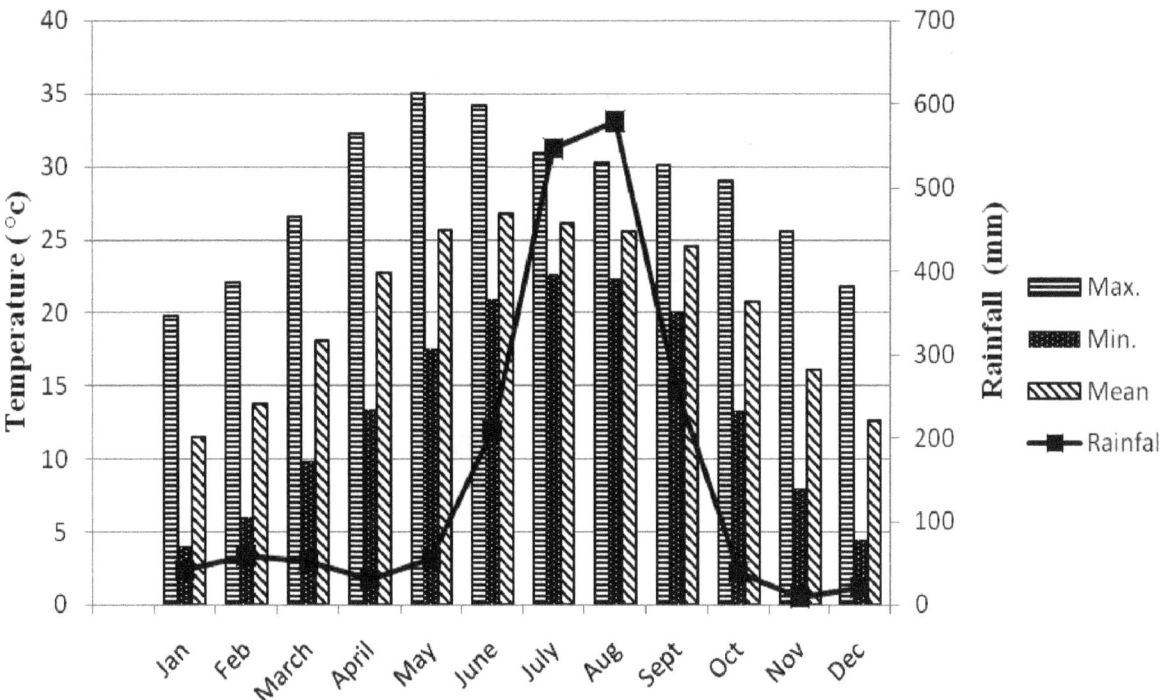

Figure 1. Ombrothermic graph of rainfall, mean, maximum and minimum temperature for 30 years (1980 to 2010).

meter.

Soil bulk density

A metal core cylinder (by core sampler) of known weight and volume was used to determine the soil bulk density (Wilde et al., 1985). Soil bulk density was determined by the following expression:

Bulk density (g cm³) = $\dfrac{W1-W2}{V}$

Where, W1= weight of cylinder + weight of soil, W2 = weight of empty cylinder, V= volume of cylinder

Biomass estimation of *Ailanthus excelsa*

The stratified tree technique method of Art and Marks (1971) was used to harvest the sample trees. Temporary sample plots (30 m × 30 m) were laid out in the plantation and the diameter at breast height (DBH at 1.3 m) of all the standing trees were recorded within the sample plots. The DBH range was divided into five different diameter classes that is, 10 to 20 cm, 20 to 30 cm, 30 to 40 cm, 40 to 50 cm and 50 to 60 cm from which 2 trees were harvested from 10 to 20 cm diametr class, 3 trees from 20 to 30 cm, 2 trees from 30 to 40 cm, 1 tree from 40 to 50 cm and 1 from 50 to 60 cm and in this way 9 representative sample trees were selected for the study.

The tree components (leaves, twigs, branches, bark, bole and roots) were separated immediately after felling and their fresh weights recorded. Samples of all tree components (100 g of each component) were selected for oven dry weight estimation and chemical analysis for C content.

The bole of each sample trees was cut into 2 m long sections (billets) for convenience of weighing.

Biomass estimation for the associated tree species

Biomass of the associated tree species (*Acacia catechu* and *Eucalyptus* hybrid) was estimated using the volumetric equations of the Forest Survey of India (FSI, 1996). Estimated volumes were multiplied by the density of the corresponding wood following the methods of Chaturvedi and Khanna (1982) to get the dry weight stem biomass. The biomass of branches and leaves were estimated using 45 and 11% of the stem biomass, respectively, as per Sharma (2003). BGB was estimated using the root-shoot ratios (R:S) of these species (FAO, 2000); Table 1). Total biomass per tree was obtained by summing AGB and BGB for each sample tree and averaging over the sample. The total biomass per ha for each of these species was esyimated by multiplying the average biomass per tree by the trees per ha for each of the species (80 trees ha⁻¹ for Acacia Catechu and 40s tree ha⁻¹ for *Eucalyptus* hybrid). C was estimated as 43% of the total biomass (Negi et al., 2003).

Biomass estimation of understory vegetation

Ten quadrats of 3 m × 3 m and 1 m × 1 m were laid out for shrubs and herbs, respectively. Complete harvesting of all shrub and herb species present in all quadrats was done; the plant materials were separated into above and below ground portions. Fresh and dry weights were measured for biomass and C was estimated according to the methodology given earlier. Biomass values were then multiplied by an expansion factor to sclae them to a one hectare area.

Estimation of litter biomass

Litter biomass was estimated by laying out ten 5 m × 5 m sample plots in the plantation. Litter samples were collected on these sam-

Table 1. Volumetric equations and root shoot ratio (R:S) used for estimation of biomass of associated tree species.

Tree specie	Volumetric equations (FSI, 1996)	R:S (FAO, 2000)
Eucalyptus hybrid	$V = 0.02894 - 0.89284\ D + 8.72416\ D^2$	0.30
Acacia catechu	$V = 0.048535 - 0.183567\sqrt{D} + 3.78725D^2$	0.25

V = volume, D = diameter.

Table 2. Moisture, Bulk density and Texture of soil at different depths under *A. excelsa* plantation.

A. excelsa plantation (cm)	Moisture (%)	BD (g cm^{-3})	Texture (Sandy loam)		
			Sand (%)	Silt (%)	Clay (%)
0-30	5.990± 0.198	1.223± 0.004	51.23 ± 0.470	26.83 ± 0.536	21.93 ± 0.133
30-60	7.075± 0.363	1.267± 0.003	51.67 ± 0.636	25.80 ± 0.851	22.53 ± 0.606
60-90	8.848± 0.203	1.283± 0.003	52.80 ± 0.208	25.37 ± 1.538	22.50 ± 0.589

ple plots and a fresh and an oven dry weight at 80°C (until a constant weight was achieved) were obtained. The litter was ground for chemical analysis to estimate C content.

Estimation of carbon in A. excelsa trees, shrubs, herbs, litter and soil organic carbon (SOC)

Samples of all three parts of *A. excelsa* trees, shrubs, herbs and litter soil were analyzed for C content using Wakley and Black's titration method (Jackson, 1967). For estimation of SOC, soil samples were taken from the surface to 90 cm depth (in three depth classes of 0 to 30 cm, 30 to 60 cm and 60 to 90 cm) from randomly selected points in the plantation area. Three replicates from each point were collected.

RESULTS AND DISCUSSION

Floristic struture

In the plantation, *A. excelsa* showed the maximum density (350 ha^{-1}) followed by *A. catechu* (80 ha^{-1}) and *Eucalyptus* hybrid (40 ha^{-1}).

Physical attributes of soil

Soil texture was observed to be sandy loam in nature, soil moisture was higher (8.848%) in the deepest layer that is, 60 to 90 cm depth, lower (5.990%) in uppermost layer that is, 0 to 30 cm depth. The trend of bulk density in soil depths was in the order 60 to 90 cm > 30 to 60 cm > 0 to 30 cm (Table 2).

Biomass of A. excelsa species

The DBH and heights of the nine sample trees varied from 15.5 cm to 55.09 cm and 8.85 m to 20.20 m, respectively. This variation in the total tree biomass which ranged from 75.04 to 759.56 kg tree^{-1}. The other tree components ranged from: bole, 40.21 to 551.48 kg; leaves, 0.32 to 12.71 kg; twigs, 0.43 to 5.10 kg;

branches, 6.71 to 52.63 kg; bark, 5.84 to 18.28 kg; and roots 21.27 to 119.36 kg.

The total biomass of *A. excelsa* trees was estimated at 126.07 t ha^{-1}, of which the AGB comprised 102.96 t ha^{-1} and the BGB comprised 23.11 t ha^{-1}. The highest percentage of total biomass was found in boles (66.94%), followed by roots (18.33%), branches (9.07%), bark (3.36%), leaves (1.59%) and twigs (0.71%). The percentage contribution to the total biomass varied among dbh classes: 10 to 20 cm, 4.52%; 21 to 30 cm, 17.93%; 31 to 40 cm, 27.98%; 41 to 50 cm, 26.79%; and 51 to 60 cm, 22.76%. More than 77% of the *A. excelsa* trees have a dbh between 31 to 60 cm.

Biomass of associate tree species

The biomass values of *Acacia catechu* and the *Eucalyptus* hybrid were estimated as product of wood density (kg/m^3) and volumes using volumetric of FSI (1996). The biomass estimated for the *Eucalyptus* hybrid was 36.15 t ha^{-1} and for *Acacia catechu* was 7.76 t ha^{-1}.

Understory biomass

The shrub species present in the plantation ecosystem were *Lantana camara*, *Justicia adhatoda*, *Murraya koenigii*, *Eucalyptus* hybrid saplings, *Syzigium cumini*, *Lemonia acidissima* and *Cassia tora*. Herb species were *Ageratum conyzoides*, *Sida cuta*, *Oxalis corniculata*, *Aerva scandens*, *Rundia pectinata*, *Cyperus esculentus*, *Oplismenus compositus*, *Parthenium hysterophorus*, *Cynodon dactylon*, *Murraya koenigii* seedlings and *Achyranthes aspera*. The AGB of shrubs was 1.027 t ha^{-1} and the BGB was 0.591 t ha^{-1}. The total shrub biomass was 1.618 t ha^{-1}. The herb biomass was 0.983 t ha^{-1} (AGB 0.705 t ha^{-1} and BGB 0.278 t ha^{-1}). The total understory biomass (shrub and herb) was estimated at 2.60 t ha^{-1}. Total litter biomass estimated as 0.98 t ha^{-1}.

Table 3. Total Biomass distribution (t ha^{-1}) among different components of *A. excelsa* plantation ecosystem.

Level	Vegetational components				Total
	Main tree species	Associated tree species	Shrubs	Herbs	
Above ground	102.96	34.01	1.027	0.705	138.702
Below ground	23.11	9.90	0.591	0.278	33.88
Total	**126.07**	**43.91**	**1.618**	**0.983**	**172.58**

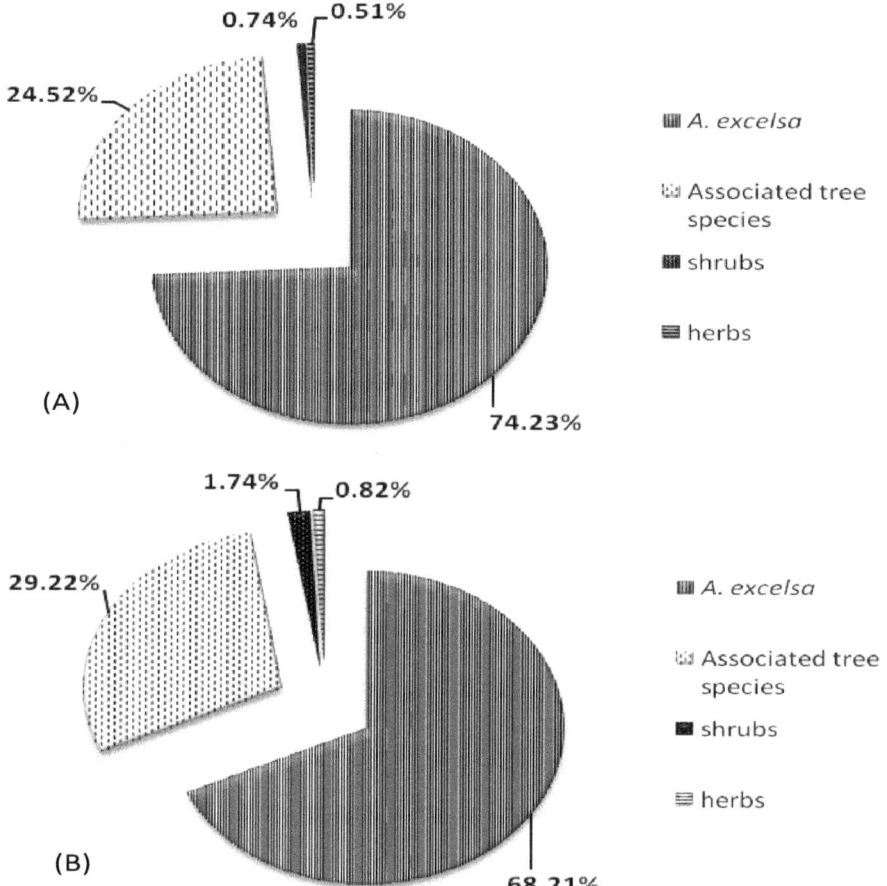

Figure 2. (a) AGB (%) contribution of different components of the ecosystem. (b) BGB (%) contribution of different components of the ecosystem.

Total biomass estimation

Total biomass of the whole ecosystem was estimated at 173.56 t ha^{-1}, which is the sum of the biomass of main tree species that is, *A. excelsa*, the biomass of associated tree species, shrub, herb biomass and litter biomass (Table 3).

The biomass contribution of the main tree species, associated tree species, shrubs and herbs to the total AGB and BGB was 72.64% for *A. excelsa* trees, 25.3% for associated tree species, 0.93% for shrubs, and 0.57% for herbs. Figure 2 (a) and (b) depicts the AGB and BGB contribution (%) separately for these components.

Carbon content / carbon pool

A. excelsa tree species

The total C content (t ha^{-1}) in the different *A. excelsa* tree components were: 40.27 (boles), 10.22 (roots), 5.36 (branches), 1.61 (bark), 0.73 (leaves), and 0.33 (twigs). The amount of C content contributed by *A. excelsa* trees was 58.52 t ha^{-1}.

Associated tree species

Total C content in associated tree species was 18.88 t ha^{-1}, of which 14.62 t ha^{-1} (77.44%) was contributed by

Table 4. Carbon stock (t ha^{-1}) in different pools of *A. excelsa* plantation ecosystem.

Parameter	Carbon Pools				Total C
	AGB C	BGB C	Litter C	SOC	
Main tree species					
Ailanthus excelsa	48.3	10.22			**58.52**
Associate species					
(1) *Acacia catechu*	2.67	0.67			3.34
(2) *Eucalyptus* hybrid	11.95	3.59			15.54
					18.88
Understory vegetation					
(1) Shrubs	0.5	0.24			0.74
(2) Herbs	0.34	0.12			0.46
					1.20
			0.35		**0.35**
				46.27	46.27
Grand total	**63.76**	**14.84**	**0.35**	**46.27**	**125.22**

AGB and 4.26 t ha^{-1} (22.56%) by BGB.

Understory vegetation

Understory vegetation (shrubs+herbs) contributed 1.20 t ha^{-1} to the C pool. For shrubs 67.56% of the C was in the above ground material and 32.43% was in the below ground material. For the herb layer 73.91% of the C was in the above ground material and 26.09% was in the below ground material. Litter contains 0.35 t ha^{-1} of total C content (Table 4).

Soil organic carbon (SOC)

Soil Organic Carbon was estimated at 46.27 t ha^{-1}.

Total carbon stock

The total C stock was determined to be 125.22 t ha^{-1} of which 46.73% was contributed by *A. excelsa* trees, 15.08% by associated tree species, 0.59% by shrubs, 0.37% by herbs, 0.28% by litter and 36.95% contibuted by soil (Table 3).

Pande et al. (1988) also the estimated biomass of *A. excelsa* of different ages in Uttar Pradesh. They have reported that contribution of bole to AGB was just over 50%, bark contributed 19.9 to 23.3%, branches contributed 9.68 to 14.5% and roots 18.1 to 25%. This study has showed a similar order of contribution of different tree components to AGB. The percent contribution of AGB to total biomass was estimated as 81.67% in the present study, which is similar to overstory biomass contribution of 81.9 and 81% reported by Nascimento and Laurance (2002) and Henry et al. (2009), respectively. However, it is less than the 92.7 to 94% of overstory contribution reported by Clark and Clark (2000).

Rana and Singh (1990) showed that the understory (shrubs+herbs) accounted for 1.5% of the total forest biomass (432.8 t ha^{-1}) in a *Pinus roxburghii* plantation located in Kumaun Himalaya of Uttarakhand. Mac Lean and Wein (1977a) found that understory biomass in *Pinus banksiana* ranged from 1 to 6% of the ecosystem biomass in old stands. The per-cent contribution of the understory to the total biomass in this study was 1.50%. Negi (1984) reported 2.3 and 0.9% understory biomass contribution to the total stand tree biomass in *Shorea robusta* (sal) forest and *Eucalyptus* hybrid plantation ecosystems, respectively.

The maximum concentration of C was found in the bole (47.2%) and the minimum concentration was in the leaves (36.08%). Similar findings have been reported by Kraenzel et al. (2003) and Negi et al. (2003) in teak plantations of Panama and India, respectively.

Conclusion

Long rotation forests have larger long term C storage in the forest biomass and product pool. Biotic interferences and changes in land use cause significant exchanges of carbon between the land and the atmosphere. The phytomass carbon pool estimates are associated with significant uncertainties due to deficiency of data, volume biomass conversion approach and the extent of the human activity on ecosystem and environment, because many ecological processes depend on the carbon cycle.

In the tropical forest the carbon in the soil is roughly equivalent to or less than the AGB due to degradation (cited from Ramachandran et al. 2007). Ravindranath et al. (1997) reported that the ratio of SOC and biomass carbon was 1.25. Kaul (2010) has given the range of this ratio between 0.7 to 2. She indicates that in the plantations, the carbon content in the soil was double the biomass carbon but not 2.5 to 3 times the biomass carbon as recorded earlier. The fact she gives that the sequestered SOC came from the original vegetation in the past before exploitation. The SOC and AGB ratio of the present study comes to be 0.72.

Biomass and productivity of *A. excelsa* plantation of 16 and 21 years at Mohand range of Shiwalik forest division of Uttarakhand have been estimated by Pande et al. (1988) and they have reported 37.62 t ha^{-1} biomass of 16 years and 31.78 t ha^{-1} of 21 years plantations. The productivity of both the plantations was 1.95 and 1.45 t ha^{-1} yr^{-1}, respectively. 126.07 t ha^{-1} biomass of the present study of *A. excelsa* species with 3.23 t ha^{-1} yr^{-1} of productivity showed a high value when compared to the study of Pande et al. (1988), which may be because of high density of *A. excelsa* trees and associate species, and more age (39 years old) of the species, which would have supported more biomass in the present study site and signifies that at this age the species shows high productivity and better C stock.

Stand level estimates of biomass according to tree components are needed when biomass productivity and litter fall by biomass components of different quality are modeled and linked to soil as Liski et al. (2002) model describes the decomposition of dead organic matter also. For these purposes it is important to be able to observe the dynamics of C stock in different tree components, such as foliage, branches, bark, stem, stump and roots according to stand age (Lehtonen et al. 2004).

ACKNOWLEDGEMENTS

The authors are thankful to Forest Research Institute (FRI), Dehradun and Uttarakhand Forest Department for logistical and financial support for conducting the study.

REFERENCES

Art HW and Marks PL (1971). A summary table of biomass and net annual primary production in forest ecosystems of the world. In: Forest biomass Studies, ed. Young H.E (ed.), pp.1-32. (16th IUFRO Congr.) Univ. Maine Press, Orono.

Black CA (1965) (ed.). Methods of soil analysis. American Society of Agronomy, Inc. Publisher. Madison, Wisconsin.

Bohm W (1979). Methods of studying root systems, Ecological Studies pp.33.

Brown S, Sathaye J, Cannel M, Kauppi P (1996). Management of forests for mitigation of greenhouse gas emissions. In: Climate Change 1995: Impacts, Adaptations, and Mitigation of Climate Change: Scientific-Technical Analyses. Watson, R.T., et al. (Eds.), Cambridge University Press, Cambridge.

Burger J, Zipper C (2002). How to restore forests on surface mined land. Virginia Cooperative Extension Publication, pp. 460-123.

Cairns MA, Brown S, Helmer EH, Baumgardner GA (1997). Root biomass allocation in the world's upland forests. *Oecologia* 111:1-11.

Chaturvedi AN, Khanna LS (1982). Forest Mensuration. International Book Distributors, Dehra Dun, India. p.408.

Clark DB, Clark DA (2000). Landscape-scale variation in forest structure and biomass in a tropical rain forest. For. Ecol. Mgmt. 137:185-198.

FAO (2000). Global Forest Resource Assessment *FAO For.* Paper No. 40. FAO, Rome.

FSI (1996). Volume Equations for Forests of India, Nepal and Bhutan. Forest Survey of India, Ministry of Environment and Forests, Dehradun.

Henry M, Tittonell P, Manlay RJ, Bernoux M, Albrecht A, Vanlauwe B (2009). Biodiversity, carbon stocks and sequestration potential in AGB in smallholder farming systems of western Kenya. Agric. Ecosyst. Environ. 129:238-252.

IPCC (Intergovernmental Panel on Climate Change) (2000). Land Use, Land-Use Change and Forestry. Cambridge: Cambridge Univ.Press (ISBN: 92-9169-114-3).

Jackson ML (1967). Soil chemical analysis. Prentice-Hall India, New Delhi.

Jackson RB, Canadell J, Ehleringer JR, Mooney HA, Sala OE, Schulze ED (1996). A global analysis of root distributions for terrestrial biomes, Oecologia 108:389-411.

Jat HS, Singh RK, Mann JS (2011). Ardu (*Ailanthus* sp) in arid ecosystem: A compatible species for combating with drought and securing livelihood security of resource poor people. Indian J. tradit. Knowledge 10(1):102-113.

Kaul M (2010). Carbon budget and carbon sequestration potential of Indian forests. Ph D Thesis, Wageningen Univ. Wagenongen, Nether Lands, 2010.

Kraenzel M, Castillo A, Moore T, Potvin C (2003). Carbon storage of harvest-age teak (*Tectona grandis*) plantations, Panama. For. Ecol. Mgmt. 173(1-3):213-225.

Lehtonen A, Makipaa R, Heikkinen J, Sievanen R, Liski J (2004). Biomass expansion factors (BEFs) for Scots pine, Norway spruce and birch according to stand age for boreal forests. For. Ecol. Mgmt. 188:211-224.

Liski J, Peruuchoud D, Karjalainen T (2002). Increasing carbon stocks in the forest soils of western Europe. For. Ecol. Mgmt. 169:168-179.

Mac Lean DA, Wein RW (1977a). Nutrient accumulation for post fire jack pine and hardwood succession patterns in New Brunswick. Can. J. For. Res. 7:562-578.

Nascimento HEM, Laurance WF (2002). Total AGB in central Amazonian rainforests: a landscape-scale study. For. Ecol. Mgmt. 168:311-321.

Negi JDS (1984). Biological productivity and cycling of nutrients in managed and man-made ecosystems. Ph. D Thesis, Garhwal University, Srinagar. p.161.

Negi JDS, Manhas RK, Chauhan PS (2003). Carbon allocation in different components of some tree species of India: A new approach for carbon estimation. Curr. Sci. 85(11):1528-1531.

Pande MC, Tandon VN, Negi M (1988). Biomass production in plantation ecosystem of *Ailanthus excelsa* at 5 different ages in Uttar Pradesh. Indian For. 114(7):362-371.

Ramachandran A, Jayakumar S, Haroon RM, Bhaskaran A, Arockiasamy DI (2007). Carbon sequestration:estimation of carbon stock in natural forests using geospatial technology in the Eastern ghats of Tamil Nadu, India. Curr. Sci. 92(3):323-331.

Rana BS, Singh RP (1990). Plant biomass and productivity estimates for central Himalayan mixed Banj oak (*Quercus leucotrichophora* A.camus)-chir pine (*Pinus roxburghii*). Indian For. 116(3):220-226.

Ravindranath NH, Somshekhar BS, Gadgil M (1997). Carbon flows in Indian forest. Climate change 35(3):297-320.

Robledo C, Forner C (2005). Adaptations of forest ecosystems and the forest sector to climate change. Forest and climate change working Paper No. 2, FAO, Rome.

Santantonio D, Hermann RK, Overton WS (1977). Root biomass studies in forest ecosystems. Pedobiologia 17:1-31.

Sharma RP (2003). Relationship between tree dimensions and biomass, sapwood area, leaf area and leaf area index in *Alnus nepalensis D.Don* in Nepal. Agricultural University of Norway (NLH), Aas.

Skog K, Nicholson G (1998). Carbon cycling through wood products: the role of wood and paper products in carbon sequestration. Forest Products J. 48:75-83.

Trexler MC, Haugen C (1994). Keeping it Green: Tropical Forestry Opportunities for Mitigating Climate Change. World Resources Institute, Washington, DC.

Watson RT, Noble IR, Bolin B, Ravindranath NH, Verado DJ, Dokken DJ (2000). Land Use, Land-Use Change, and Forestry. Cambridge University Press, Cambridge.

Wilde SA, Corey, RB, Iyer JG, Voigt GK (1985). Soil and plant analysis for tree culture. Oxford and IBH publishing co., New Delhi.

16

Influence of Zn stresses on growth and physiology in Khus-khus (*Vetiveria zizanoides* Nash.) and its essential sesquiterpene oil(s), in relation to roots diameter circumferential positions

A. MISRA*, N. K. SRIVASTAVA and A. K. SRIVASTAVA

Central Institute of Medicinal and Aromatic Plants, P. O. CIMAP, Kukrail Picnic Spot Road Lucknow – 226015, India.

Culturing Khus in controlled glasshouse condition in sand culture techniques for maximum essential monoterpene oil(s) was found (0.21%) in young developed middle position circumferences of roots. At middle position of leaf, net photosynthetic rate and contents of chlorophyll (Chl) were affected. The maximum peroxidase activity was obtained at the middle position of leaf and roots circumferences area, with the maximum production of biomolecule of khusimol and khusinol at 250 mg Zn/L. The maximum monoterpene oil(s) (0.21%) was also found at middle position of developed roots. However, the relative contents of khusimol and Khusinol varied at different circumference area of positions. As a result of different root developmental positions, the contents of Fe, Mn, Zn and Cu were smaller in quantity in Vetiver. Their maximum contents were observed at middle positions of developed roots. Thus, the value addition of essential monoterpene oil(s) seems to be used for commercial exploitation at large scales, to be for the collection of developed roots position at middle levels (roots area of 5th positions developments) and for better quality of total essential oil of Khus.

Key words: Chlorophyll, dry mass, leaf area, net photosynthetic rate, plant height, saccharides, Zn.

INTRODUCTION

Khus-khus (*Vetiveria zizanoides* Nash.) is an aromatic grass of the family Poaceae and it is the only source of one of the most important essential monoterpene oil(s) called the oil of Khus. It is commonly known as Khus oil. It is distinctly different from the horticultural khus, which are basically used for soil erosion and have no commercial usage in perfumery industries (Douglas, 1969). *V. zizanoides Nash* widely grown cultivars improved varity' Gulabi' (Rajeshwar and Bhatacharya, 1992) are commonly cultivated in India. Steam distillation of root biomass of Khus-Khus yields Khusimol and Khusinol rich monoterpene Vetiver oil(s), extensively used for perfuming soaps and cosmetics and in

aromatherapy to which they impart a pronounced and lasting Vetiver oil peculiar odour. It is also largely used in flavouring soft drinks products and other pharmaceutical preparations, and also the plant is used in soil erosion as the roots are having soil binding properties. Apart from this, it is widely used in phytochelatin properties and in phytoremediation of heavy metals extraction from the heavily polluted environment.

Zn is an essential micronutrient and acts as a phytochelatin that acts either as a metal component of various enzymes or as a functional, structural, or regulatory cofactor, and is thus, associated with saccharide metabolism, photosynthesis and protein synthesis (Marschner, 1986). Zn deficiency reduces plant growth and inhibits photosynthesis in many plants including forest trees (Dell and Wilson, 1985), fiber crops (Ohki, 1986), rice (Ajay and Rathore, 1985) and spinach (Randal and Bouma, 1973). Zn retards the activity of carbon metabolism enzymes such as carbonic anhydrase

*Corresponding author. E-mail: amisracimap@yahoo.co.in or a.misra@cimap.res.in.

(Ohki, 1976; Ohki, 1978), ribulose 1,5-bisphosphate carboxylase/oxygenase and fructose-1,6-bisphosphate (Marschner, 1986). Zn, Se and Cr are antioxidants scavenging free radicals. Zn stimulates the removal of free radicals (Chakmak and Engles, 1999).

Essential oil biosynthesis in geranium is strongly influenced by Zn acquisition and the stresses caused by Zn on nutrition and growth. Zn is involved in carbon assimilation, saccharide accumulation, free radical removal, antioxidant enzymes, carbon utilization in terpene biosynthesis and the overall growth of the plants. The requirement of Zn for Japanese mint and its limitations imposed on photosynthetic carbon metabolism and translocation in relation to essential oil accumulation in mint were shown by Misra and Sharma (1991), whereas antioxidants enzymes for free radical quenching in geranium have not been fully documented.

In the present paper, we report on the role of Zn as a stimulant of quenching of free radicals through Zn affected antioxidant enzyme activity. Simultaneously, photosynthetic efficiency in terms of net photosynthetic rate (P_N), content of chlorophyll (Chl), leaf fresh and dry mass, leaf area, Zn content in plant shoot biomass and oil yield were also determined.

MATERIALS AND METHODS

Plant tips (12.5 to 15.0 cm) with 3 to 4 leaves of *V. zizanoides* L. genotype of diploid – Gulabi were obtained from the farm nursery of the Central Institute of Medicinal and Aromatic Plants (CIMAP), Lucknow, India. Uniform slips were initially planted in 10 000 cm³ earthen pots filled with purified silica sand (Agrawala and Sharma, 1961) for the development of roots. After 15 days, rooted cuttings were transferred to 2 500 cm³ pots. The salts used in nutrient solution of Hoagland and Arnon (1952) were purified for Zn (Hewitt, 1952). The nutrient solution was used in the experiment except Fe which was supplied as Fe- ethylenediaminetetra acetic acid (EDTA). Three pots each of Zn treatments ranging from 0.0 to 1.0 g (Zn) m⁻³ were maintained in controlled glasshouse condition at ambient temperature (30±5°C) and irradiance (800 to 1 000 µmol m⁻² s⁻¹). The nutrient solution in each treatment was added at alternate days. With onset of deficiency and toxicity (after 20 days), growth and detailed physiological, and biochemical data characteristics were determined. P_N was measured using a computerized portable photosynthesis system *Li-COR 6000* (LiCOR, USA) (Srivastava, 1991). Chl amount in 80% acetone extracts from 3rd leaf was determined spectrophotometrically on *Pye Unicam PU8610* according to Arnon (1949). Leaf fresh and shoot dry mass, and area (area meter *Li-3000*) were also recorded. For tissue element analysis, 1 g dried leaf samples were digested with 1 M HCl at 60°C for 24 h. Aliquot samples of the clear digest were diluted with water (10 cm³) and analyzed for Zn by atomic absorption spectrophotometer (*Pye Unicam SP 2800*) (Misra and Sharma, 1991). Antioxidant reactive peroxidase enzyme activity was estimated as described in Sharon et al. (1966). Two gram (2 g) of freshly chopped leaves at 3rd position was homogenized with 5 cm³ of 0.1 M phosphate buffer (pH 6.8). Each treatment was replicated 3 times and assayed by sodium dodecyl sulfate polyacrylamide gel electrophoresis (SDS-PAGE).

Vetiver oil was estimated by steam distillation of 100 g freshly plucked roots area from circumferences (Root area #1 to 7, that is, outer area (ar.#1:1.5 cms, then the b following ar.2 to 7 in descending orders), in an apparatus of Clevenger (1928). Khusimol

and Khusinol, and other associated oil contents were determined by gas liquid chromatography (*Perkin-Elmer* model *3920 B*). The stainless steel column was packed with 10% carbowax (20 meshes) on *Chromosorb WNAW*. Injector and detector temperature were maintained at 200°C. The flow of H_2 was 0.47 cm s⁻¹; data processing for area % was done on a *Hewlett- Packard* integrator model *HP-33*.

RESULTS AND DISCUSSION

The fresh and dry biomasses increased with increase in the supply of Zn (Table 1). Maximum fresh and dry biomass, and leaf area were observed at $Zn_{0.250}$. Plant height was maximum at $Zn_{0.500}$. $Zn_{1.000}$ was toxic to all growth parameters. The Chl content increased up to $Zn_{0.250}$ and then decreased. The maximum P_N was found at $Zn_{0.250}$; at this Zn supply also the saccharide content was the highest. Zn deficiency and toxicity inhibited P_N in cotton (Ohki, 1976), peppermint (Srivastava et al., 1997), soybean (Ohki, 1978) and sweet mint (Misra et al., 2003). A decrease in Chl content represents a decline in photochemical capacity of leaf at deficient Zn supply (Ohki, 1976).

Maxima of peroxidase activity were observed at $Zn_{0.250}$. The Zn deficient and toxic cultured plants revealed lesser peroxidase activity with lesser peroxidase isoenzyme band profiles. In Japanese, mint similar report was given for Mn nutrition (Misra, 1996). The maximum of monoterpene oil(s) was found at $Zn_{0.250}$. However, relative contents of citronellol, geraniol, linalool and nerol varied at different Zn treatments. As a result of different Zn supply, the contents of Fe, Mn, Zn and Cu were smaller in shoots ofgeranium. Their maximum contents were observed at $Zn_{0.250}$. Statistical analysis showed a positive significant association between Zn content in leaf and P_N (γ, 0.924 ≤ p, 0.5%) and between P_N and content of saccharides (γ, 0.879 ≤ p, 0.05%). However, Zn content in leaf was negatively correlated with Chl a/b ratio. P_N showed a positive significant association with leaf fresh mass (γ, 791 ≤ p, 0.05%), leaf dry mass (γ, 692 ≤ p, 0.05%), leaf area and total monoterpene oil(s) (γ, 0.721 ≤ p, 0.01). A positive significant correlation was also observedbetween saccharides and total oil (γ, 0.695 ≤ p, 0.01%). A quadratic trend was observed for all these characters which were comparable in +Zn than in plants grown at Zn deficit or much higher Zn supply.

We found that optimum supply of Zn is $Zn_{0.250}$. Utilization of metabolites from primary photosynthetic process in secondary metabolism regulates monoterpene production (Gershezon and Croteu, 1991). Thus, a close relation between photosynthesis, photorespiration and terpenoid synthesis exists in essential monoterpene oil(s) bearing plants (Maffei and Codignola, 1990). Moreover, the actively growing leaves require a larger supply of an antioxidants stimulator Zn, in association with greater supply of photosynthates. Since essential oil biosynthesis occurs in these rapidly growing leaves, the initial growth period would require a greater supply of photosynthates

Table 1. Effect of root positions on parameters of V. zizanoides.

Growth attribute	Root ar. #1 0.00	Root ar. #2 0.01	Root ar. #3 0.100	Root ar. #4 0.200	Root ar. #5 0.250	Root ar. #6 1.0	Root ar. #7 1.5 mg Zn/L	LSD At 5%	LSD at 1%
Plant height [cm]	57.0	58.0	61.0*	62.5**	63.4**	64.1**	59.0	2.5	4.1
No. of branches	9	10*	13**	18**	10*	10*	8	1.1	3.2
Fresh mass [g plant^{-1}]	218.8	238.6*	224.8	252.1**	282.5**	215.5**	196.2	11.1	16.3
Dry mass [g plant^{-1}]	14.11	16.33**	16.81*	17.37**	19.36**	18.46**	15.85	2.10	3.30
Leaf area [cm^2]	8.2	12.1*	25.2**	39.1**	40.3**	37.2**	11.2	3.5	6.2
Chl a [g kg^{-1} (FM)]	0.68	0.79*	0.94**	1.35**	1.48**	1.01**	0.82*	0.11	0.15
Chl b [g kg^{-1} (FM)]	0.50	0.56	0.61*	0.69**	0.79**	0.40	0.29	0.08	0.12
Chl a/b	1.36	1.41	1.54	1.96	1.87	2.53	2.83	-	-
P_N [µg(CO$_2$) m^{-2} s^{-1}]	0.15	0.19*	0.75**	0.76**	0.82**	0.71**	0.42**	0.03	0.06
Saccharides [µg (CH$_2$O) m^{-2} s^{-1}]	0.102	0.129	0.510	0.516	0.558	0.483	0.286	-	-
Oil %	0.15	0.16	0.17*	0.19	0.21**	0.16	0.15	0.02	0.04
Khusimol [% of total oil]	0.21	0.27**	0.29**	0.32**	0.25**	0.18**	0.17**	0.01	0.02
Khusinol [% of total oil]	0.09	0.09	0.10**	0.11**	0.07**	0.12**	0.10**	0.01	0.01
Roots tissue concentrations									
Fe [mg kg^{-1}]	98	112	142*	249**	537**	419**	312*	21	42
Mn [mg kg^{-1}]	26	37**	41**	57**	98**	62**	53**	9	11
Zn [mg kg^{-1}]	12	19*	34**	45**	64**	41**	36**	7	9
Cu [mg kg^{-1}]	7	9	11**	11	12**	7	5	3	5

Chl: Chlorophyll; P_N, net photosynthetic rate; oil amounts in % of total oil. *, ** Values are significant at P, 0.05 or P, 0.01 levels, respectively.

and energy.

Conclusion

The Zn at 0.250 mg/L application and at # 5 diameter circumference position has given the maximum total oil % as well as the maximum khusimol and khusinol oil contents. Therefore, 0.250 Zn mg/L application is more useful for better growth of Khus-khus and maximum oil production.

REFERENCES

Agrawala SC, Sharma CP (1961).The standardization of sand and water culture technique for the study of macro and micronutrients (trace) element deficiencies under Indian conditions. Curr. Sci., 40: 424-428.

Ajay, Rathore VS (1985). Effect of Zn^{2+} stress in rice (Oryza sativa cv. Manhar) on growth and photosynthetic processes. Photosynthetica, 31: 571-584.

Arnon DI (1949). Copper enzymes in isolated chloroplasts. Polyphenoloxidase in Beta vulgaris. Plant Physiol., 24: 1-15.

Chakmak I, Engels C (1999). Role of mineral nutrients in photosynthesis and yield formation – In: Ringel, Z. (ed.): Mineral Nutrition of Crops. Haworth Press, New York, pp. 141-168.

Clevenger JF (1928). Apparatus for determination of essential oils. J. Am. Pharmac. Assoc., 17: 346.

Dell B, Wilson SA (1985). Effect of Zn supply on growth of three species of Eucalyptus seedlings and wheat. Plant Soil, 88: 377-384.

Douglas JS (1969). Essential oil crops and their uses. World Crops, 21: 49-54.

Hewitt EJ (1952). Sand and water culture methods used in the study of plant nutrition. Commonwealth Bureau bot. Plantation Crops Tech. Commun., 22: 405-439.

Hoagland DR, Arnon DI (1952). The water culture method for growing plants without soil. Calif. Agr. Exp. Stat. Circ., 347: 1-32.

Maffei M, Codignola A (1990). Photosynthesis, photorespiration and herbicide effect on terpene production in peppermint (Mentha piperita L.). J. Essent. Oil Res., 2: 275-286.

Marschner H (1986). Function of mineral nutrients. Micronutrients. – In: Mineral Nutrition of Higher Plants. Academic Press, New York, pp. 269-300.

Misra A (1996). Genotypic variation of manganese toxicity and tolerance of Japanese mint. J. Herbs Spices Medic. Plants, 4: 3-13.

Misra A, Sharma S (1991). Zn concentration for essential oil yield and menthol concentration of Japanese mint. Fertilizer Res., 29: 261-265.

Misra A, Srivastava NK, Sharma S (2003). Role of an

antioxidant on net photosynthetic rate, carbon partitioning and oil accumulation in sweet mint. Agrotropica, 15: 59-74.

Ohki K (1976). Effect of zinc nutrition on photosynthesis and carbonic anhydrase activity in cotton. Physiol. Plant, 38: 300-304.

Ohki K (1978). Zinc concentration in soybean as related to growth, photosynthesis, and carbonic anhydrase activity. Crop Sci., 18: 79-82.

Rajeshwar-Rao BR, Bhatacharya AK (1992). History and botanical nomenclature of rose scented geranium cultivars grown in India. – Indian Perfumer, 36: 155-161.

Sharon LM, Kay E, Lew JY (1966). Peroxidase isoenzymes from horse raddish roots 1-1 Isolation and physical properties. J. Biol. Chem., 241: 2166-2172.

Srivastava NK, Misra A (1991). Effect of thr triacontanol formulation" Miraculon" on photosynthesis, growth, nutrient uptake and essential oil yield of lemon grass (Cymbopogon flexuosus steud, Watts). Plant Growth Regul., 10: 57-63.

Srivastava NK, Misra A, Sharma S (1997). Effect of Zn deficiency on net photosynthetic rate, ^{14}C partitioning, and oil accumulation in leaves of peppermint. Photosynthetica, 33: 71-79.

Comparison of the trees regeneration at different distances from Alang Dareh forest roads considering tourist pressure

Aidin Parsakhoo , Mohammad Hadi Moayeri and Majid Poursadeghi

Department of Forestry, Faculty of Forest Sciences, Gorgan University of Agricultural Sciences and Natural Resources, Gorgan, Iran.

Natural regeneration is the most important factor in survival and sustaining forest parks. This study was conducted to compare the regeneration frequency of trees species at distances of 15, 40 and 80 mfrom roads considering tourist pressure in Alang Dareh forest park. Results show that number of high seedling in low tourist pressure area was more than that of area with severe tourist pressure. Moreover, total number of seedlings in low tourist pressure area was more than that of area with severe tourist pressure. Lowest number of seedlings was recorded at distance of 80 m from road edge because of the tourist density and consequently soil compaction.

Key words: Tourist pressure, regeneration, forest road, Alang Dareh park.

INTRODUCTION

Changes to regeneration conditions are considered by forest park managers to be an important impact of tourism use and consequences of soil compaction (Good, 1995). Reduction of regeneration is a well-documented negative effects of tourism use (Good and Grenier, 1994). Decompaction using subsurface tilling, grazing and blocking compacted area after regeneration or plantation are preferred site-preparation treatments by many public and private park managers (Shestak and Busse, 2005). Beside, ground cover by slash and plant residuals is said to decrease soil compaction by providing a pressure absorbing layer, lowering the net ground pressure of passing equipment.

Siikamäki (2009) in a study in Paanajärvi National Park, Republic of Karelia, Russia indicated that high compaction

of the soil mineral horizons lead to critical status for normal growth and development of the root system. Beside it was concluded that the most important factors in soil resistance to trampling are moisture and type of the plant community. Natural regeneration is the most important factor in survival and sustaining forest parks. Therefore, study of the regeneration condition in a forest park can be useful to predict ecosystem future and apply forestry programs for improving forest stands. Despite Alang Dareh Forest Park's size and status, it has not been closely studied, and thus little information is available about floristic condition of forest to develop ecology-based management tools. The objective of this study was to compare the regeneration frequency of trees species at different distances from roads considering population

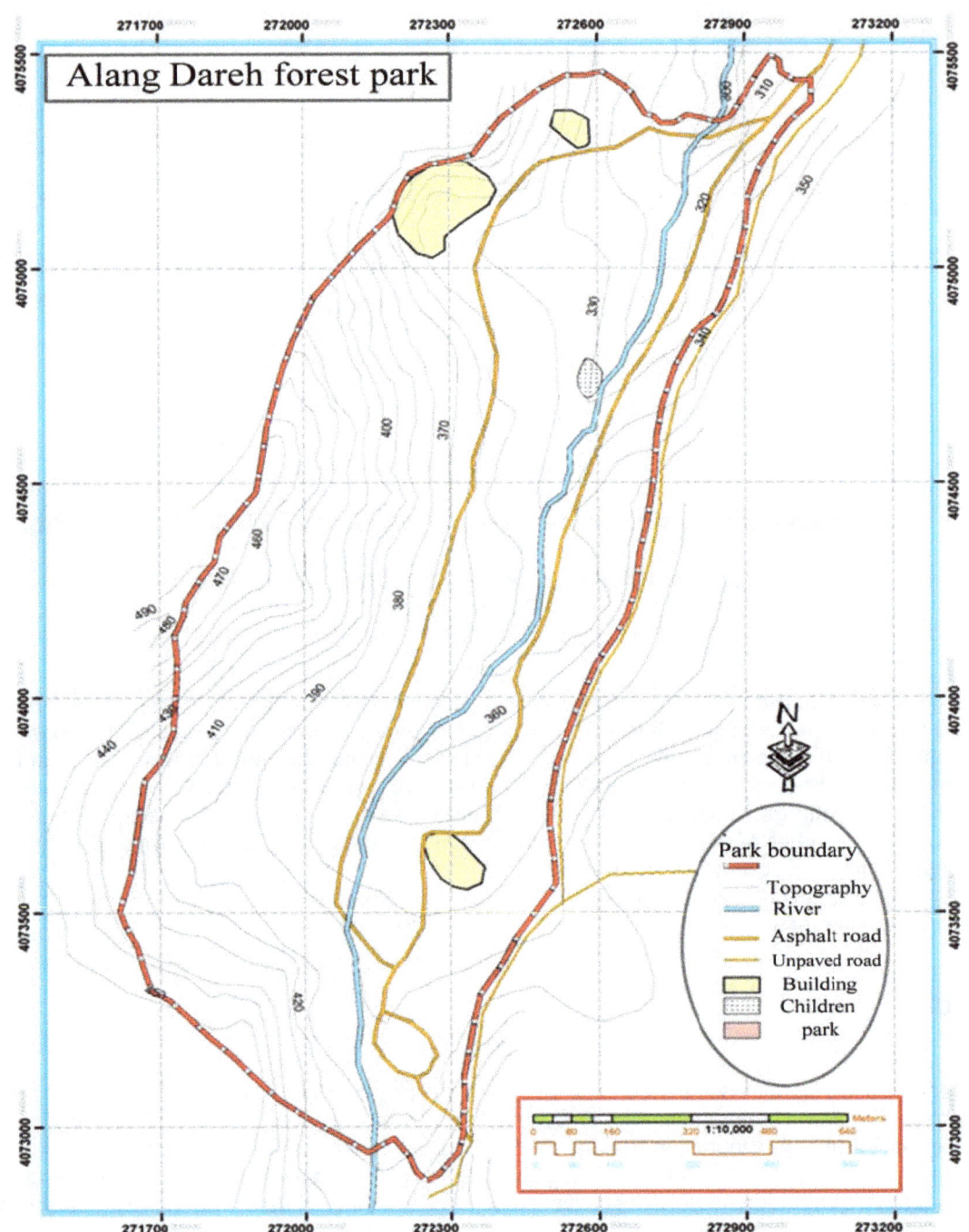

Figure 1. The geographical position of the study area.

density in Alang Dareh forest park.

MATERIALS AND METHODS

Study site

Alang Dareh forest with an area of 185 hectares Alang Dareh is a forest 5 km away from Gorgan city in the south west on the way to Naharkhoran (36°47′43″ N and 54°26′44″ E). The climate of the region is moderate, moist and mid-moist. The bedrock of this forest is green schist and sedimentary loess stone with altitude ranging from 300 to more than 400 m above sea level. The forest is mixed deciduous which has been established on brown forest soil. The mean annual precipitation is 837 mm which the highest is occurred in autumn (Figure 1). The Ambrotermic curve of the study area can be shown as Figure 2. Floristic composition of the park are *Oplismenus undulatifolia*, *Oplismenus* sp., *Carex silvatica*, *Viola* spp., *Juncus* sp., *Euphorbia* spp., *Agropyrum* sp., *Convolvulus* sp., *Parrotia persica*, *Carpinus betulus* and *Quercus castaneifolia*.

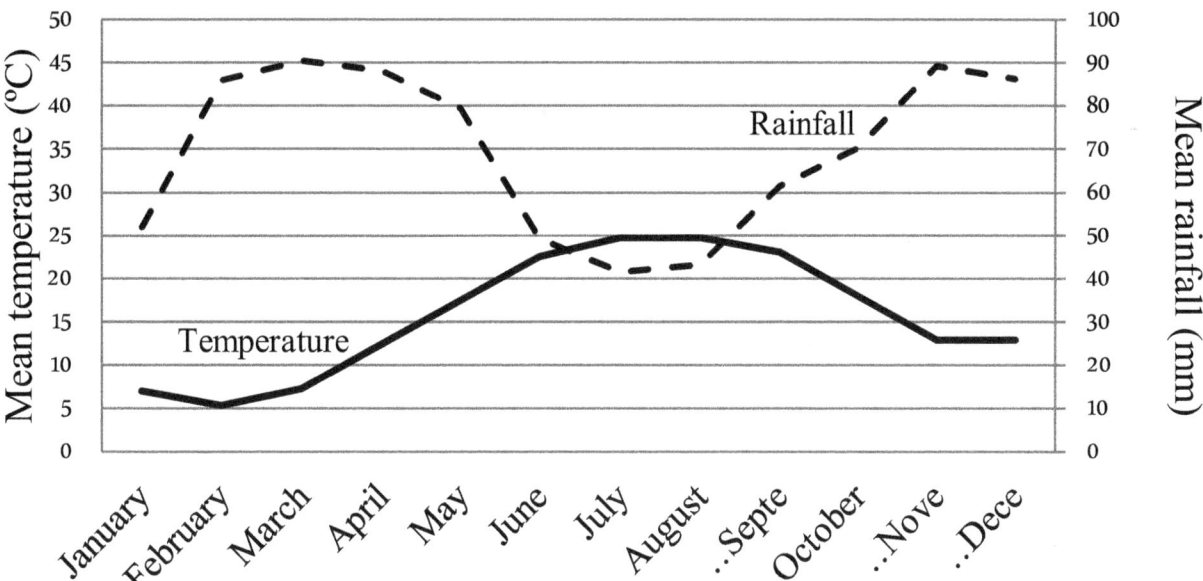

Figure 2. Ambrotermic curve of the Alang Dareh forest park.

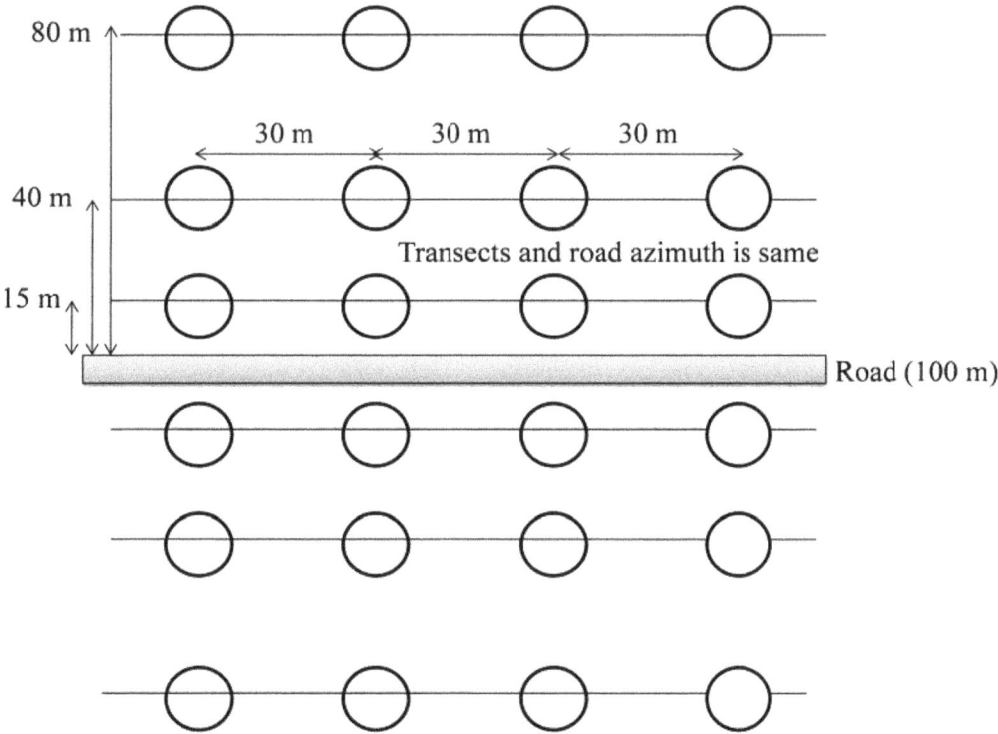

Figure 3. Sampling design in study area.

Sampling design and field survey

In this study two road segment each of with a length of 100 m was selected in Alang Dareh forest park. It was attempted to select roads with similar condition considering altitude, slope gradient, slope direction, vegetation type and soil. Road segment one is located in a region with high population density in park and another is located in a region with low population density. Three transects were established at each sides of the road and at distances of 15, 40 and 80 m. On each transect, four plots with an area of 100 m² and radius of 5.7 m was systematic randomly selected. The distance of plots to each other was 30 m (Figure 3). Trees regeneration with

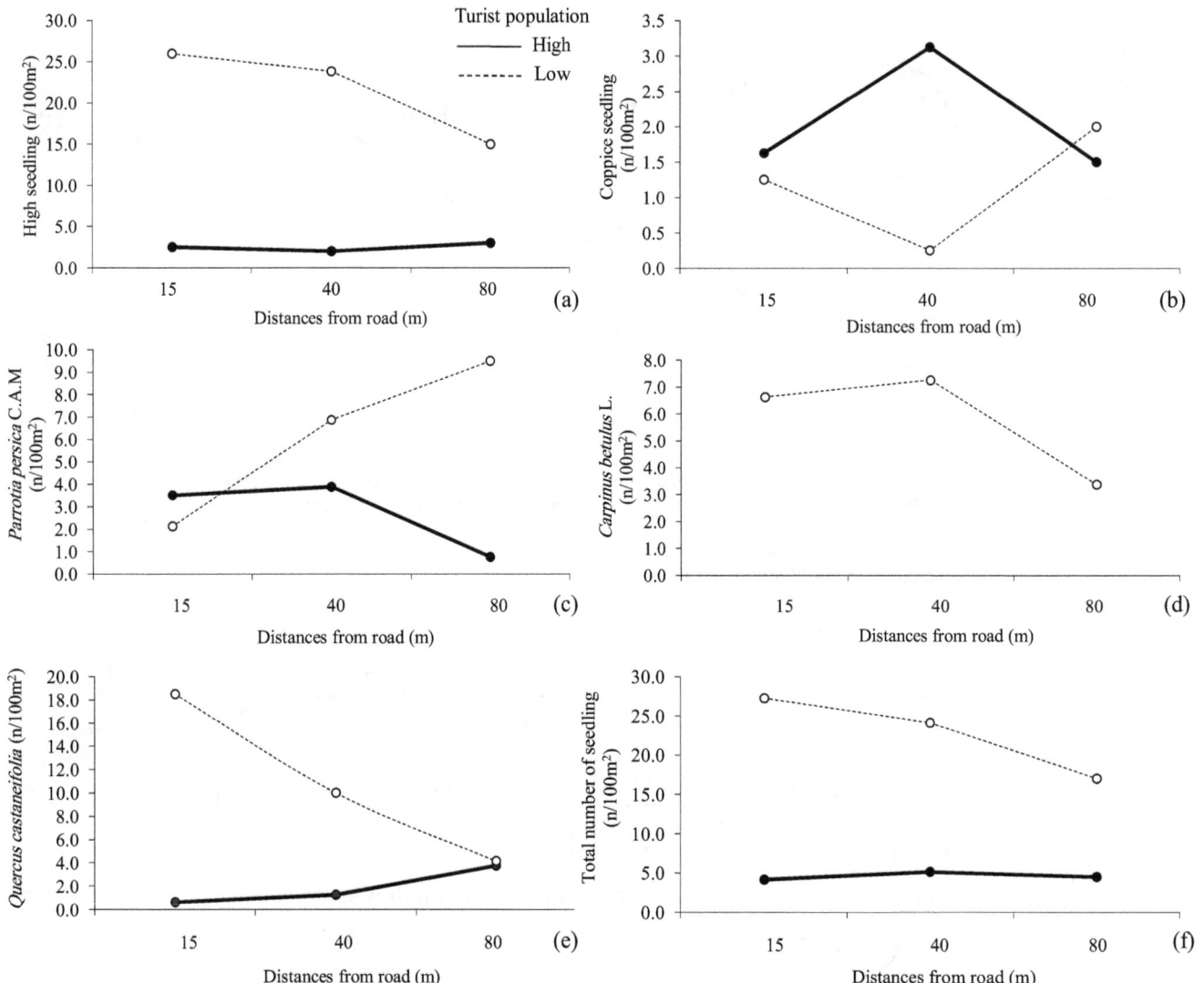

Figure 4. Regeneration trend at different distances from road edge considering population density.

diameter at breast height (DBH) less than 7.5 cm and height of less and more than 1.30 m were counted separately for each species within the plots. Graphs were designed in Excel software.

RESULTS AND DISCUSSION

Results of the study in high density population area indicates stationary trend at different distances from road concerning high seedling frequency (Figure 4a). Low and high tourist population showed that there was no obvious trend in coppice seedling frequency at the different distances from road edge due to the human interference (Figure 4b). The frequency of different species including *Parrotia persica* (Figure 4c), *Carpinus betulus* (Figure 4d) and *Quercus castaneifolia* (Figure 4e) are shown. Lowest number of seedlings grew at the distance of 80 m from road edge, because tourist density and consequently soil compaction was high in this zone. Indeed number of seedling

decreased with increasing distances from road edge in low tourist pressure area. In severe tourist pressure area there was no significant differences among distances in term of seedlings frequency (Figure 3f). Glaeser (2006) conducted a research within a Forest Park, in Queens County, New York, to document the current floristic composition and structure of the woodland community. His findings about the disturbance patterns, the decline in traditional dominant tree species, the abundance of pioneer tree species and the continued colonization by *Phellodendron amurense* may be the signs of structural change throughout the park.

Number of high seedling in low density population area was more than that of high density population area (Figure 5a), while number of coppice seedling in low density population area was less than that of high density population area (Figure 5b). Low density population area has been dominated by *P. persica* (Figure 5c), *C. betulus* (Figure 5d)

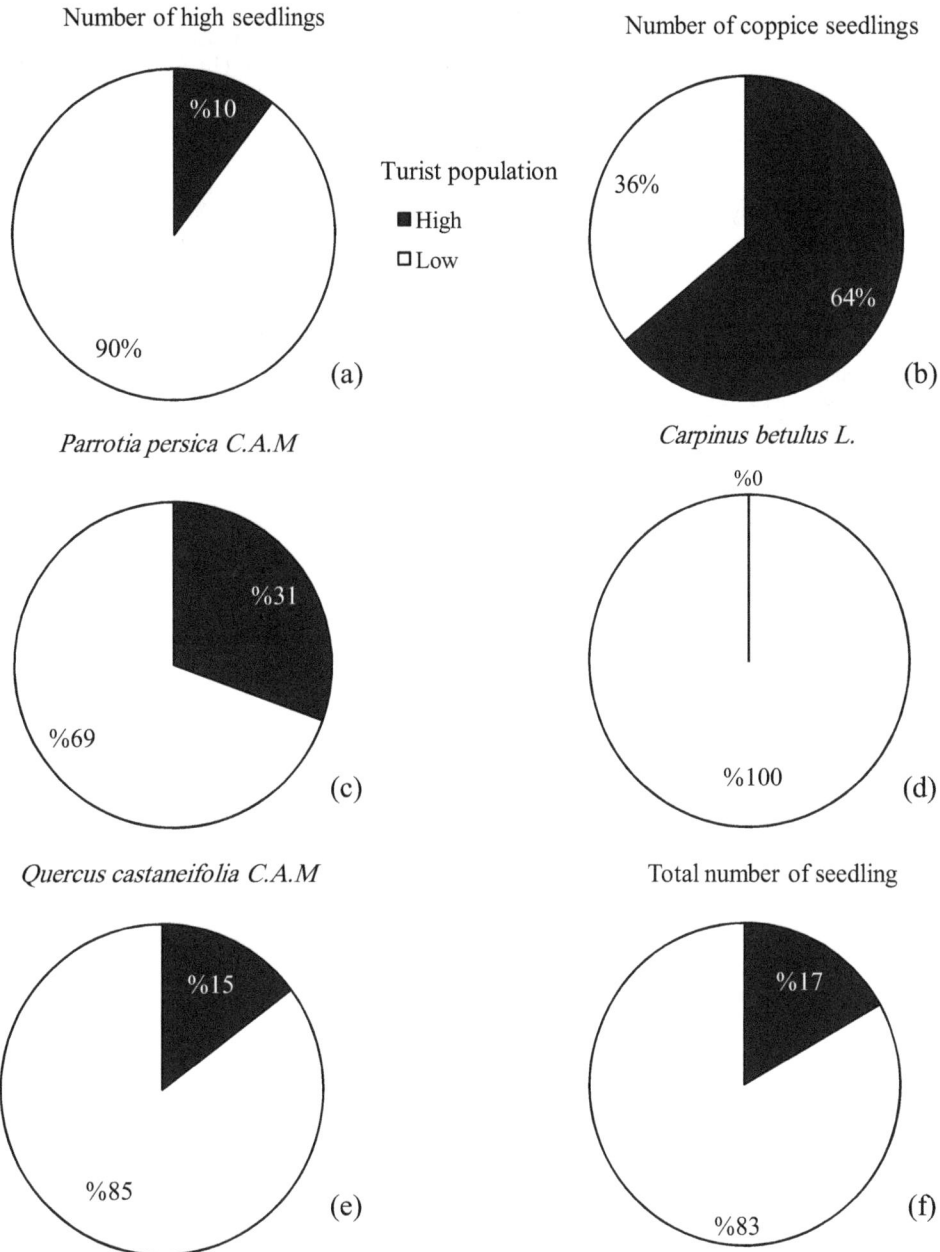

Figure 5. Regeneration in low and high density population area.

and *Q. castaneifolia* (Figure 5e). Total number of seedlings in low density population area was more than that of high density population area (Figure 5f). The loss of regeneration cover and severe compaction and erosion of soil due to human activities resulted in a forest park without restoration (Trn et al., 2006). In forest park, the diversity of ecosystems species and wild gene should be maintained. Native species and populations are highly sensitive to human and domestic animal disturbance (Bhandari, 1999). Increasing the amount of tourism use within an area usually results in increased disturbance to vegetation (Whinam et al., 1994; Cooper et al., 2007).

Conclusions

Study about regeneration, and how it is affected by trampling, etc., considering a conservation area is necessary to develop management recommendations. It was concluded that the number of high seedling in low density population area was more than that of high density population area. Moreover, total number of seedlings in low density population area was more than that of high density population area. Lowest number of seedlings was recorded at distance of 80 m from road edge because of the tourist density and consequently soil compaction. It is necessary to carry out much more research studies in completion of the study.

Conflict of Interests

The author(s) have not declared any conflict of interests.

REFERENCES

Bhandari M (1999). Tourism Raised Problems in Masai Mara National Park Narok, Kenya APEC, Nepal. 14p.

Cooper C, De Lacy T, Jago L (2007). Impacts of recreation and tourism on plants in protected areas in Australia. Catherine Pickering and Wendy Hill Technical Reports, University of Queensland,

Glaeser CW (2006). The Floristic Composition and Community Structure of the Forest Park Woodland, Queens County, New York. Urban Habitats. Urban Habitats 4:102-126.

Good RB (1995). Ecologically sustainable development in the Australian Alps. Mountain Res. Dev. 15:251-258.

Good RB, Grenier P (1994). Some environmental impacts of recreation in the Australian Alps. Australian Parks & Recreation (Summer). pp. 20-26.

Shestak CJ, Busse MD (2005). Compaction Alters Physical but Not Biological Indices of Soil Health. Soil Sci. Soc. Am. J. 69:236-246.

Siikamaki P (2009). Research and monitoring of sustainability of nature-based tourism and recreational use of nature in Oulanka and Paanajarvi National Parks. Oulanka Research Station University of Oulu,

Trn A, Rautio J, Norokorpi Y, Tolvanen A (2006). Revegetation after short-term trampling at subalpine heath vegetation. Ann. Bot. Fenn. 43:129-138.

Whinam J, Cannel E, Kirkpatrick J, Comfort JB (1994). Studies on the potential impact of recreational horseriding on some alpine environments of the central plateau, Tasmania. J. Environ. Manage. 40:103-117.

18

Density and distribution of bongos (*Tragelaphus eurycerus*) in a high forest zone in Ghana

Kwaku Brako Dakwa[1] , Kweku Ansah Monney[1] and Daniel Attuquayefio[2]

[1]Department of Entomology and Wildlife, School of Biological Sciences.University of Cape Coast, Cape Coast, Ghana.
[2]Department of Animal Biology and Conservation Science, University of Ghana, Legon, Accra, Ghana.

This research was undertaken at Kakum Conservation Area (KCA) in the Central Region of Ghana, from October 2011 to September 2012. The aim was to determine the population density and factors affecting distribution of bongos *(Tragelaphus eurycerus)* for management planning and conservation of the bongo as well as tourism promotion. The methodology involved a field study of sampled plots that represented three habitat types, namely closed forest, open forest and thickets and habitat classification based on canopy coverage and locations of these habitats, whether marginal or deep inside the forest within each of the nine ranges. It was observed that encounters with bongos in KCA were more likely to be during early hours of the day, from 05.00 to 07.00 h GMT and later in the day, from 17.30 to 18.00 h GMT. The usual location was in their preferred thickets at four out of the nine ranges of KCA, and their distribution was not affected by seasonality or habitat utilization. About 5.3 bongos/km^2 currently occupy the KCA, which can be said to be currently under severe pressure as evidenced by the presence of hunting tools and human activities all over. The results of Pearson's correlation coefficient regarding bongo densities and water availability suggested that sources of water affected the distribution of the bongos in the KCA since more bongos were encountered closer to water sources. This underscores the importance of sources of water in the KCA for the conservation of the bongos, and the need to ensure adequate protection of the rivers and rivulets in KCA and off-reserve areas. These results have implications for the formulation of adaptive management plans that would protect the secretive, charismatic and largest antelopes in the KCA, thereby promoting tourism.

Key words: Population density, distribution, bongos, secretive, forest margins, Kakum Conservation Area, hunting pressure, water availability, tourism.

INTRODUCTION

The bongo (*Tragelaphus eurycerus* Ogilby, 1837) is the largest social forest-dwelling antelope in Africa, with geographical distribution within three discontinuous parts: East, Central and West (Bosley, 2003) (Figure 1). The species has been classified as Low Risk or Near Threatened with extirpations occurring in some African countries such as Benin, Togo and Uganda (IUCN, 2002). The species inhabits tropical jungles with dense undergrowth up to altitude of 4000 m in Ghana, with exacerbated loss of habitat for mammals due to agriculture and

Figure 1. Distribution of bongos in Africa (Bosley, 2003).

deforestation. As expanding human populations compete with mammals for habitat, few forests including Kakum Conservation Area (KCA) remain for the bongo and; the future of bongos depends entirely on protected areas. Proper management of protected areas is thus very important and requires useful information from research studies as guide to the implementation of management schemes, specifically for the conservation of species, and more so for those endangered or near threatened such as the bongo. The bongo is a spectacular species with a relatively high touristic value. Yet, very few studies have been undertaken on wild bongos (Hillman, 1986; Hillman and Gwynne, 1987; Klaus-Hugi et al., 2000) with most information coming from captive populations in zoos. In the KCA, the bongo is second to the elephant in terms of size of the large mammal species, and its range in West Africa is limited as compared to elephants. Whilst the threatened status of elephants and some primates like the western chimpanzee *(Pan troglodytes)* and Miss Waldron's red colobus monkey *(Procolobus badius waldron)* has been given wide publicity (Oates et al., 1997), little is known about the bongos (East, 1990). Hiking expeditions for bongo sighting at the KCA have not been successful in many cases, even though this charismatic mammal would be interesting to view. In this study, the factors affecting the density and distribution of the bongo in KCA were assessed for management planning and action towards the conservation of the

species, as well as tourism promotion. The study also investigates effects of water availability, habitat utilization and hunting pressure on the distribution of bongos in the study area.

Study area

KCA is located in a fragmented moist evergreen high forest zone in Southern Ghana (Figure 2), and consists of the Kakum National Park and its adjacent Assin Attandanso Resource Reserve, and occupies a 366-km² land area. Both areas were originally Forest Reserves but were legally gazetted in 1992 as wildlife conservation areas under the Wildlife Reserves Regulations (LI 1525). This transferred administrative jurisdiction to the then Wildlife Department, following recommendations based on an initial faunal survey (Hawthorne and Abu-Juam, 1993; Nchanji, 1994). The general climatic conditions of the country characterized by bimodal rainfall and two dry seasons (Durand and Skubich, 1982) prevail in the park. A heavy rainy season from April to July is followed by a light dry season from August to September. A light rainy season from October to early December is then followed by a heavy dry harmattan season from December to March (Kouadio et al., 2008). The fauna may concentrate in and around the few water spots available in the park during the dry harmattan from December to March.

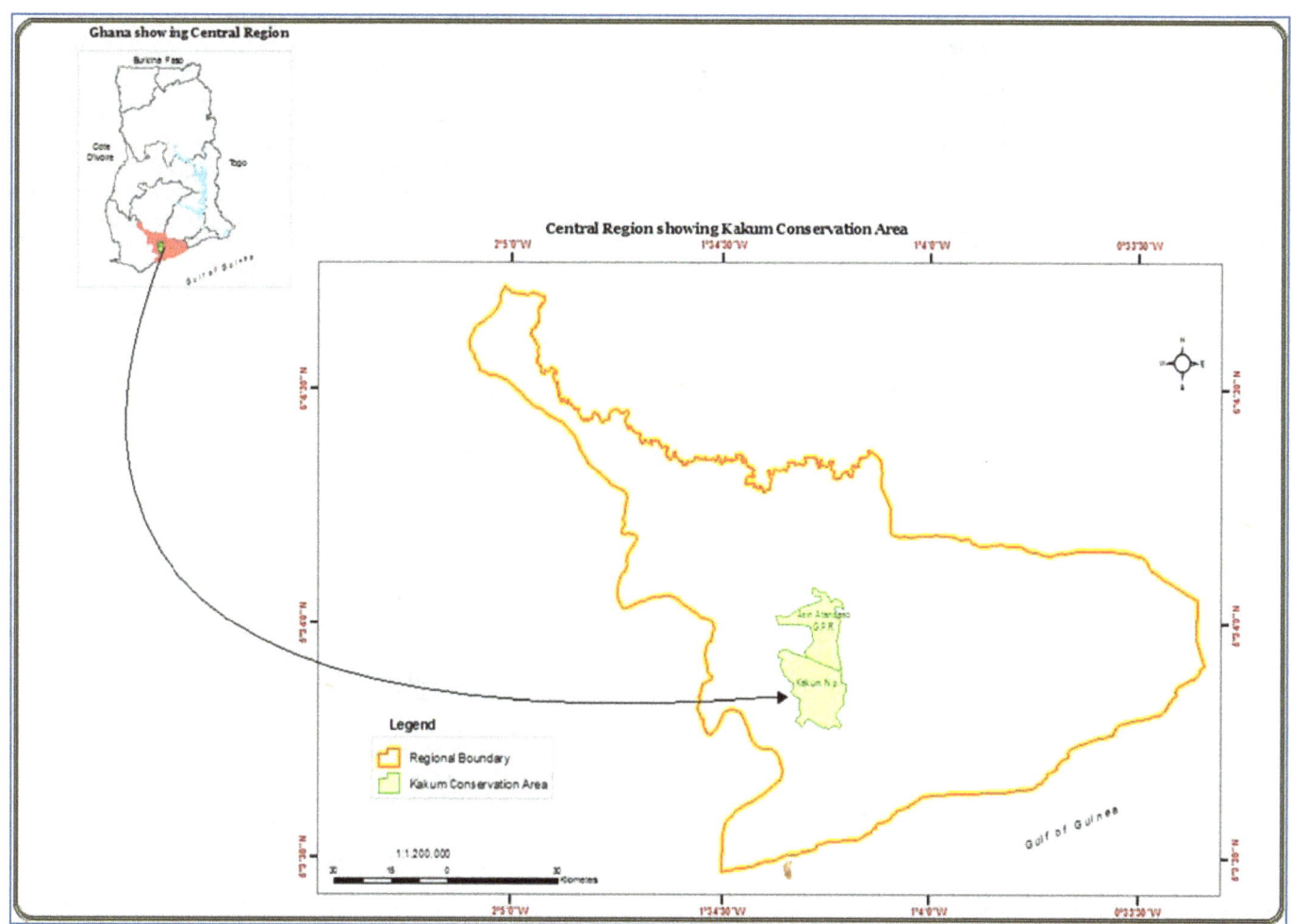

Figure 2. Location map of Kakum Conservation Area in the Central Region, Ghana.

The average annual rainfall is about 1600 mm (Forestry Commission, 2007). The average relative humidity is about 80% throughout the year while temperature ranges from 18.2 to 32.1°C. The terrain is flat to slightly undulating with an elevation of between 15 to 250 m above sea level (asl) (Forestry Commission, 2007). Most of the elevations occur in the south-western portion of the park. Light south westerly winds blow over the area almost throughout the year. The KCA is surrounded by about 52 local communities with a population of about 40,000 people who are mainly peasant farmers cultivating various food and cash crops, often close to the park boundaries (Monney et al., 2010).

About 105 species of vascular plants (Wildlife Department, 1996), 69 species of mammals (Yeboah, 1996) and about 266 species of birds (Dowsett-Lemaire and Dowsett, 2005) have so far been identified in KCA. Mammals include the potto (*Perodicticus potto*), Demidoff'sgalago (*Galagoides demidoff*), bongo, African forest elephant (*Loxodonta cyclotis*), and leopard (*Panthera pardus*). Many herpetofaunal species (Yeboah,

1996; Monney eta l., 2011) and a great number and diversity of butterflies (at least 405 species) (Larsen, 1994, 1995) have been recorded in the KCA, which, for effective patrol and monitoring is divided into nine ranges, namely Abrafo, Kruwa, Briscoe II, Adiembra, Homaho, Aboabo, Afiaso, Antwikwaa and Mfuom. Field staff are deployed from their camps adjacent to their ranges, and tourists led by tour guides use traditional routes in the park for hiking.

MATERIALS AND METHODS

Habitat classification

This study was undertaken from October 2011 to September 2012 using the nine ranges of KCA as study blocks, and camping at some vantage points from 04:00 to 08:00 h GMT and 16:00 to 18:30 h GMT. This became necessary because feasibility studies failed to sight the animal during the day to confirm reports by the staff. The study relied on a field study of sampled plots which were representative of three habitat types (closed forest, open forest and thickets) within each range. Habitat types were classified according to canopy coverage (Wiafe et al., 2010). In the closed forest, light penetration to the forest floor was less than 25%, and tree canopy coverage was more than 75%. In the open forest, light penetration to forest floor was more than 25% with tree canopy coverage less than 75%. In the thickets, light penetration was less than 25% and the canopy consisted of underbrush with coverage of more than 75%.

Sample plots, herd sizes and sighting times

To equalize sampling effort, two 200 m square plots were studied in different locations at each habitat type in each range, one at forest margin and another deep in the forest, and these locations were at least 1 km apart. In all 54 plots were surveyed over the period of study and each one was surveyed by eight people working in pairs and each pair taking charge of a portion of the plot to increase efforts. GPS coordinates at the centre of each plot were recorded. At each range plot surveys were conducted in each of eight months including both rainy and dry months and; from hideouts, including tree tops, hill tops and observation platforms, the number of bongos sighted, herd sizes and sighting times were recorded. Binoculars were used to facilitate viewing where necessary.

Mean bongo densities

Bongo densities were estimated by counting the number of individuals of bongo in each plot as follows: (1). The number of individuals of bongo in any plot divided by the plot area gave the bongos' plot density; (2). The number of individuals of bongo in the same habitat type were summed up and the result divided by the total area of all the plots in that habitat type to give the bongos' density for a specific habitat type; (3). The number of individuals of bongo in each habitat location were summed up and the result divided by the total area of all the plots in the same habitat location to give the bongos' density for a specific habitat location; (4). The number of individuals of bongo in each range were summed up and the result divided by the total area of all the plots in the same range to give the bongos' density for a specific range and; (5). The number of individuals of bongo in all plots in the study area were summed up and the result divided by the total area of all the plots to give a bongo density in the study area. As surveys were replicated eight times all densities were divided by eight to give mean densities.

Population densities and distribution of bongos

Distribution of bongos was measured in terms of the presence and absence of bongos, and their population densities in survey plots, in the different habitat types and their locations, and ranges of the Park during both rainy and dry seasons.

Habitat use

There was also daytime searching for signs of the presence of the bongos in each plot. The presence of bongo spoors (scats, footprints, etc.) was used as evidence of their presence. The degree of habitat use by the bongos was measured by signs of bongos' presence or absence, coded as follows: 0 = no sign of presence; 1 = signs of presence (footprints, dung), but no evidence of browsing; 2 = signs of presence and < 50% browsing; 3 = signs of presence and ≥ 50% browsing of the area.

The codes scored in each plot in the respective habitat types were ranked (1st for habitat that had the highest, 2nd for the next and 3rd for habitat that recorded the least number).

Water availability and hunting pressure

To find out whether water availability affected the distribution of bongos in KCA, the distance of each plot from the nearest source of available water was recorded using the nearest-features extension method in ArcView GIS (v 3.2), based on the GPS coordinates of the plots and geospatial data on the parks water bodies obtained from the Centre for Remote Sensing and Geographic Information System (CERSGIS), Accra. A correlation between distances of plots from water and the bongos' plot densities was then determined. Hunting pressure on bongos was measured by counting any sign of hunting activity in each plot, notably traps, spent cartridges, poachers' camping sites and footprints. Each tool or activity sighted was recorded as 1 and removed from the study area. Correlation between bongos' density and hunting pressure was determined.

Analysis of data

We used IBM's SPSS version 16.0 to calculate descriptive statistics including mean densities and their standard errors to analyze all data. To assess habitat use of bongos at KCA, Levene's test of homogeneity of variance (Zar, 2010) was used to test the null hypothesis that population variances were equal. A two-factor ANOVA was conducted twice to evaluate the: (i) seasonal differences in bongo densities with habitat type (closed forest at Park margins, closed forest deep inside the Park, open forest at margins, open forest deep inside, thickets at margins or thickets deep inside), and (ii) seasonal differences in bongo densities among the nine study ranges (Aboabo, Abrafo, Adiembra, Afiaso, Antwikwaa, Briscoe II, Homaho, Mfuom and Kruwa). The data was transformed using the log (base 10) function in order to convert it into a normally-distributed one. Where differences were statistically significant, a post-hoc analysis of the variances by either non-parametric Games-Howell multiple comparisons or parametric Tukey's HSD multiple pairwise comparisons test (Kleinbaum et al., 1988) was conducted. Descriptive statistics of ANOVA were used to evaluate the population densities as a function of the distribution of the bongos in the various habitat types and locations, and the ranges in the Park in both rainy and dry seasons. Descriptive statistics of ANOVA were used also to assess the differences in habitat use in the three habitat types and in the two different locations of habitat types and; Chi-square was used to test for the significance of the differences. In order to determine the association between bongo densities and water availability or hunting pressure, total bongo densities for habitats in all ranges for both rainy and dry seasons were log-transformed to obtain a linear relationship and also to meet the assumption of normality. A bivariate correlation between the density of bongos and distances from sources of water or hunting pressure was computed and Pearson's correlation coefficient (Cohen and Cohen, 1975) was calculated.

RESULTS

Sighting times and herd sizes

We observed bongos during early hours of the day, from 05:00 to 07:00 h GMT and in the evening between 16:00 and 18:30 h GMT (Figure 3) meaning that this species is likely crepuscular. There was no significant difference between morning and evening periods of encounter with the bongo (t=0.7806, p=0.4575). Of all the bongo herds encountered throughout the study, herd sizes ranged from one to eight individuals, with two as the modal size,

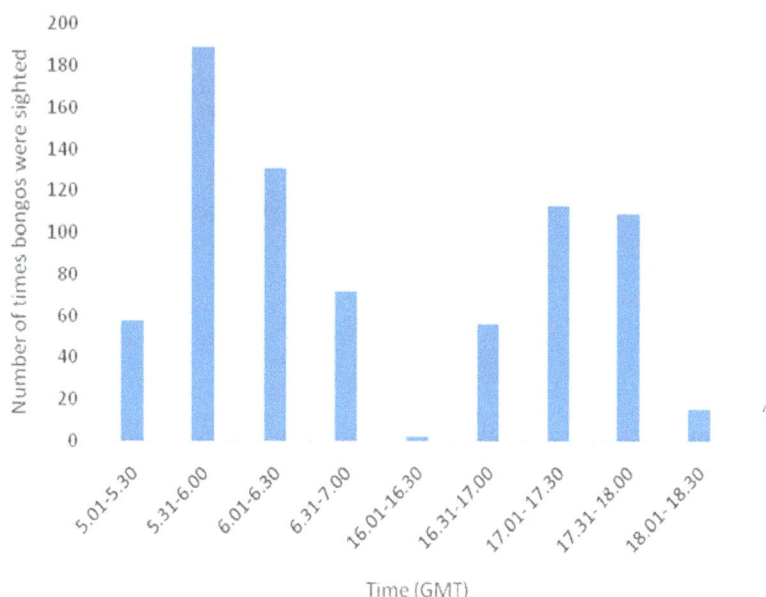

Figure 3. Time of encounter with bongos in the Kakum Conservation Area

though sizes as high as 15 individuals have been reported by field staff.

Mean bongo densities

ANOVA for the various combinations of the factors (season and habitat) and the dependent variable (bongo densities) indicated that in both the rainy and dry seasons, the highest bongo densities were recorded in thickets at the Park margins (Table 1). The mean was 0.9219 (σ=0.59056) per 100,000 m^2 in the rainy season and 1.0005 (σ=0.54398) per 100,000 m^2 in the dry season. The next highest bongo density was also recorded in the thickets deep inside the Park. The mean values in the rainy and dry seasons were 0.6002 (σ = 0.59056) and 0.6213 (σ = 0.55254) respectively per 100,000 m^2. At the margin's closed forests, means of 0.4451 (σ = 0.1387) and 0.4370 (σ = 0.10544) were recorded for the rainy and dry seasons respectively per 100,000 m^2 estimates while deep inside closed forests were 0.3574 (σ = 0.22486) and 0.4863 (σ = 0.7325), respectively per 100,000 m^2. Also, mean bongo density for deep inside open forests was higher (μ/100,000 m^2 = 0.3548; σ = 0.44697) in the dry season than in the rainy season (μ/100,000 m^2 = 0.1995; σ = 0.41140). At margin's open forest, mean bongo density of 0.4478/100,000 m^2 (σ = 0.23199) was recorded during the rainy season, and 0.3028/100,000 m^2 (σ = 0.37707) during the dry season.

Overall, mean bongo population density per 100,000 m^2 was 0.5252 (σ = 0.45819) ranging from μ = 0.5495 (σ = 0.44841) in the dry season to μ = 0.5009 (σ = 0.46807) in the rainy season. The bongo population density per 100,000 m^2 was highest (μ = 0.9612; σ = 0.58670) in the thickets at the margins and lowest (μ = 0.2616; σ =

0.40872) in the open forests deep inside. The test for homogeneity of variance was highly significant (Levene'stest statistic = 3.820; $p < 0.05$). This indicates variances were not equal across groups, and therefore an assumption of ANOVA is violated. Games-Howell's post-hoc analysis which is free of assumptions of normality indicated a significant difference in bongo population densities between the open forests deep inside and the margins thickets in the Park. The estimated marginal mean for margins thickets was 0.961 ± 0.123, while that for open forests deep inside was 0.277 ± 0.138. It could therefore be concluded that margins thickets have the highest bongo density in the park.

Descriptive statistics of ANOVA for the independent variables (season and range) and the dependent variable (bongo densities) revealed interesting results. While the population densities were zero for three ranges, representing 33% of all the ranges or plots surveyed in the park in both rainy and dry seasons, others showed different results for the different seasons. At Abrafo, the population density of the bongo community during the rainy season was very low, with a mean of 0.02258 (σ = 0.15051), but shot up slightly to a mean of 0.0587 (σ = 0.10167) in the dry season. A similar trend was recorded at Kruwa, with mean 0.1910 (σ = 0.35236) in the rainy season and 0.894 (σ = 0.45189) in the dry season. Three of the four remaining sites recorded marginal increases in population density from the rainy to the dry season. At Adiembra, the mean population densities for the rainy and dry seasons were 0.6191 (σ = 0.38158) and 0.7131 (σ = 0.35374) respectively. Values for Afiaso were 0.7765 (σ = 0.36413) and 0.8112 (0.35088), while those for Antwikwaa were 0.6926 (σ = 0.37644) and 0.6919 (σ =0.35788). Mfuom recorded 0.7095 (σ = 0.38268) and

Table 1. Means, standard deviations, and the number of observations (N) of the factors affecting bongo densities (response).

Season	Habitat Type	Mean	Standard Deviation	N
	CF-M	0.4451	0.13872	5
	CF-D	0.3574	0.22486	5
	OF-M	0.4478	0.23199	6
Rainy	OF-D	0.1995	0.41140	6
	TH-M	0.9219	0.67651	6
	TH-D	0.6002	0.59056	6
	Total	0.5009	0.46807	34
	CF-M	0.4370	0.10544	5
	CF-D	0.4863	0.07325	5
	OF-M	0.3028	0.37707	6
Dry	OF-D	0.3548	0.44697	4
	TH-M	1.0005	0.54398	6
	TH-D	0.6213	0.55254	6
	Total	0.5495	0.44841	32
	CF-M	0.4410	0.11624	10
	CF-D	0.4218	0.17166	10
	OF-M	0.3753	0.30793	12
Total	OF-D	0.2616	0.40872	10
	TH-M	0.9612	0.58670	12
	TH-D	0.6108	0.54536	12
	Total	0.5244	0.45577	66

CF-M = closed forest margin; CF-D = closed forest deep; OF-M = open forest margin; OF-D open forest deep; TH-M = thickets margin and; TH-D = thickets deep.

0.7145 (σ = 0.34998). Overall, the highest bongo population density was recorded during the dry season at Afiaso (mean = 0.7938; σ = 0.34141) and the lowest at Aboabo, Briscoe II and Homaho (all recording zero in both seasons) followed by Abrafo during the rainy season (μ = 0.01039; σ = 0.19466). The means recorded at four stations, namely Adiembra, Afiaso, Antwikwaa and Mfuom were always much higher than at Abrafo and Kruwa, and there was not much difference in the mean bongo population densities between Abrafo and Kruwa. Again, it could be said that the population distributions of bongos in the various ranges were not uniform. They were absent from Aboabo, Briscoe II and Homaho, low in Abrafo and Kruwa and relatively high in Adiembra, Afiaso, Antwikwaa and Mfuom.

Population density and distribution of bongos

Bongos were present in six of the nine ranges and 36 of the 54 sample plots (representing 67% of all the ranges surveyed) in both seasons. They were absent from all the 18 plots in three ranges, namely Aboabo, Briscoe II and Homaho in both seasons. Bongos were present in both seasons in the same habitat type or absent in both seasons from the same habitat type, but never present in one habitat type during one season and absent during the other.

The 2*6 factorial ANOVA to determine the possible interaction between the distribution of the bongos in various habitats of the study area and the season of the year on the population densities of the bongos also revealed interesting results for all possible combinations of the analyses. There was no significant interaction between seasons and population distribution of bongos in the various habitats $\{F (5, 54) = 0.181, p \gg 0.05,$ Table 2$\}$. From the partial ETA computed the interaction effect only accounted for 1.7% of the total variance in bongo densities between season and distribution of bongos in the habitats (Table 2). The profile plots of interaction (Figure 4) gives a pictorial representation of the interaction. It is observed that the lines are almost all parallel. The main factor (season) was not significant $\{F (1, 54) = 0.132, p \gg 0.05,$ Table 2$\}$, only accounting for 0.2% of the variations in the population densities of the bongos. However, the habitat main effect was found to be significant $\{F (5, 54) = 0.005, p < 0.05,$ Table 2$\}$. This means that bongo population densities for the various habitats were different and Games-Howell's post-hoc analysis (Table 3) revealed the difference existed between open forests deep inside (with low population distribution) and margins thickets (with high population

Table 2. Test of subject effects for season and habitat type.

Source	df	Mean square	F	Significance	Partial Eta squared
Season	1	0.024	0.132	0.717	0.002
Habitat	5	0.681	3.750	0.005	0.258
Season x Habitat	5	0.033	0.181	0.968	0.017
Error	54				

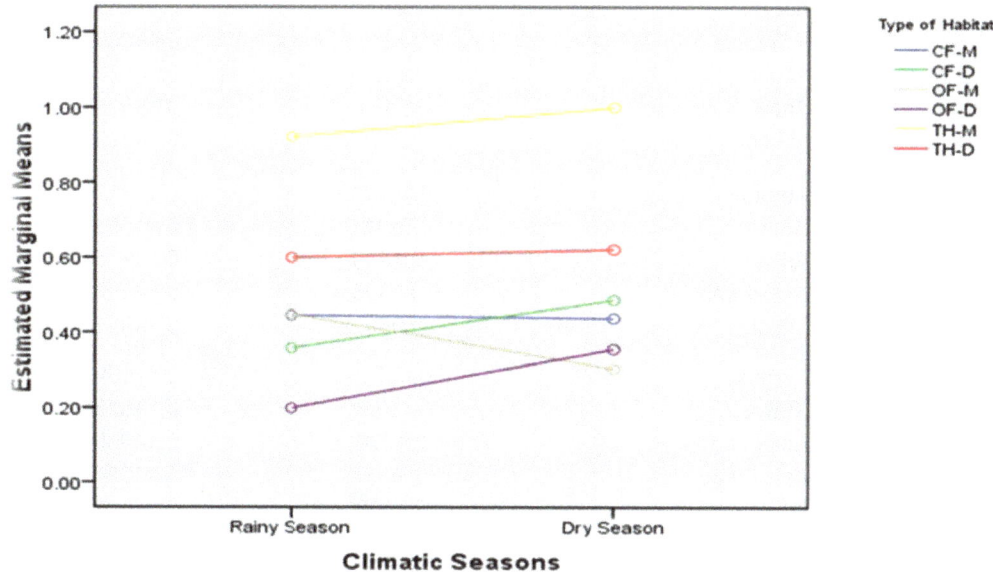

Figure 4. Comparing estimated marginal means of bongo densities in different types of habitats in the rainy and dry seasons during the study.

distribution). Bongos' population densities however did not differ between all other possible pairings of habitats (Table 3).

Also, there was no significant interaction between season and distribution of bongos in the various ranges of the park {F (5, 54) = 0.278, p>>0.05, Table 4}. Only 2.5% of the total variance in the bongos' densities was accounted for by the interaction between season and range of occurrence of the bongos in the park. The main effect by the season on the population distribution of the bongos in the ranges was also not significant {F (1, 54) = 0.562, p >>0.05, Table 4} and the season main effect accounted for less than 1% of the total variance in the bongos' densities. However, there was a highly significant {F (5, 54) = 9.591, p <<0.05, Table 4} main effect by the range factor. In other words, the bongos' population densities were statistically different across the different ranges of the bongos in the park. As the ratio of the highest to least recorded population densities was 7:1, variances across groups were expected to be small, which was confirmed by the test of homogeneity of variance {F (11, 54) = 1.090, p>>0.05}. However, Tukey's post-hoc analysis revealed significant differences (Table

5) in bongos' population densities between Abrafo and each of four ranges namely Adiembra, Afiaso, Antwikwaa, and Mfuom, but not Kruwa and; also between Kruwa and each of the four ranges. Between pairs of Adiembra, Afiaso, Antwikwaa and Mfuom, the differences were not significant (Table 5). This means that population densities varied across the ranges, but with Abrafo and Kruwa having similarly low densities and Adiembra, Afiaso, Antwikwaa and Mfuom similarly high densities. Bongos were absent from Homaho, Briscoe II and Aboabo.

In summary, the population densities of the bongos were not dependent on the time of climatic season (rainy or dry). On the other hand, the population densities of the bongos depended on the habitat in which they lived, particularly in four of the nine ranges of the park.

Habitat use

The results of cross-tabulation among degrees of habitat use are presented in Table 6. The figure in each cell indicates the number of times that degree of use was assigned in that habitat. For example, 24 in cell 1 implies

Table 3. Tukey's HSD post-hoc multiple comparisons of the various levels of the factor range.

Range(I)	Range(J)	Mean Difference (I-J)	Std. Error	Sig.
Abrafo	Kruwa	-0.2441	0.17039	0.707
	Adiembra	-0.7657*	0.17322	0.001
	Afiaso	-0.8977*	0.17039	0.000
	Antwikwaa	-0.7961*	0.17039	0.000
	Mfuom	-0.8158*	0.17039	0.000
Kruwa	Abrafo	0.2441	0.17039	0.707
	Adiembra	-0.5216*	0.14955	0.012
	Afiaso	-0.6536*	0.14626	0.001
	Antwikwaa	-0.5521*	0.14626	0.005
	Mfuom	-0.5718*	0.14626	0.003
Adiembra	Abrafo	0.7657*	0.17322	0.001
	Kruwa	0.5216*	0.14955	0.012
	Afiaso	-0.1320	0.14955	0.949
	Antwikwaa	-0.0305	0.14955	1.000
	Mfuom	-0.0502	0.14955	0.999
Afiaso	Abrafo	0.8977*	0.17039	0.000
	Kruwa	0.6536*	0.14626	0.001
	Adiembra	0.1320	0.14955	0.949
	Antwikwaa	0.1015	0.14626	0.982
	Mfuom	0.0818	0.14626	0.993
Antwikwaa	Abrafo	0.7961*	0.17039	0.000
	Kruwa	0.5521*	0.14626	0.005
	Adiembra	0.0305	0.14955	1.000
	Afiaso	-0.1015	0.14626	0.982
	Mfuom	-0.0197	0.14626	1.000
Mfuom	Abrafo	0.8158*	0.17039	0.000
	Kruwa	0.5718*	0.14626	0.003
	Adiembra	0.0502	0.14955	0.999
	Afiaso	-0.0818	0.14626	0.993
	Antwikwaa	0.0197	0.14626	1.000

*The mean difference is significant at the 0.05 level.

Table 4. Test of subject effects for season and ranges.

Source	df	F	Significance	Partial Eta squared
Season	1	0.340	0.562	0.006
Range	5	9.591	0.000	0.470
Season * Range	5	0.278	0.923	0.025
Error	54			

that in the closed forest, the degree of use of that habitat assigned a "no use" was coded 24 times. It appears that the degree of use coded 'no use', 'low use' or 'moderate use' was always least in the thickets while 'high use' was highest in the thickets (Table 6). In other words, it appears that the bongos used the thickets more than other habitat types. However, the trend was not clear between the closed and open forests and as Chi-square statistic at an alpha level of 0.05 was not significant (χ = 3.2121, df = 3, p = 0.36006), none of them could be said

Table 5. Mean differences of pair-wise comparisons of the types of habitats and their significance.

Habitats	CF-M	CF-D	OF-M	OF-D	TH-M	TH-D
CF-M						
CF-D	-0.01919					
OF-M	-0.06573	-0.04654				
OF-D	-0.17939	-0.16021	-0.11367			
TH-M	0.52016	0.53935	0.58589	0.69956**		
TH-D	0.16971	0.18890	0.23544	0.34911	-0.35045	

CF-M: closed forest margins; CF-D: deep inside closed forest; OF-M: open forest margin; OF-D: deep inside open forest; TH-M: thicket margins; TH-D: deep inside thicket. **Mean difference is significant at 0.05 by Games-Howell multiple comparisons test.

Table 6. Cross-tabulation between degree of habitat use and habitat type.

Degree of habitat use	Habitat type		
	Closed forest	Open forest	Thicket
No use	24	20	20
Low	5	6	3
Moderate	6	5	3
High	1	5	10

Table 7. Cross-tabulation between degree of habitat use and habitat location.

Degree of habitat use	Habitat location	
	Margin	Interior
No use	32	32
Low	4	10
Moderate	10	4
High	8	8

to be used more than the other. Also, in the case of habitat locations, scores for both forest margin and deep forest were the same for 'no use' and 'high use' and it appeared that margins were used more than deep forests judging from the results of low and moderate uses (Table 7). However, the difference in the degree of use between forest margins and deep forests was not significant (χ = 5.143, df = 3, p = 0.162).

Water availability and hunting pressure

Pearson's correlation coefficient for bongo densities and water availability was -0.468 and this was statistically significant (p = 0.005), suggesting a moderate and inverse correlation between bongo density and water availability. Thus bongo densities were lower when water was scarcer or farther away from bongo locations, and this affected bongo distribution in the KCA, with bongo standing

to occur in areas closer to water sources.

A correlation coefficient of -0.267 suggested an inverse relationship between bongo densities and hunting pressure that could also suggest that higher hunting pressure reduced bongo densities and vice-versa; but as the correlation was found to be not significant at an alpha level of 0.05 (p = 0.127) hunting pressure could not therefore be said to have any effect on the distribution of bongos at KCA. Evidence of hunting activities included spent cartridges, traps of different types, poachers' camps, and matchboxes and reports from field staff confirmed that poaching was rampant with the use of dogs, guns and traps at all ranges and habitats.

DISCUSSION

Mammals of the tropical moist forest are not easy to see and count, as in the case of bongos, which are particularly secretive, making it difficult to encounter especially during daytime. Considering the total survey effort in this study, however, the results could be considered reliable. Spinage (1986) and Estes (1991) described bongos as nocturnal and Hillman (1986) observed most activity of the species from dusk to early morning; but Bosley (2003) described bongos as diurnal. This study found virtually no direct activity during the day, but there was no opportunity to obtain evidence of night activity since there were no surveys at night. It appearsthat the bongos in KCA are active in low light during theday.

Apart from direct encounter with the bongos, critical examination of footprints and feeding activities further support the hypothesis that intense activity occurred during the early and late hours of the day, rest by lying under dense cover during high light in the day and sleep at night. The explanation could be that poachers return home in the early and late hours after night and day duty respectively. Also leopards, the historical predators of calves of bongos (An Ultimate Ungulate Fact Sheet, 2004), are exclusively nocturnal and may find it difficult to locate the bongos when they are asleep at their hideouts under dense cover in the night. This may also account for the higher use of thickets by bongos in both forest margins and forests deep inside the park than other habitat types, as observed in this study. The thickets normally comprise slow and low-growing regenerating plants used as hideouts for the bongos as well as food sources, unlike the primeval or less-disturbed areas in the interior parts of the reserve where leaves and twigs of tall trees cannot be reached for consumption.

East (1990) reported crop-raids by bongos and though this study did not investigate field staff's reports on crop raids, it is suggested that location of bongos near forest margins, and therefore crop fields just after the boundary of the conservation area sometimes, is one possible benefit which may account for the use of thickets at forest margins by bongos. Dense thickets might offer good hiding places for bongos to raid nearby crop farms bordering the KCA. As there was a significant difference in densities among the ranges of occurrences of bongos, other factors than chance may account for the distribution of bongos in the ranges. For instance, the chance of encountering bongos at KCA is high at the margins of Afiaso, Adiembra, Antwikwaa and Mfuom ranges perhaps because of the abundance of thickets in these areas, which are re-growths of extensively logged forests.

There were very few bongo encounters at the Kruwa and Abrafo ranges and no encounters at Briscoe II, Homaho and Aboabo. These five ranges had evidence of severe human interference in the form of direct poaching and noise due to increased human populations or visitor influx. For example, Aboabo shares boundary with the Park; Kruwa and Briscoe II harbour the most notorious hunters, according to Park Management and; Abrafo experiences noise due to regular influx of visitors to the canopy walkway. For the purpose of bongo viewing, observation platforms would be more useful if they were erected at Afiaso, Antwikwaa, Mfuom and Adiembra near the forest margins as tourists have failed to view bongos from existing platforms at Briscoe II and Abrafo (Monney and Dakwa, 2014).

The results also indicated that water sources were necessary for the distribution of the bongos, since even thickets were avoided if they were farther away from water sources. This underscores the importance of conserving water bodies (rivers and rivulets) in the KCA and off-reserve.

Large herd sizes of up to 15 have been recorded in field reports and; elsewhere, Klaus-Hugi et al. (2000) encountered 10-20 bongo herd sizes in the Dzanga National Park, Central African Republic. This study did not however record herd sizes larger than eight. Even though large herds may split temporarily (Klaus-Hugi et al., 2000) or permanently, it is possible that threats to bongos in KCA in the form of poaching and predation may have reduced the herd sizes. There was no significant difference between bongo densities and hunting pressure in the various habitat types, habitat locations and ranges, suggesting that the mammals were equally exposed to hunting pressure, which could not therefore account for the distribution of the bongos. It is noteworthy that factors including illegal hunting and predation affect the abundance and distribution of bongos as they do to other mammals, notably elephants. However, Ottow et al. (1996) reported that bongo populations in a predominantly secondary forest in Bangassou in the Central African Republic were stable even though there was hunting pressure.

This study was not extensive enough to find evidence of reducing bongo populations in the KCA, yet patrol staff reported reducing bongo encounter rates. Estes (1991) reported a drastic decline of some isolated bongo populations in Africa and in Kenya, bongo populations are declining throughout their range as a result of over-hunting, habitat loss and rising exploitation through safari hunting and have been nearly extirpated (Kingdon, 1997; East, 1999). The field staff reported active hunting with guns, traps and dogs, inside KCA, as evidenced by the spent cartridges and traps (wire snares and gin traps) found all over the reserve in this study. There was however no evidence of predation in this study, even though field staff reported that pythons (Python sebae) and leopards preyed on young bongos.

The elusive nature of bongos, coupled with difficulties in detecting over-exploitation of bongos, makes more reliable population estimates difficult, leading to 'sudden' drops in bongo numbers. This study estimated bongo populations at approximately 0.53 bongos per 100,000 m^2 (5.3/km^2) within an area of about 360 km^2 at the KCA. Estimated bongo density in the about 150 km^2 Dzanga National Park in the Central African Republic was 0.25/km^2 (Klaus-Hugi et al., 2000). This is an indication that bongo population densities in Africa's protected areas are low, and that there is need for institution of measures to ensure the adequate protection of bongos (East, 1990). The results of this study suggest that the bongo population at KCA is currently under severe pressure.

Protected area management requires information about species distribution, trends in species population densities and knowledge about the impact of potential threats on the population, such as hunting pressure (Carrillo et al., 2000) and logging (Frumhoff, 1995). Also, wildlife monitoring is essential for assessing the success of implemented management actions such as law enforcement strategies and the establishment of research and tourist sites (Hockings et al., 2006). The results of this study have implications for the formulation of adaptive management plans to protect the secretive, charismatic and largest antelope in the Kakum Conservation Area.

Conflict of Interests

The author(s) have not declared any conflict of interests.

ACKNOWLEDGEMENTS

The authors are grateful to the management and staff of Kakum Conservation Area, as well as Ms. Rose Maku Sackey, Mr Samuel Kusi Ampofo and Mr. Kwaku Frimpong for assisting in field data collection.

REFERENCES

An Ultimate Ungulate Fact Sheet (2004). *Tragelaphuseurycerus*, Bongo. Proc. Zool. Soc. Lond.1836:120 [1837]. West Africa.

Bosley LF (2003). International Studbook for Bongo Antelope (*tragelaphus eurycerus isaaci*, Vol. XVIII, Fort Worth Zoo, Fort Worth, TX.

Carrillo E, Wong G, Cuaro´n AD (2000). Monitoring mammal populations in Costa Rican protected areas under different hunting restrictions.Conserv. Biol. 14:1580-1591.

Cohen J, Cohen P (1975).Applied Multiple Regression/Correlation Analysis for the Behavioral Sciences,1stEdition, Lawrence Erlbaum Associates, Mahwah (2ndEdition, 1983; 3rd Edition, with West, S.J. & Aiken, L.S., 2003).

Dowsett-Lemaire F, Dowsett RJ (2005). Ornithological surveys in Kakum National Park. WDSP Report No. 50-s. Unpublished.

Durand JR, Skubich M (1982). Lagoons of Ivory Coast. Aquaculture 27: 211-250

East R (1990). Antelope's global survey and regional action plans. Part 3.West and Central Africa.171 pp.

East R (1999). African Antelope Database 1998.Occasional Paper of the IUCN Species Survival Commission No. 21.IUCN, Gland, Switzerland and Cambridge, U.K.

Estes D (1991). The Behavior Guide to African Mammals. University of California Press,London.

Forestry Commission (2007). Monitoring staff deployment, patrol effort, illegal activity and wildlife trends to facilitate adaptive and performance management in nine protected areas in Ghana. Second Evaluation, Accra, p. 55.

Frumhoff P (1995). Conserving wildlife in tropical forests managed for timber. Bioscience 45:456-464.

Hawthorne W, Abu-Juam M (1993). Forest protection in Ghana.Unpublished Report, ODA and Forest Inventory and Management Project Planning Branch, Forestry Department, Kumasi.

Hillman JC, Gwynne MD (1987). Feeding of the bongo antelope *Tragelaphuseurycerus*(Ogilby, 1837) in southwest Sudan. Mammalia 51:1.

Hillman JC (1986). Aspects of the biology of the bongo antelope *Tragelaphuseurycerus* (Ogilby 1837) in South West Sudan.Biol.Conserv. 38(3):255-272.

Hockings M, Stolton S, Leverington F, Dudley N, Courrau J (2006).Evaluating Effectiveness: A Framework for Assessing Management Effectiveness of Protected Areas,2ndEdn.IUCN, Gland, Switzerland and Cambridge, UK, xiv + 105 pp.

International Union for Conservation of Nature and Natural Resources (IUCN) (2002).IUCN Red List of Threatened Species. Available online at http://www.redlist.org/

Kingdon J (1997). Field Guide to African Mammals. Academic Press, London.

Klaus-Hugi, C, Klaus, G, Schmid, B (2000). Movement patterns and home range of the bongo *(Tragelaphuseurycerus)* in the rain forest of the Dzanga National Park, Central African Republic. Afr. J.Ecol. 38: 53-61.

Kleinbaum DG, Kupper LL, Muller KE (1988). Applied regression analysis and other multivariable methods.2nd Ed. Duxbury Press in Pacific Grove.

Kouadio KN, Diomande D, Ouattara A, Koné YJM, Gourène G (2008). Taxonomic diversity and structure of benthic macroinvertebrates in Aby Lagoon (Ivory Coast, West Aftrica). Pak. J. Biol. Sci. 11:2224-2230.

Larsen TB (1994). *Diopeteskakumi*, a new hairstreak from Kakum National Park, Ghana.Trop.Lepid. 5:83-84. (WA 5).

Larsen TB (1995). Butterflies in Kakum National Park, Ghana. Part 1: Papilionidae, Pieridae&Lycaenidae. Bulletin of the Amateur Entomological Society, 54:3-8, Part 2: Nymphalidae, Hesperiidae, *ibid*. 54:43-46. (WA 11).

Monney KA, Dakwa KB (2014). Prospects and potentials of the Kakum Conservation Area. J. Ecol. Nat. Environ. 6: 140-153.

Monney KA, Dakwa KB, Wiafe ED(2010). Assessment of crop raiding situation by elephants (*Loxodontaafricanacyclotis*) in farms around Kakum Conservation Area, Ghana. Int. J. Biodiver. Conserv. 2(9): 243-249.

Monney KA, Darkey ML, Dakwa KB (2011). Diversity and distribution of amphibians in Kakum National Park and its surroundings.Int.J.Biodivers.Conserv.3(8):358-366.

Nchanji AC (1994). Preliminary survey of the forest elephant, *Loxodontaafricanacyclotis* crop damage situation around the Kakum National Park. Conservation International, Cape Coast, Ghana.Unpublished.

Oates JF, Struhsaker TT, Whitesides GH (1997). Extinction faces Ghana's red colobus monkey and other locally endemic subspecies. Primate Conservation 17:138-144.

Ottow K, Jean-Bosco K, Bloom A (1996). A survey of the BangassouForest.Gnusletter 16:23-25.

Spinage CA (1986). Natural History of Antelopes.London: Croom Helm. 224pp.

Wiafe ED, Dakwa KB, Yeboah S (2010). Assemblages of avian communities in forest elephant *(Loxodontacyclotis)* range in Ghana. Pachyderm 48: 41- 47.

Yeboah S (1996). Small mammals and the herpetofauna of the Kakum National Park.Facing the Storm, Proceedings of the Kakum Conservation Area Research Colloquium)Consvervation International, Washington D.C. pp. 37-45.

Zar JH (2010). BiostatisticalAnalysis.Prentice Hall.944 pp.

19

Regulation of usages and dependency on indigenous fruits (IFs) for livelihoods sustenance of rural households: A case study of the Ivindo National Park (INP), Gabon

Mikolo Yobo Christian[1] and Kasumi I. T. O.[2]

[1]Department of Bioengineering Science, Graduate School of Bioagricultural Sciences, Division of International Cooperation in Agricultural Science Laboratory of Project Development, Nagoya University, Furo-cho, Chikusa-ku, Nagoya, 464-8601, Japan.
[2]International Cooperation Center for Agricultural Education, Nagoya University, Furo-cho, Chikusa-ku, Nagoya, 464-8601, Japan.

The dependency of many rural people on restricted access and use of natural resources of national parks for livelihoods sustenance is poorly acknowledged and detailed surveys clarifying usages and dependency on forest resources by local people are often lacking, especially for regulations and laws improvement purposes. A semi-structured questionnaire was administered to six villages of 252 households (152) in close and (100) far areas following about 80% sampling intensity coupled with focus groups' discussions, to clarify usages and the dependency of rural people on indigenous fruits' species around the Ivindo National Park in Gabon. The results of the study revealed that these forest products collected represent an important component of the household livelihoods as source of food and income generation. Almost all the people, 250 (99.2%) reported harvesting all the six forest products in both locations of the park. Among the harvested products, three out of the six were considered as most popular such as *Coula edulis, Dacriodes buettneri* and *Irvingia gabonensis* while the others three were perceived as less popular ones, for example *Baillonella toxisperma, Gambeya lacourtiana* and *Trichoscypha abut*. In addition, purpose of forest products harvesting were both directed to household consumption, 250 (100.0%) and income generation, 88 (75.2%). Moreover, two out of the three most popular fruit species are sold at higher price per unit including *C. edulis* and *I. gabonensis*. Since the trends on usage were different mainly by ethnic group, distance and residential period, therefore it is necessary to be flexible when designing future rules and regulations on resources utilization of the Ivindo National Park that ensure livelihood of rural people in the meanwhile.

Key words: Usages, dependency, indigenous fruits species, regulation, Ivindo National Park, Gabon.

INTRODUCTION

National parks are one of the typical and worldwide approaches for protection and sustainable management of natural resources. However, difficulties in keeping the balance between protection and resource utilization by rural people for their livelihood have been reported from many countries such as Indonesia, India, Malaysia, Nepal and Zimbabwe (Wells et al., 1999; Agrawal, 2001; Lynam et al., 2007; Spiteri and Nepal, 2008; Frost and Bond, 2006). According to past studies and experiences, natural resources management by preventing rural people from using resources for their livelihood tend to fail (Beltrán, 2000; Bawa et al., 2007; Naughton-Treves et al., 2005). Responding to the lessons learned, natural resources management especially in developing countries have been gradually shifting from protection by prohibiting usages of resources to sustainable utilization by rural people including participatory natural resources management which will provide resources for basic human needs (Beltrán, 2000; Agrawal and Ostrom, 2006; Hayes and Ostrom, 2005; Naughton-Treves et al., 2005). However, there are still many countries falling behind the trend such as the Gabon.

The republic of Gabon is located on the Atlantic coast of Central Africa, and covers a total area of 267,667 km^2 with 1.50 millions of people (UNDP, 2010). The country with an equatorial climate is partly covered by the Congo Basin, the second largest tropical forest after the Amazon Basin. Necessity of its protection and sustainable management of its valuable biodiversity has been gradually recognized after Nobel Prize winner Wangari Maathai became roving ambassador for its protection and sustainable management. More than 80% of the country is recognized as rich diversified forests with more than 6,500 plant species, 3,020 mammal species and 617 bird species (Blaser et al., 2011). Thanks to its abundant natural resources endowment, the economy of the country is largely dependent on natural resources especially exporting oil, timber and manganese. With per capita gross national income (GNI) of USD 7,370, Gabon is classified as middle income country in the world, or one of the highest among African countries (AFDB, 2011). On the other hand, agriculture accounts for only 4% of gross domestic product (GDP) of the country, and nearly 85% of foods are imported due to an undeveloped agriculture and its manufacturing (AFDB, 2011).

The undiversified economy appears to be a cause of unstable economy by fluctuating international price of oil, reduction and loss of forest resources and biodiversity by exporting timbers, and acceleration of rural poverty by restricting access and usage of forest resources such as fruits, nuts, tree leafs for wrapping, medicinal plants, construction material and wild animal for meats (AfDBG, 2011), especially inside national parks. Additionally, since agricultural sector of the country is very weak, people in rural area tend to depend more on collection of natural resources rather than production. It means that forest resources are the important lifeline for livelihood of local

people as well as the economy of the country. The recent national development strategy of the country therefore prioritizes conservation of natural environment while seeking to develop competitive manufacturing industries and services sector, and exporting raw timber was already prohibited by law in 2010 (AfDBG, 2011).

In Gabon, protection of natural resources has started since colonial period with Lopé reserve establishment in 1946 followed by the Ipassa Makokou Biosphere Reserve in 1979 and more recently with a network of 13 national parks established throughout the country covering nearly 2.9 million (11%) ha of total land area (Blaser et al., 2011) with some of them representing extensions of the previous biosphere reserves. The main objectives of establishing such parks were strict biodiversity conservation and ecotourism development for the most part (Gabonese Republic, 2001, 2007). Rules and regulations of the national parks, as a part of natural resources management of the country, are existing under the decree on Customary Rights Law of 2004 (Gabonese Republic, 2004), Forestry Code of 2001 (Gabonese Republic, 2001), and the National Parks Law of 2007 (Gabonese Republic, 2007). Access and use of resources are strictly prohibited in the core area by the National Park Law of 2007, regulated in the buffer zone and let free of use in the transition area.

However, these laws and regulations have not fully considered the livelihoods of rural people even though most park areas used to be utilized by them. As a result, firewood is the only forest product allowed to be collected from the national parks to sustain the livelihoods of rural people, other forest products even non-timber forest products (NTFPs) including nuts and fruits from indigenous trees, known as indigenous fruits (IFs), have been prohibited from use inside the park, regulated in the buffer zone, and let free of use outside by the above mentioned laws and regulations on forest and national parks. The harvesting, utilization and marketing of indigenous fruit and nuts have been central to the livelihoods of majority of rural communities throughout Africa (Akinnifesi et al., 2007; Leakey et al., 2005).

In Gabon, it is also recognized as one of the important traditional resources for rural people and that restrictions on usages may have enormous negative impact on their livelihood. Although buffer zone is available in all national parks of Gabon, rules of resources use by local people for their livelihood are not clearly mentioned by current laws and regulations. Setting up restrictive measures without securing livelihoods of rural people could threaten their lives as well as biodiversity and natural resources as the other countries have experienced. It is therefore urgently necessary to set clear rules and regulations by concerning livelihood of local people as well as sustainable forest resources management.

However, only a limited numbers of quantitative studies based on field survey have been conducted to understand reality of usages and dependency on forest resources by local people, which is urgently necessary for the improvement of regulations and laws on national park management that ensure livelihood of rural people. This study therefore aims to clarify the current situation of natural resources utilization and dependency by rural people near the national park. The Ivindo National Park (INP), one of the oldest protection areas in Gabon, is selected as a case study of typical national park of Gabon.

MATERIALS AND METHODS

Study area

The study was carried out in communities around the Ivindo National Park (INP) in the province of Ogooué-Ivindo, north-eastern Gabon, about 620 km from Libreville, main capital city of Gabon. This area is located in Central African region (0° 23'-0° 33'N, 0° 42'-12° 49E) (Figure 1). The population of the area is about 15,000 people (IRET/CENAREST, 2003). The poor development of roads in this area makes difficult commercial exchanges between towns and other parts of the country coupled with a poor development of agricultural and tourism industries (Lescuyer, 2006). Therefore, people in this region probably need to depend more on natural resources for their livelihood as compare to the urban area. The region of Makokou is characterized by an equatorial climate, marked by a high humidity, middle high rainfall of 1,700 mm, temperature averaging 24°C year round with four distinguished seasons: small dry (from December to February), rainy (from March to May), dry (from June to August) and small rainy (from September to November) (IRET/CENAREST, 2003).

The forest of the area is known as dense evergreen and humid type (Cabalé, 1978), and has characteristics of the Guineo-Congolese forests (White, 1992) of rich fauna and floristic composition. According to the existing report, about 1,200 floral species have been inventoried in this area as total (IRET/CENAREST, 2003). Among valuable timber and non-timber forest products commonly encountered around the study area include *Scorodophleus zenkeri*, *Santiria trimera*, *Coula edulis*, *Anonidium mannii*, *Afrostyrax lepidophyllus*, *Baillonella toxisperma*, *Dacriodes buettneri*, *Irvingia gabonensis* and *Coula edulis*. The park hosts also a large variety of wildlife species including mammals, birds species, etc. (Vander Weghe, 2006).

The current area of the park formerly known as Ipassa Biosphere Reserve of 10,000 ha was established in 1979. The reserve area was then extended to form the actual Ivindo National Park covering area of 300,000 ha today. The park is composed of three main areas including transition area, buffer zone and central or core area. Access and use of resources are strictly prohibited in the core area, regulated in the buffer zone and permitted in the transition area, as well as the other national parks in Gabon as an adopted type of management approach. The population in this area consists of many ethnic groups including the Bantus and the Baka Pygmies who live near the park area. The Bantus break up into small groups including Fang, Kwélé and Kota. The Fang and Kota are the main dominant ethnic groups with a small number of migrants' people in the area (IRET/CENAREST, 2003).

According to the several existing reports, rural people are using the areas of the park for their livelihoods activities such as slash and burn agriculture, hunting, fishing, gathering resources and

unsustainable forest resources utilization through illegal access especially by people who live close to the park has been suspected (Okouyi-Okouyi, 2006; Lescuyer, 2006; Sassen and Wan, 2006). Although logging operations inside of national parks have been banned for conservation purposes, several important species including IFs appear to be threatened or vulnerable. For instance, multiple use plant species such as Moabi (*Baillonella toxisperma*) has been included as Red List of Threatened Species under International Union for Conservation of Nature and Natural Resources (IUCN) due to overexploitation by logging companies and rural people unsustainable harvesting of its fruits or seeds for oil making (Sassen and Wan, 2006; White, 1998). A past study on small number of households carried out mainly in Loaloa, the closest village of this area revealed that forest products including indigenous forest products near the Ivindo National Park are valuable sources of food and revenue for livelihoods of rural people living nearby (Sassen and Wan, 2006). However, usages and dependency of rural people on indigenous fruits based on detailed survey is not identified yet for the improvement of regulations on national park management.

Data collection and analysis

In this study, several preliminary surveys were conducted near the Ivindo National Park to select appropriate study area by identifying migration history, social structures, popular fruit trees and general resource usage through workshops for key informants such as leader or village chiefs. As a result of the preliminary surveys, a total of six villages, three villages each, close (less than 3 km from the park gate) and far (more than 3 km from the park gate) were selected as target area for this study. This study attempted to access 80% of all the existing households in each village for semi-structured interviews based on questionnaire form. A well conducted semi-structured interview contributes to yield of an appropriate relationship between researcher and the respondents (Longhurst, 2003; Whiting, 2008).

In order to clarify resources usages and dependency on indigenous fruits species for livelihoods sustenance of rural households around the park, questions on socio-economic status, resources utilization tendency and awareness and knowledge on the park were prepared and pretested before the survey. The first part of the interview consisted of the questions on socio-economic status such as academic background, employment status and residential period. The second part of was questions on name of harvested fruits species, amount of harvesting, amount of selling and income by selling the fruits were asked to identify tendency of resources utilization in the study area. This study focused on six popular indigenous fruit species, Coula edulis, Inrvingia gabonensis, Dacriodès buettneri, Gambeya lacourtiana, Trichoscypha abut and Baillonella toxisperma, according to the results of preliminary survey of the key informants. Final part of the interview was awareness and knowledge on the national parks such as its boundary, protection status and issues affecting its development, roles of national park's staff, and available laws or regulations on resources utilization. Tendency and characteristics of resource utilization and people's dependency were analyzed by distance, socio-economic status and awareness on the park. SPSS (17.0) was used for all the statistical tests for comparative analysis.

RESULTS AND DISCUSSION

Respondents and their socio-economic status

As a result of field survey, 79.8% (260) of all households in the target villages were visited for interview, and which

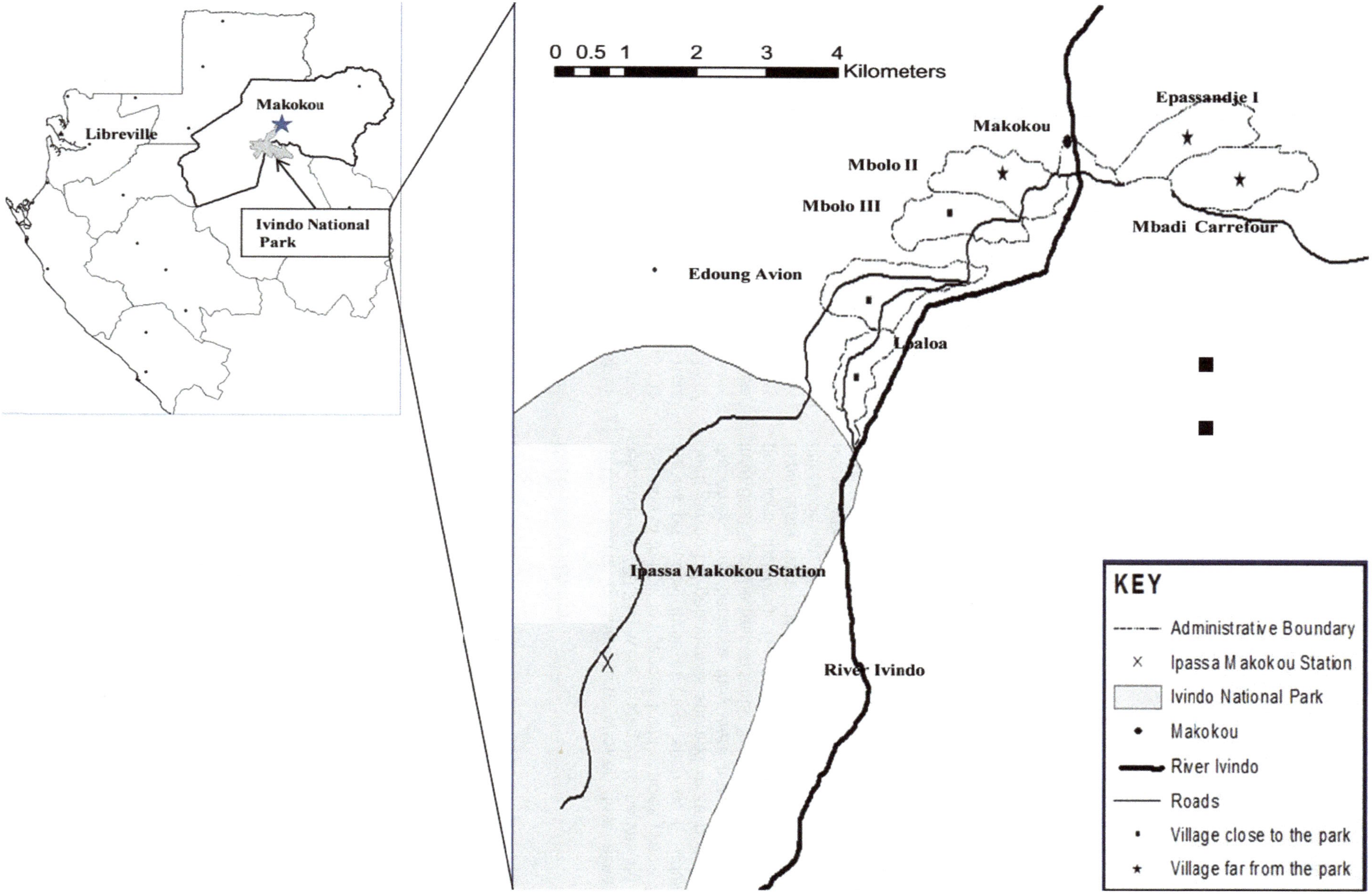

Figure 1. The study area around the Ivindo National Park in Makokou, north east of Gabon.

Table 1. Results of sampling.

-	Target villages	Existing HHs	Sampled HHs (%)		Valid response (%)	
	A	60	50	(83.3)	50	(100)
Close	B	56	51	(91.1)	50	(98.0)
	C	70	58	(82.9)	52	(89.7)
-	Sub total	186	159	(85.5)	152	(95.6)
	D	80	51	(63.8)	51	(100)
Far	E	20	15	(75.0)	15	(100)
	F	40	35	(87.5)	34	(97.1)
	Sub total	140	101	(72.1)	100	(99.0)
Total		326	260	(79.8)	252	(96.9)

consists of 159 households in close area and 101 households in far area, (Table 1). Among all respondents, 95.6% (152) in close area and 99.0% (100) in far area, a total of 252 (96.9%) households, were accepted as valid responses. The number of respondents was considered as sufficient to analyze tendency and characteristics of resources usage and dependency of this area were obtained.

Table 2 shows the socio-economic status of respondents in the study area. As a result of sampling, this study succeeded in collecting nearly half (42.5%) of the female respondents among 252 responses. Since Kota (57.5%) and Fang (27.8%) groups represent majority of the total respondents, the samples of this study were presumed to reflect real social structure of the study area. According to the sampled data, almost all the Fang households were located in close area and there was only one Fang households in far area. Therefore, it is considered that Fang households are concentrated only in close area while Kota households spread into the both close and far villages. Education level of respondents was considered as relatively low since more than half (50.8%) of the respondents have received only primary education. Education level of respondents found in close area was slightly higher than that of far area with regards to the proportion of people who have reached both primary (52.6%) and secondary (42.8%) education than in far area.

Regardless of the slightly higher education status of close respondents, their unemployment rate was lower than that of far area. Generally, education status and employment stats tend to have positive correlation. However, the study area did not show such typical trend. The employment status in this area may be influenced by geographical conditions because areas near the city such as three villages in the far area may have more job opportunities than rural area of close area. The results also indicated that most people in both close and far area have migrated about thirty years ago from the other areas. Average residential period is slightly higher in the

close area of 28.8 years than the far area of 25.6 years therefore, migrants might start to occupy near the current national park at the beginning, and gradually expand their residential area to the farther area.

Resource utilization

The results of the study showed that almost all respondents (99.2%) were engaged in harvesting of at last one of the six indigenous fruits species (Table 3). Of the six produces, three of them such as *C. edulis*, *D. buettneri* and *I. gabonensis* seemed more popular than the other three because they were harvested by more than 80.0% of the respondents. By contrast, other three produces including *G. lacourtiana*, *T. abut* and *B. toxisperma* were identified as less popular ones with utilization of 65.2, 48.4 and 45.2% of respondents, respectively. Seasonality of harvesting the fruits was observed since the more popular species were harvested during the dry season and more often while the less popular ones were harvested in the big dry season and less often. The seasonal nature of such forest based activities refers to the fact that resource users harvest them only at certain periods of the year (Timko et al., 2010). Consequently, out of these given periods, resource users tend to depend on other forest products to meet their households' livelihoods needs.

Additionally, trends for harvesting were different by species because average harvesting amount and frequency of less popular species for both consuming and selling purpose were only half of the popular species. According to these results, availability of more popular species assumed to be higher than the less popular ones, and it may influence the resources utilization trend.

With regards to purpose of fruits' utilization, all the respondents were consuming them while 75.2% of respondents were selling at least one of the six indigenous fruits therefore; more respondents were engaged in consumption than sales of these produces. However,

Table 2. Social structure of sampled area.

-		Close: 152 HHs (%)		Far: 100HHs (%)		Total: 252 HHs (%)	
			Distance				
Gender	Male	87	(57.2)	58	(58.0)	145	(57.5)
	Female	65	(42.8)	42	(42.0)	107	(42.5)
Ethnicity	Kota	62	(40.8)	83	(83.0)	145	(57.5)
	Fang	69	(45.4)	1	(1.0)	70	(27.8)
	Kouele	7	(4.6)	1	(1.0)	8	(3.2)
	Sacke	2	(1.3)	2	(2.0)	4	(1.6)
	Ossamaye	8	(5.3)	7	(7.0)	15	(6.0)
	Massango	4	(2.6)	6	(6.0)	10	(4.0)
Education	None	5	(3.3)	11	(11.0)	16	(6.3)
	Primary	80	(52.6)	48	(48.0)	128	(50.8)
	Secondary	65	(42.8)	41	(41.0)	106	(42.1)
	University	2	(1.3)	0	(0.0)	2	(0.8)
Employment status	Employed	104	(68.4)	80	(80.0)	184	(73.0)
	Unemployed	48	(31.6)	20	(20.0)	68	(27.0)
Residential period (yrs)	Average period	29.2		25.6		27.8	
Size of household	Average number	9.0		8.8		8.9	

average harvesting amount and frequency of each six species for selling were more than that of consumption as mentioned above. It means that indigenous fruits harvesting for selling purpose is considered as major usage and that large amount of resources have been utilized even by fewer respondents. In addition, the purpose of indigenous fruits' use depended on species since the number of respondents was more than 100 in popular species while it was less than 30 in less popular ones. This tendency may be influenced by price as well as resources' availability because both the number of harvesting respondents and selling price per kg of less popular species were smaller than that of more popular species. Thus, both selling and consuming purpose were considered as main purpose of fruits usage in the study area.

Similar results have been stressed by Awe et al (2011) who have revealed that objectives of NTFPs gathering is to meet households sustenance's needs since almost (98%) of rural people collect and use NTFPs as source of food in Kogi State (Nigeria). In the case of Pachmarhi Biosphere Reserve (India), Kala (2011) showed that out of a total of 46 tree species gathered from the wild by local people, 41% of them are used as source of food with trees used for medicine purposes representing fifty percent of response in terms of usage. These results indicate the importance of indigenous fruits in sustaining the livelihoods of people engaged in their harvesting as source of income generation and food (Awe et al., 2011).

Moreover, the popularity of these forest products is also revealed by their mean market price per unit since *I. gabonensis* and *C. edulis* represent the two species

fetching higher market price per FCFA out of the six species in the study area, 500 and 300 FCFA, respectively while *G. lacourtiana, T. abut* and *Baillonella toxisperma* fruits species averaged 200 FCFA each. Thus, mean price of forest products may reflect the importance or direct use values that the respondents have for the resources in terms of consumption and income generation from sale. Given this importance, uncontrolled price (demand) driving resources supply may have serious implications on forest resources management, livelihoods sustainability and conservation goals (Duchelle et al., 2011; Saha and Sundriyal, 2012).

Resource use and socio-economic status of the respondents

Table 4 shows amounts of resources harvested (T), consumed (C) and sold (S) for each of six indigenous fruits species and the total with regards to socio-economic status of the respondents such as ethnic group, family size and residential period. According to the results, all ethnic groups were involved in harvesting some of the six indigenous fruits species for both consumption and selling purposes, however, each ethnic group showed different trends by purpose and species. As total harvesting amount of the six produces, Ossamaye was the largest in harvest in average amount followed by Fang. For consuming purpose, the average harvesting amount was the largest for Massango then followed by Ossamaye while the Kouélé group had the largest harvested amount in average for selling purpose

Table 3. Resource utilisation in the study area.

Species	Collection No. of HHs (%)[1]	Collection Season[2]	Collection Frequency[3]	Purpose (n=250)	No. of HHs (%)	Amount (kg) /season[2] Mean±SD	Income (FCFA[4]) /season[2] Mean±SD	Mean price (FCFA)/ kg
All species	250 (99.2)			Selling	188 (75.2)	21.4 ± 1.2	7,397.3± 397.2	326.5
				Consuming	250 (100)	21.0±0.7		
Coula edulis	230 (92.0)	Small dry	2	Selling	152 (66.1)	9.1±0.6	2,715.8 ±168.6	300.0
				Consuming	230 (100)	6.1±0.3		
Irvingia gabonensis	227 (90.8)	Small dry	2	Selling	145 (63.9)	10.3±0.5	5,162.1 ±270.2	500.0
				Consuming	227 (100)	6.8±0.3		
Dacriodes buettneri	211 (84.4)	Small dry	2	Selling	111 (52.6)	7.9±0.7	1,589.2 ±135.9	200.0
				Consuming	211 (100)	5.2±0.3		
Gambeya lacourtiana	163 (65.2)	Big dry	1	Selling	30 (18.4)	4.7±0.5	946.7 ±105.5	200.0
				Consuming	162 (99.4)	3.0±0.2		
Trichoscypha abut	121 (48.4)	Big dry	1	Selling	10 (8.3)	6.5±2.0	1,310.0 ±403.4	200.0
				Consuming	119 (98.4)	2.8±0.2		
Baillonella toxisperma	113 (45.2)	Big dry	1	Selling	16 (14.2)	3.6±0.6	725.0 ±118.1	200.0
				Consuming	113 (100)	3.2±0.2		

1. N=252; 2. There are following four seasons in the study area, Small dry: December to February and from March to May and Big dry: June to August and from September to November. 3. Harvesting frequency per season; 4. Local currency of Gabon. This study calculated 1 FCFA = 655.957 Euro.

mainly. From the perspective of species, although three popular ones, C. edulis, D. buettneri and I. gabonensis were harvested for both consuming and selling purposes by all the six ethnic groups, however, Kouélé, Sacké and Massango were not involved at all in selling the less popular species of G. lacourtiana, T. abut and B. toxisperma. The less popular species were harvested for consumption and selling purposes by Kota group, the dominant ethnic group of far area.

Although, there is no data to show the availability of each of the less popular fruits species in this study, however, if those species were available in far area, Kouélé Sacké and Massango groups may have different customs from other ethnic groups since they were not involved at all in harvesting of those species. As another possibility, Kota people may be visiting near the national park for harvesting if the species are not available in far area. Highly significant relationships were found among the six ethnic groups for total harvested amount (p=0.00), consumed (p=0.00), sold (p=0.00).

of all the six indigenous fruits species at 5% significant level, as a result of Kruskal-Wallis test. In addition, significant relationships among six ethnic groups were also found with regards to total harvested amount, consumed and sold amount of the three more popular species at 5% significant level, except for sold amount of I. gabonensis and consumed amount of D. buettneri. However, no significant relationships among the six ethnic groups were found for any of total, consumed and sold amounts of the less popular

Table 4. Amounts of resources used, income and socio-economic status of the respondents (N=250).

Variables	Contents	All species				Coula edulis (CE)				Irvingia gabonensis (IG)			
		Consumed	Sold	Total harvested amount	Total income	Consumed	Sold	Total harvested amount	Total income	Consumed	Sold	Total harvested amount	Total income
Ethnicity	Kota	18.3	18.3	32.0	6253.3	5.1	6.9	9.5	2082.4	6.0	9.3	11.3	4653.8
	Fang	23.9	27.3	45.0	9570.4	7.5	12.0	16.1	3600.0	8.1	12.0	17.4	6010.4
	Kouele	18.9	31.5	42.5	11250.0	9.7	23.3	25.2	6975.0	5.6	13.6	15.3	6800.0
	Sacke	22.5	13.3	32.5	4500.0	6.3	5.0	8.8	1500.0	5.0	5.0	8.8	2500.0
	Ossamaye	28.3	23.6	45.7	7990.9	7.9	8.9	12.7	2662.5	7.5	10.7	13.9	5333.3
	Massango	29.6	14.0	40.4	4300.0	6.1	8.6	12.8	2571.4	12.9	7.5	15.0	3750.0
	KW*5	10.4	10.4	10.9	13.6	24.0	24.0	21.4	24.0	9.8	2.7	16.2	2.7
	df*6	2	2	2	2	2	2	2	2	2	2	2	2
	P-value	0.00**	0.00**	0.00**	0**	0.00**	0.00**	0.00**	0**	0.00**	0.26	0.00	0.26
HH size	Rs*7	0.07	0.06	0.13	0.04	0.02	0.10	0.09	0.09	-0.05	-0.06	0.06	-0.06
	P-value	0.24	0.40	0.03 *	0.54	0.76	0.23	0.16	0.23	0.50	0.46	0.40	0.46
Residence duration	Rs	0.16	0.21	0.21	0.23	0.12	0.28	0.22	0.27	0.12	0.27	0.20	0.27
	P-value	0.01**	0.00**	0.00**	0**	0.06	0.00**	0.00**	0**	0.06	0.00	0.00	0

*4 MW U: Mann Whitney U test; *5 KW: Kruskal Wallist test; *6 df: degree of freedom; *7 Rs: Spearmann correlation coefficient.

Table 4. Contd.

Variables	Contents	Gambeya lacourtiana (GL)				Dacriodes buettneri (DB)			
		Consumed	Sold	Total harvested amount	Total income	Consumed	Sold	Total harvested amount	Total income
Ethnicity	Kota	3.0	4.9	4.3	976.0	4.6	6.1	8.0	1214.5
	Fang	3.0	4.0	3.2	800.0	6.2	11.6	11.8	2325.9
	Kouele	1.5	0.0	1.5	0.0	4.6	14.0	10.2	2800.0
	Sacke	3.5	0.0	3.5	0.0	4.0	7.5	7.8	1500.0
	Ossamaye	3.1	4.0	3.8	800.0	6.9	10.4	12.4	2075.0
	Massango	4.2	0.0	4.2	0.0	5.8	7.7	8.3	1533.3
	KW*5	2.7	0.0	4.8	0.0	0.8	12.6	0.1	12.6
	df*6	2	1	2	1	2	2	2	2
	P-value	0.26	0.85	0.09	0.84	0.68	0.00	0.95	0.00 **
HH size	Rs*7	0.03	-0.11	0.10	-0.10	0.06	0.03	0.11	0.03
	P-value	0.68	0.57	0.22	0.57	0.39	0.72	0.11	0.72
Residence duration	Rs	0.00	-0.37	-0.07	-0.37	0.17	0.19	0.16	0.18
	P-value	0.99	0.04 *	0.39	0.04*	0.02	0.05	0.02	0.05*

indigenous fruits species.

The differences found may be due to ethnic groups' customs depending on usages. Therefore, resources usages tend to be influenced by ethnicity of the respondents as an illustration of their alimentary behavior, especially for the most popular ones. This result is in line with the study conducted by Ozanne et al. (2014) in Central Guyana which has revealed that variation in indigenous forest resource's use among communities could be attributed to socio-cultural drivers in terms of consumptive behavior. This means that ethnic groups have acquired complex knowledge on their environment that allow them to face challenges related to food security for example. As a result, ethnic groups' knowledge needs to be taken into account when management of the park resources in the country.

For all of the six indigenous fruits species, a significant correlation (Rs= 0.03) was only found between household size and total harvested amounts of all the six fruits species at 5% significance level through Spearman Correlation Coefficient. No significant correlations were found between household size and total amounts of the six indigenous fruits species consumed, sold on one hand and between household size and total harvested, consumed and sold amounts of the most and less popular species on the other hand. In addition, no significant relationships were also found among household size and income generated from each of the six indigenous fruits species, the most and the less popular species. These results imply that the size of the household tend to have a more direct influence on amounts of resources harvested (accessed) rather than usages (consumption and sale). This has probably to do with respondents' ability to mobilize their household labor to extract forest resources in time of needs. This result is in line with the study carried out by Ding et al. (2012) who showed that households' size represents one of the determinants of energy's consumption in a semi-arid rural area of northwest China. Consequently, family size may be a relevant variable to be taken into account to regulate resources usages in the study.

For all the six indigenous fruits species, significant correlations were also found between residence duration and total amounts harvested (Rs = 0.01), sold (Rs= 0.00), consumed (Rs= 0.01) of all the six fruits species at 1% significance levels, using Spearman Correlation Coefficient. Significant correlations were found only between residence duration and total amounts of D. buettneri harvested; sold and consumed at 1 or 5% significant levels, except for consumed amounts of C. edulis and I. gabonensis. In addition, significant correlations were also found between residence duration and total income of all the six indigenous fruits species (Rs = 0.00), all three of the most popular species C. edulis (Rs = 0.01), I. gabonensis (Rs = 0.00), D. buettneri (Rs = 0.05), and only one of the three less popular species G. lacourtiana (Rs = 0.04). Regarding the less popular species as well,

significant correlations were found only between residence duration and amounts of B. toxisperma harvested (Rs = 0.05), consumed (Rs = 0.01) and amounts of G. lacourtiana sold (Rs = 0.04) at 1 or 5% significant levels. On the contrary, no significant correlations were found at all between residence duration and total amounts of T. abut harvested, sold and consumed. As a result, residence duration in terms of respondents' knowledge (affinity) with natural resources or experience (market involvement) seemed to drive respondents' usages and dependency on the resources. Residence duration may also represent a key variable that needs to be taken into account while designing resource usages.

Resource use and distance

Total amounts of indigenous fruits used (harvested (T), consumed (C), sold (S) and income (I) gained) all tend to vary according to respondents' distance (proximity) to the park (Table 4). Regarding all the six indigenous fruits, respondents closer to the park harvest larger amounts of all of the six produces than farther ones. Resources harvested were more importantly directed towards sales than households' consumption, especially for respondents closer than farther from the park. In addition, respondents closer to the park also harvest larger amounts of each of the three most popular fruits species in comparison with the ones living further away from the park. The three most popular harvested fruits were all directed more importantly for sale (income generation) than households' consumption. Similar trends were also observed for the less popular fruits species, except for harvested amounts of G. lacourtiana and B. toxisperma. These results imply that sale (income generation) represents one of the most important usage of the resources by respondents living close to the park (Table 5). However, respondents living farther away from the park have also managed to enter an activity dominated by respondents living close to the park as a result certainly of the importance of the resources to them.

Since significant differences were found between total amounts of all the six indigenous fruits' species harvested (P= 0.00), consumed (P= 0.00), and sold (P= 0.05) and distance at 1% significance level as a result of Mann Whitney U test. Significant differences were also found between distance and amounts of each of the more popular indigenous fruits' species such as C. edulis consumed and sold, amount of D. buettneri sold at 1% significance level while no significant differences were found between distances and mean amounts of I. gabonensis consumed and sold, and mean amount of D. buettneri consumed. On a contrary, a significant relationship was only found between distance and mean amount of T. abut consumed. Lastly, significant differences were found between distance and mean income all of the six indigenous fruits' species (P = 0.00),

Table 5. Amounts of resources used, income and respondents distance from the park (N=250).

Variables	Contents	All species				Coula edulis (CE)			
		Consumed	Sold	Total harvested amount	Total income	Consumed	Sold	Total harvested amount	Total income
Distance	Close	22.83	24.84	42.07	5369.44	6.76	11.13	14.01	1860.94
	Far	18.23	15.82	29.62	2597.50	5.19	6.20	9.37	1584.50
	MW U*4	5952.00	2767.50	5342.50		######	1584.50	4926.00	
	P-value	0.05	0.00	0.00**	0.00**	0.00**	0.00**	0.00**	0.00**

*4 MW U: Mann Whitney U test; *5 KW: Kruskal Wallist test; *6 df: degree of freedom; *7 Rs: Spearmann correlation coefficient.

Table 5. Contd.

Variables	Contents	Irvingia gabonensis (IG)				Dacriodes buettneri (DB)			
		Consumed	Sold	Total harvested amount	Total income	Consumed	Sold	Total harvested amount	Total income
Distance	Close	7.38	11.00	15.47	4411.11	5.78	9.98	10.91	1034.04
	Far	6.01	8.82	10.21	######	4.45	5.17	7.24	796.00
	MW U*4	5720.50	1976.00	4357.00		######	796.00	4883.00	
	P-value	0.29	0.23	0.00**	0.22	0.09	0.00 **	0.23	0.00

*4 MW U: Mann Whitney U test; *5 KW: Kruskal Wallist test; *6 df: degree of freedom; *7 Rs: Spearmann correlation coefficient.

Table 6. Respondents' awareness on the Ivindo National Park.

Questions on awareness	Aware		Unaware	
	Respondents	(%)	Respondents	(%)
Date of laws & regulations establishment	95	(38.0)	155	(62.0)
Boundaries of the Ivindo National Park	42	(16.8)	208	(83.2)
Any problems about the park	74	(29.6)	176	(70.4)
Villages visited by-National Park' staff	64	(25.6)	186	(74.4)
Protection status of the INP	233	(93.2)	17	(6.8)
Issues of elephants destroying agricultural fields	189	(75.6)	61	(24.4)

N=250.

mean income of two of the three most popular species including C. edulis (P = 0.00), and D. buettneri (P = 0.00) as well as with one of the three less popular species mainly T. abut (P = 0.04) at 1 or 5% significance levels via Mann Whitney U tests. These results may mean that spatial proximity plays a crucial role in driving people's access and use of the resources, especially the most popular ones. Scholars such as Timko et al. (2010) and Yemiru et al. (2010) have all stressed that physical location (distance) has a potential impact on people's ability to access and use forest resources and marketplaces. On the contrary, when resources are so valuable to the people distance does not matter since local people can walk long distance to collect the needed resources. Inappropriate policies and legal and/or tradi-

tional institutions arrangements that restrict or enable people to access forest and marketplaces may also yield illegal access and use of the resources (Laird et al., 2009, Timko et al., 2010). Subsequently, encroachments of forested areas are among the common challenges faced by forest and land managers (Biswas and Choudhury, 2007, Laudati, 2010, Balilla et al., 2012).

Respondents' awareness level on information on the park

Table 6 further stresses respondents' awareness on information about the park. Respondents' awareness on information about the park tends to vary with regards to questions asked as an indication of their levels of expectations or knowledge. Since almost all respondents, 233 (93.2%) were aware of the protection status of the park therefore it can be assumed that awareness campaigns carried out by relevant institutions of the park prior to its establishment have contributed to raise respondents' knowledge on protection status of the park. Given that 189 (75.6%) of them were also aware of issues of elephants destroying agricultural fields therefore mean that wildlife damages caused by *Loxodonta africana* (especially) represents one of the major concerns affecting the livelihoods of people in Gabon at large and the study in particular.Given that most of the respondents were unaware of each of the following questions: (i) the date upon which the laws and regulations of the park have been established (62.0%), (ii) the boundaries of the Ivindo National Park (83.2%), (iii) any problem about the park (70.4%) and (iv) whether villagers have been visited by national park staff (74.4%). These results contradict not only the previous results but also contribute to raise several questions. Awareness campaigns previously carried out by park authorities prior to the establishment of this park appeared not to be effective in raising respondents' awareness on the park and revealing "physical" boundaries of the park. In addition, communicational issues seemed to exist between park authorities and local people since they were talking less about issues affecting their daily livelihoods in general, probably due to retliation from park authorities (Sassen and Wan, 2006). Alleviting the previous issues will more likely contribute to a proper management of the park and increase people's awareness on the park through more targeted awareness campaigns and communicational approaches as suggested by Katel and Schmidt-Vogt, (2011) in the case of Jigme Singye Wangchuck National Park in Bhutan.

Resource use and respondents' awareness level

Table 7 shows relationships between resources use and respondents' awareness level on the previously asked questions (six) about the park. Amounts of resources use (species) and purposes all tend to vary according to respondents' awareness. For all of the six indigenous fruits' species, aware respondents tend to harvest (sale and consume) larger amounts of resources than unaware ones, especially with regards to all the six questions asked. Thus, resources accessed are more importantly directed for sale and household consumption for the most part. Significant relationships were found between all the six harvested indigenous fruits species (sold, consumed) and some of the questions asked including boundary of the park, its protection status, issues affecting its development, roles of national park's staff, and respondents awareness on available laws or regulations on resources utilization. These results imply that awareness on information on the park including restriction of access does not prevent people from accessing and using these forest products of the park for meeting their

households' needs in terms of sale (income generation) and consumption. Thus, the current encroachments (entering the park) observed by surrounding communities may raise some concerns about the effectiveness of the management (land tenure) of the park resources by national park authorities (Sassen and Wan, 2006). Proper management or interventions will more likely contribute to enhance respondents' awareness on the park while reducing their dependency in terms of access and usage (sale and consumption) of forest resources of the park as suggested by various scholars (Blouch, 2010; Khan and Bhagwat, 2010, Van der Ploeg et al., 2011, Vedeld et al., 2012; Gandiwa et al., 2013; Gandiwa et al., 2014).

For the most popular indigenous fruits species, more knowledgeable respondents on questions asked were also harvesting larger amounts of C. edulis and D. buettneri, except for I. gabonensis. Two of the most popular harvested species (C. edulis and D. buettneri) were more importantly directed towards selling (income generation) and consuming by more knowledgeable respondents, except for I. gabonensis sold and consumed for the most part. These results mean that purposes of most popular species used are for meeting household needs in terms of income and consumption as already mentioned in the previous sections.

On the contrary, respondents well-informed about the six questions asked were harvesting lesser amounts of G. lacourtiana and T. abut, except for B. toxisperma. Two of those lesser popular harvested species (G. lacourtiana and T. abut) were more importantly directed towards both income generation and households consumption, especially for respondents well-informed about the following questions: i) the date upon which laws and regulations of the park have been established, ii) issues affecting the park, iii) boundaries of the park, and iv) visit of villages by park's staff and its protection status. Consequently, respondents' awareness on national parks status does not prevent them from entering and making use of forest resources of the park, even the less popular ones. Raising people awareness may therefore drive well informed people to obey restriction of access and use of resources, however, proper alternatives have also to be provided to the affected people including a greater management and improved governance over natural resources access and use (Campbell et al., 2013).

Since significant relationships were found between total amount harvested, consumed, sold (income generation) for all the six indigenous fruits' species and respondents awareness on each of the following question: i) date of laws and regulations establishment, ii) issues of elephants, iii) boundaries of the park, and iv) visit by the park staff to villages therefore management of forest resources based on restriction of access and use of resources may not stop people from accessing and using resources of the park. Thus, reducing well informed people's dependence on resources use may call for providing alternative livelihoods opportunities as stressed

Table 7. Relationship between amounts and income gained from IFs and awareness on INP (N=250).

Variable		All species				Coula edulis (CE)				Irvingia gabonensis (IG)			
		C¹	S²	T³	TI⁴	C¹	S²	T³	TI⁴	C¹	S²	T³	TI⁴
Date of Laws and regulations establishment	Aware	25.4	23.1	41.7	8217.9	7.1	10.4	13.5	3105.6	8.0	12.6	14.8	6295.9
	Unaware	18.3	20.4	34.3	6943.0	5.5	8.3	11.2	2501.0	6.0	9.2	12.4	4583.3
	MW U*⁵	5262.0	3948.0	6208.5	3885.5	4693.5	2202.0	5382.0	2202.0	5156.5	1928.0	5987.5	1928.0
	P-value	0.00**	0.76	0.04**	0.64	0.00**	0.08	0.70	0.08	0.02**	0.07	0.67	0.07
Issues of the park	Aware	23.0	25.6	41.4	8520.8	7.1	11.2	14.0	3358.5	6.4	10.2	13.3	5102.3
	Unaware	20.1	19.7	35.3	6956.3	5.7	8.3	11.3	2478.4	7.0	10.4	13.4	5188.1
	MW U*⁵	5726.5	2961.0	5885.0	2974.5	4552.5	1692.5	4765.5	1692.5	4852.0	2069.5	5167.5	2069.5
	P-value	0.13	0.06	0.23	0.72	0.04*	0.01**	0.15*	0.12*	0.31	0.45	0.82	0.50
Issues of elephants	Aware	22.0	23.4	40.1	8196.6	7.1	10.0	13.1	2992.2	7.1	10.8	14.6	5418.0
	Unaware	17.8	14.2	27.7	4619.0	5.8	6.1	9.0	1825.0	5.8	7.6	9.2	3804.3
	MW U*⁵	4696.0	1763.0	3728.5	1697.0	3814.0	1233.5	3558.0	1233.5	4049.5	1055.0	3114.5	1055.0
	P-value	0.03*	0.00**	0.00**	0.00**	0.01*	0.00**	0.02*	0.00**	0.25	0.05	0.00**	0.05
Boundaries of the INP	Aware	23.6	28.5	44.7	10109.7	6.6	12.4	15.5	3717.9	7.2	11.3	15.2	5650.0
	Unaware	20.5	20.0	35.6	6861.8	6.0	8.3	11.4	2489.5	6.8	10.1	12.9	5034.8
	MW U*⁵	3708.0	1490.0	3336.0	1515.5	3274.0	1142.0	2968.0	1142.0	3859.5	1614.0	3532.0	1614.0
	P-value	0.12	0.00**	0.00**	0.00**	0.19	0.00**	0.04	0.00**	0.94	0.58	0.35	0.58
Visit of INP park staffs to villages	Aware	25.6	25.8	46.2	9086.3	7.6	12.0	16.0	3600.0	8.9	12.2	17.5	6122.0
	Unaware	19.4	19.7	33.9	6768.6	5.6	8.0	10.7	2389.2	6.1	9.6	11.9	4783.7
	MW U*⁵	4498.5	2354.0	3997.0	2423.5	3528.0	1360.0	3368.5	1360.0	1518.0	3366.5	3532.0	1518.0
	P-value	0.00**	0.00**	0.00**	0.00**	0.00**	0.00**	0.00**	0.00**	0.00**	0.00**	0.00**	0.00**
Protection status of the INP	Aware	21.1	21.6	37.5	7468.9	6.1	9.0	12.1	2691.6	6.9	10.4	13.5	5215.3
	Unaware	19.7	17.7	31.2	6245.5	6.1	10.3	11.9	3100.0	6.5	8.5	11.8	4250.0
	MW U*⁵	1845.0	821.0	1714.0	842.5	1586.5	488.0	1655.0	488.0	1309.0	444.0	1322.0	444.0
	P-value	0.63	0.38	0.35	0.45	0.59	0.21	0.82	0.21	0.69	0.35	0.75	0.35

1:Amount for consumed, 2:Amount for sold, 3:Total harvested amount, 4:Total income, 5:Statistica value of U for Mann Whitney U test.

by Campbell et al. (2013) in the case of the Karimunjawa National Park in Indonesia.

CONCLUSION AND RECOMMENDATIONS

Indigenous fruits species harvested are important sources of food and income generation which are contributing to sustain the livelihoods of many rural people. In the study, almost all respondents (99.2%) were engaged in harvesting at least one of the six forest products. Out of the six indigenous fruits species harvested, *C. edulis*, *D. buettneri* and *I. gabonensis* seemed to be the more popular species as compared to *G. lacourtiana*, *T. abut* and *B. toxisperma* considered as less popular as a result of the proportion of people engaged in their harvesting, sale and consumption amounts and because of their

Table 7. Contd.

Variable		Dacriodes buettneri (DB)				Gambeya lacourtiana (GL)				Trichoscypha abut (TA)			
		C[1]	S[2]	T[3]	TI[4]	C[1]	S[2]	T[3]	TI[4]	C[1]	S[2]	T[3]	TI[4]
Date of Laws and regulations establishment	Aware	5.7	9.3	9.7	1852.9	3.4	3.9	3.7	771.4	3.0	1.8	3.1	350.0
	Unaware	5.0	7.4	9.2	1472.7	2.8	5.0	4.0	1000.0	2.7	9.7	3.6	1950.0
	MW U[5]	4866.0	1018.0	5058.0	1018.0	2887.0	56.0	3169.5	56.0	1757.5	5.0	1782.5	3.5
	P-value	0.43	0.05	0.08	0.05	0.23	0.21	0.74	0.21	0.97	0.12	0.82	0.06
Issues of the park	Aware	6.5	10.3	11.7	2066.7	3.4	4.3	4.2	866.7	3.4	8.8	4.6	1760.0
	Unaware	4.7	6.9	8.4	1387.2	2.9	4.9	3.7	981.0	2.6	4.2	2.9	860.0
	MW U[5]	4291.0	1062.0	4446.5	1062.0	2476.5	81.5	2456.0	81.5	1356.0	6.0	1341.0	7.0
	P-value	0.24	0.13	0.46	0.13	0.32	0.53	0.23	0.53	0.70	0.15	0.40	0.23
Issues of elephants	Aware	5.4	8.2	9.9	1646.5	3.1	5.0	4.1	1000.0	2.9	7.8	3.5	1550.0
	Unaware	4.9	7.0	7.9	1392.0	2.7	3.7	3.2	733.3	2.8	1.5	2.9	350.0
	MW U[5]	3985.0	644.0	3652.5	644.0	2189.5	52.5	2213.0	52.5	1123.0	3.0	1163.5	4.5
	P-value	0.28	0.00**	0.06	0.00**	0.31	0.29	0.33	0.29	0.32	0.17	0.37	0.35
Boundaries of the INP	Aware	5.6	8.4	10.9	1680.0	3.2	5.3	3.8	1050.0	3.2	2.0	3.3	400.0
	Unaware	5.2	7.8	9.1	1569.2	3.0	4.7	3.9	930.8	2.8	7.0	3.4	1411.1
	MW U[5]	2520.5	738.5	2437.0	738.5	1878.5	44.0	1824.0	44.0	1147.0	3.5	1161.5	3.0
	P-value	0.37	0.17	0.25	0.17	0.50	0.61	0.24	0.60	0.67	0.71	0.63	0.60
Visit of INP park staffs to villages	Aware	6.0	9.2	12.1	1843.8	3.2	6.3	3.6	1266.7	3.4	0.0	3.3	0.0
	Unaware	5.0	7.4	8.6	1486.1	3.0	4.6	4.0	911.1	2.6	6.5	3.4	1310.0
	MW U[5]	3502.5	905.5	3090.0	905.5	2316.5	26.0	2234.5	26.0	1518.0	1553.5	3744.0	0.0
	P-value	0.00**	0.00**	0.00**	0.00**	0.00**	0.00**	0.00**	0.00**	0.00**	0.00**	0.00**	0.0
Protection status of the INP	Aware	5.2	8.0	9.6	1605.6	3.1	4.9	4.0	971.4	2.9	6.5	3.5	1310.0
	Unaware	5.4	5.8	6.9	1150.0	2.2	3.0	2.6	600.0	2.1	0.0	2.0	0.0
	MW U[5]	1405.5	177.5	1325.5	177.5	886.0	14.0	916.5	14.0	337.5	0.0	328.5	0.0
	P-value	0.76	0.54	0.52	0.54	0.35	0.22	0.44	0.22	0.23	0.0	0.06	0.0

1:Amount for consumed, 2:Amount for sold, 3:Total harvested amount, 4:Total income, 5:Statistica value of U for Mann Whitney U test.

marketability. Although, harvested indigenous fruits species were directed to households' consumption and income generation, however, selling purpose is considered as major usage because it may influence resources management of the park.

Resources are accessed in a seasonal basis with the more popular species being harvested during the dry season and more often while the less popular ones were harvested in the big dry season and less often. Regarding species usage, harvested amounts and frequency of more popular

Table 7. Contd.

Variable		Baillonella toxisperma (BT)			
		C[1]	S[2]	T[3]	TI[4]
Date of Laws and regulations establishment	Aware	3.7	3.8	4.1	766.7
	Unaware	2.7	3.5	3.3	700.0
	MW U[*5]	1342.5	13.5	1399.0	13.5
	P-value	0.14	0.06	0.25	0.06
Issues of the park	Aware	3.0	3.3	3.6	657.1
	Unaware	3.3	3.9	3.8	777.8
	MW U[*5]	1431.5	27.5	1393.5	27.5
	P-value	0.65	0.65	0.50	0.66
Issues of elephants	Aware	3.1	3.8	3.7	757.1
	Unaware	3.4	2.5	3.6	500.0
	MW U[*5]	646.0	10.5	655.0	10.5
	P-value	0.16	0.56	0.19	0.56**
Boundaries of the INP	Aware	2.9	2.7	3.1	533.3
	Unaware	3.3	3.8	3.8	769.2
	MW U[*5]	986.0	17.0	955.0	17.0
	P-value	0.23	0.72	0.15	0.72**
Visit of INP park staffs to villages	Aware	3.3	3.3	3.6	666.7
	Unaware	3.1	3.7	3.7	738.5
	MW U[*5]	1365.5	18.0	1356.5	18.0
	P-value	0.00**	0.00**	0.00**	0.00**
Protection status of the INP	Aware	3.3	3.8	3.8	757.1
	Unaware	2.2	2.5	2.7	500.0
	MW U[*5]	420.0	10.5	464.5	10.5
	P-value	0.32	0.55	0.61	0.55

1:Amount for consumed, 2:Amount for sold, 3:Total harvested amount, 4:Total income, 5:Statistica value of U for Mann Whitney U test.

species for both consuming and selling purpose were twice of the less species as a result of certainly of resources availability. The latter result may influence the future management of the resources based on utilization by rural people if not properly taken into account. Although, total amounts of all the six indigenous fruits species, each of the more and less popular species harvested (T), consumed (C) and sold (S) all tended to vary with regards to socio-economic status of the respondents including ethnic group, family size, residential period, distance and awareness on information on the park to some extent, therefore it is necessary to be flexible when designing future rules and regulations on resources utilization of the park. Recommendations on potential regulation of resources utilization of the park are drawn in Table 8. Further studies need to focus on designing future rules and regulations on resources utilization by amounts, distance and seasons for both rural livelihoods and management of the Ivindo National Park.

Conflict of Interests

The author(s) have not declared any conflict of interests.

ACKNOWLEDGEMENTS

This study was carried out on the basis of the financial assistance from Japanese International Cooperation Association (JICA) for which the authors are highly thankfully. They are also grateful to the Directors of the Tropical Ecology Research Institute (IRET) of the National Centre of Research in Science and Technology (CENAREST), for providing necessary facilities and guidance. They also thank the staff at the Ipassa Makokou's station for their friendly assistance, tolerance and helpfulness throughout the study period and intro-ducing the authors to the local administration, village chiefs and local people. They would like to express their

Table 8. Recommendations on potential regulation of resources utilization of the park.

Regulation by			Inside	Buffer zone	Outside
Species	Most used	For selling	P to P	R to R	A to A
		For livelihood	P to R	R to A	A to A
	Less used	For selling	P to P	R to R	A to A
		For livelihood	P to R	R to A	A to A
By distance	More than 3 km		P to P	R to A	A to A
	Less than 3 km		P to R	R to A	A to A
Season			P to P	R to A	A to A

thanks to researchers and friends for their continuous help. They are particularly grateful to Dr. Donald Midoko Iponga and numerous anonymous reviewers for their advice and valuable comments to improve the quality of this manuscript. They are also particularly thankful to their parents and family for their unstoppable support throughout the course of this study.

REFERENCES

African Development Bank Group (AFDBG) (2011). Republic of Gabon. Country Strategy Paper 2011-2015. Operations Department Centre Region – ORCE. August 2011. Online accessed 14th July 2014. Web site. www.afdb.org/Documents/Operations/Gabon.pdf.

Akinnifesi FK, Ajayi OC, Sileshi G, Kadzere I, Akinnifesi AI (2007). Domesticating and commercializing indigenous fruit and nut tree crops for food security and income generation in Sub-Saharan Africa. In Paper presented at the New Crops International Symposium (Vol. 3, p. 4).

Agrawal A (2001). State formation in community spaces?: Decentralization of control over forests in the Kumaon Himalaya, India. J. Asian Stud. 60:9-41.

Agrawal A, Ostrom E (2006). Political Science and Conservation Biology: a Dialog of the Deaf. Conserv. Biol. 20:681-682.

Awe F, Osadebe CO, Imoagene E, Fashina AY, Eniola TS, Adeleke EO (2011). Assessment of rural households' objectives for gathering non-timber forest products (NTFPs) in Kogi State, Nigeria. Afr. J. Environ. Sci. Technol. 5(2):143-148.

Balilla VS, Anwar-McHenry J, McHenry MP, Parkinson RM, Banal DT (2012). Aeta Magbukún of Mariveles: traditional Indigenous forest resource use practices and the sustainable economic development challenge in remote Philippine regions. J. Sust. For. 31(7):687-709.

BAWA KS, JOSEPH G, SETTY S (2007). Poverty, biodiversity and institutions in forestagriculture ecotones in the Western Ghats and Eastern Himalaya ranges of India. Agric. Ecosys. Environ.121:287-295.

Beltrán JE (2000). Indigenous and Traditional Peoples and Protected Areas: Principles, Guidelines and Case Studies. IUCN, Gland, Switzerland and Cambridge, UK and WWF International, Gland, Switzerland. xi + 133pp.

Biswas SR, Choudhury JK (2007). Forests and Forest Management Practices in Bangladesh: The Question of Sustainability 1. Int. For. Review. 9(2):627-640.

Blaser J, Sarre A, Poore D, Johnson S (2011). Status of Tropical Forest Management 2011. ITTO Technical Series No 38. International Tropical Timber Organization, Yokohama, Japan.

Blouch RA (2010). Zoning for people within Indonesia's Kerinci Seblat National Park. J. Sust. For. 29(2-4):432-450.

Caballé G. (1978) Essai sur la Géographie forestière du Gabon. Adansonia 17(4):425-440.

Campbell SJ, Kartawijaya T, Yulianto I, Prasetia R, Clifton J (2013). Co-management approaches and incentives improve management effectiveness in the Karimunjawa National Park, Indonesia. Marine Policy. 41:72-79.

Corblin A (2006). Economie et perceptions des pratiques villageoises dans le Parc National de l'Ivindo (Gabon). Mémoire de Master 1, Ingénierie en écologie et gestion de la biodiversité, Université de Montpellier II.

Ding W, Niu H, Chen J, Du J, Wu Y (2012). Influence of household biogas digester use on household energy consumption in a semi-arid rural region of northwest China. Appl. Energy. 97:16-23.

Duchelle AE, Cronkleton P, Kainer KA, Guanacoma G, Gezan S (2011). Resource theft in tropical forest communities: implications for non-timber management, livelihoods, and conservation. Ecology and Society, 16(4).

Frost GH, Bond I (2006) CAMPFIRE and the payment for environmental services. International Institute for Environment and Development, London.

Gabonese Republic (2007). Law No003/2007 related to National Parks based on the Forest Code in the Gabonese Republic. Directorate of official Publications. Libreville.

Gabonese Republic (2001). Law No. Law No16/01 of 31st December 2001 on the Forestry Code in the Gabonese Republic.

Gabonese Republic (2004). Decree No000692 /PR/MEFEPEPN setting up the conditions for exercising customary use rights on forest, wildlife, hunting and fishing livelihoods activities.

Gandiwa E, Heitkönig IM, Lokhorst AM, Prins HH, Leeuwis C (2013). CAMPFIRE and human-wildlife conflicts in local communities bordering northern Gonarezhou National Park, Zimbabwe. Ecol. Soc. 18(4):7.

Gandiwa E, Zisadza-Gandiwa P, Muboko N, Libombo E, Mashapa C, Gwazani R (2014). Local People's Knowledge and Perceptions of Wildlife Conservation in Southeastern Zimbabwe. J. Environ. Prot. 5(06):475.

Hayes TM, Ostrom E (2005) Conserving the world's forests: are protected areas the only way?. Indiana Law Review, 38: 595-617.

IRET/CENAREST (2003) (draft). Station de recherche d'Ipassa, Makokou, Gabon. Presentation et publications (1962-2003). IRET/CENAREST, Libreville, Gabon.

Kala CP (2011). Reserve of India Indigenous uses and sustainable harvesting of trees by local people in the Pachmarhi Biosphere. Int. J. Med. Arom. Plants. 1(2):153-161.

Katel ON, Schmidt-Vogt D (2011). Use of Forest Resources by Residents of Jigme Singye Wangchuck National Park, Bhutan: Practices and Perceptions in a Context of Constraints. Mount. Res. Dev. 31(4):325-333.

Khan MS, Bhagwat SA (2010). Protected areas: a resource or constraint for local people? A study at Chitral Gol National Park, North-West Frontier Province, Pakistan. Mount. Res. Dev. 30(1):14-24.

Laird SA, Wynberg R, McLain RJ. (2010) (eds). Wild Product Governance: Finding policies that work for non-timber forest products. Earthscan, London.

Laudati AA (2010). The encroaching forest: Struggles over land and resources on the boundary of Bwindi Impenetrable National Park, Uganda. Society and Natural Resources. 23(8):776-789.

Leakey RRB, Tchoundjeu Z, Schreckenberg K, Shackleton S, Shackleton C (2005). Agroforestry Tree Products (AFTPs): Targeting Poverty Reduction and Enhanced Livelihoods. Int. J. Agric. Sust. 3:1-23.

Lescuyer G (2006). L'évaluation économique du Parc National de l'Ivindo au Gabon : une estimation des bénéfices attendus de la conservation de la nature en Afrique centrale. Rapport final, CIRAD Forêt, UPR36. Montpellier. 56 p.

Longhurst R (2003). Semi-structured interviews and focus groups. Key methods in geography. 117-132.

LYNAM AJ, LAIDLAW R, WAN NOORDIN WS, ELAGUPILLAY S, BENNETT EL (2007). Assessing the conservation status of the tiger Panthera tigris at priority sites in Peninsular Malaysia. Oryx, 41:454-462.

Naughton-Treves L, Holland MB, Brandon K (2005). The role of protected areas in conserving biodiversity and sustaining local livelihoods. Annual Review of Environment and Resources 30:219-252.

Okouyi-Okouyi J (2006). Savoirs locaux et outils modernes cynégétiques: développement de la filière commerciale de viande de brousse à Makokou (Gabon). Dissertation doctorale , Université d'Orléans, Orléans, France, pp165.

Ozanne CM, Cabral C, Shaw PJ (2014). Variation in Indigenous Forest Resource Use in Central Guyana. PloS one. 9(7):e102952.

Saha D, Sundriyal RC (2012). Utilization of non-timber forest products in humid tropics: Implications for management and livelihood. For. Policy Econ. 14(1):28-40.

Sassen M, Wan M (2006). Biodiversity and local priorities in a community near the Ivindo National Park, Makokou, Gabon. Report, CIFOR.

Spiteri A, Nepal S (2008). Evaluating Local Benefits from Conservation in Nepal's Annapurna Conservation Area. Environ. Manag. 42:391-401.

Timko JA, Waeber PO, Kozak RA (2010). The socio-economic contribution of non-timber forest products to rural livelihoods in Sub-Saharan Africa: knowledge gaps and new directions. Int. For. Review, 12(3):284-294.

UNDP (2010). Gabon: Country profile of human development indicators. Downloaded on 14th July 2014. Web site http://hdrstats.undp.org/en/countries/profiles/GAB.html. .

Vander W (JP) (2006). Ivindo et Mwagna: eaux noires, forêts vierges et baïs. Book, Wildlife Conservation Society, Libreville, Gabon.

van der Ploeg J, Cauilan - Cureg M, van Weerd M, De Groot WT (2011). Assessing the effectiveness of environmental education: mobilizing public support for Philippine crocodile conservation. Conservation Letters. 4(4):313-323.

Vedeld P, Jumane A, Wapalila G, Songorwa A (2012). Protected areas, poverty and conflicts: A livelihood case study of Mikumi National Park, Tanzania. Forest Policy and Economics. 21:20-31.

Viano M (2005). Caractérisation des activités de la population de Makokou à l'intérieur du Parc National de l'Ivindo. Mémoire de fin d'études, DES « Gestion des ressources animales et végétales en milieux tropicaux », Université de Liège, Belgique.

Wells M, Guggenheim S, Khan A, Wardojo W, Jepson P (1999). Investing in biodiversity: A review of Indonesia's integrated conservation and development projects. Washington, DC: World Bank.

White L (1992) Vegetation history and logging disturbance: effects on rain forest mammals in the Lopé Reserve, Gabon (with special emphasis on elephants and apes). Ph.D. thesis, University of Edinburgh, 230 p.

White L (1998). Baillonella toxisperma. The IUCN Red List of Threatened Species. Version 2014.2. <www.iucnredlist.org>. Downloaded on 30th October 2014.

Yemiru T, Roos A, Campbell BM, Bohlin F (2010). Forest incomes and poverty alleviation under participatory forest management in the Bale Highlands, Southern Ethiopia. Int. For. Review. 12(1):66-77.

20

Ethnic-based diversity and distribution of enset (*Ensete ventricosum*) clones in southern Ethiopia

Z. Yemataw[1], H. Mohamed[2], M. Diro[3], T. Addis[4] and G. Blomme[5]

[1]Southern Agricultural Research Institute, Areka Agricultural Research Center, P. O. Box 79, Areka, Ethiopia.
[2]Hawassa University, Awassa College of Agriculture, P. O. Box 05, Hawassa, Ethiopia.
[3]Capacity building for scaling up of evidence-based best practices in agricultural production in Ethiopia (CASCAPE), Addis Ababa, Ethiopia.
[4]Southern Agricultural Research Institute, Awassa Agricultural Research Center, P. O. Box 06, Hawassa, Ethiopia.
[5]Bioversity International Uganda Office, P. O. Box 24384, Kampala, Uganda.

Enset cultivation in southern and south-western Ethiopia is practiced mainly in densely populated areas. A survey covering 280 farm households and seven districts was conducted in seven zones of southern Ethiopia with the main objective of assessing the diversity and distribution of enset clones. Interviews using structured and semi-structured questionnaires were conducted to generate data. A total of 218 enset clones were recorded in the surveyed areas. The number of clones cultivated on individual farms ranged from two to 26 (mean of 8.9 ± 0.9). The highest richness of enset was recorded in Hadiya (59 clones) whereas the lowest was in Sidama zone (30); the mean richness being 39.7 ± 3.8 clones per zone. Exchange of clones among farmers in different ethnic groups in enset growing regions revealed that strong cultural and linguistic similarities exist between zones. Farmers reported that clones such as Gena and Mazia are replacing previously grown clones due to their resistance to *Xanthomonas* wilt. Several enset clones previously known by farmers have disappeared in recent years due to disease, extended drought and wild animals, pointing to genetic erosion and the necessity of genetic conservation.

Key words: Abundance, Gurage, Kembata, Mazia, richness, Wolaita.

INTRODUCTION

The genus *Ensete* belongs to the order Schistaminae and Musaceae family and comprises several species that grow in Africa and Asia (Bezuneh, 1984). Wild *Ensete ventricosum* can be found in Africa from the Ethiopian highlands to Malawi. However, domesticated enset (*E. ventricosum*) Welw. (Cheesman) is only cultivated in Ethiopia

The Ethiopian highlands are a center of genetic diversity for enset, tef, sorghum, barley and finger millet (Engels and Hawkes, 1991). The enset farming system supports over 15 million people with food, fiber, medicine and animal feed (Brandt et al., 1997).

The major food types obtained from enset are *kocho*, *bulla* and *amicho*. *Kocho* is fermented starch obtained from decorticated (scraped) leaf sheaths and grated corms. *Bulla* is a liquid which is obtained when leaf sheaths and corms are pulverized; the liquid containing starch is squeezed out from scraped leaf sheathes and grated corm and the resultant starch allowed to concentrate into white powder. *Amicho* is boiled enset corm/rhizome pieces that are prepared and consumed in a similar manner to other root and tuber crops (Brandt et al., 1997).

Reports of landrace diversity in enset are numerous. Alemu and Sandford (1991) reported names of 99 enset clones in the North Omo area, while Shigeta (1990) listed 78 vernacular names of cultivated enset clones in the Ari region of southern Ethiopia. Negash (2001) reported that farmers maintain and enrich the diversity of enset, and select or classify clones for various uses. Tesfaye (2002) indicated that enset landraces are not evenly distributed across the region mainly due to altitude variations. Tsegaye (2002) reported that numerous enset clones were identified in each region and the variations in the number of clones were attributed to a combination of socio-cultural and agroecological factors. Furthermore, Birmeta (2004) reported that the observed genetic diversity in cultivated enset in a particular area appears to be related to the extent of enset cultivation and the culture and distribution pattern of the different ethnic groups.

Some limited work has been done to evaluate, analyze and document clonal identity. Clonal names reported in the literature are associated with only limited phenotypic data provided by farmers (Shigeta, 1990). In enset, molecular characterization of clones has been done using amplified fragment length polymorphism (Negash, 2001; Negash et al., 2002; Tsegaye, 2002) and random amplified polymerphic DNA (RAPD) techniques (Birmeta, 2004). These earlier studies of enset diversity were limited to one or a few ethnic groups or a specific and limited growing region. However, a study encompassing many enset growing regions and ethnic groups has previously not been carried out, although knowledge on the level of morphological diversity of enset across a large number of ethnic groups or a large geographical area is important to assess the number of enset clones in the country and in the same time to develop a strategy for better genetic diversity conservation. Therefore, the objective of this study was to investigate farm level diversity and distribution of enset clones in seven (out of 16) major enset production areas in southern Ethiopia.

MATERIALS AND METHODS

The study area

The Southern Nations, Nationalities and Peoples' Regional State (SNNPRS) has a total area of 117,506 km², with altitudes ranging from 378 to 4,207 m above sea level (masl) (Abebe, 2005). The study was conducted in seven administrative zones: Wolaita, Kembata, Hadiya, Sidama, Gamo Gofa, Gurage and Dawro. One

district was selected in each zone. The selection was based on the prominence of enset cultivation and information about enset distribution obtained from the Departments of Agriculture of the respective zones.

Sampling and data collection

A household-level survey covering the seven zones was conducted from August 2008 to February 2009. In each zone, two peasant associations (PAs) (PA; this is the lowest tier of civil administration, equivalent to a village) were selected. Twenty households were randomly selected from each PA, giving a total of 280 households across the seven zones. Farmers were asked to name and describe each enset clone present on their farm.

In order to quantify on-farm genetic diversity, in all the directly monitored farms, a participatory zigzag sampling in diagonal direction of the plot was made in all 280 enset farms. All encountered clones were counted and discussions were made with farmers. For further verification of the clones, sample plants were taken from each clone to Areka research center for further on station assessment of selected quantitative and qualitative traits (Yemataw et al., 201)

Data analysis

Simpson (1949) and Shannon and Weaver (1949) diversity indices are the two most widely used measures of heterogeneity (Magurran, 1988). Both of them were calculated for all the zones. Simpson's index (D) measures the probability that two individuals, randomly selected from a sample, belong to the same category (Simpson, 1949) and hence, as D increases, diversity decreases. This is neither intuitive nor logical, so to get over this problem, D is often subtracted from 1 to give Simpson's Index of Diversity (1 – D). The value of this index ranges between 0 and 1; the greater the value, the greater the sample diversity. The index was computed for all the zones and all the clones using the function:

Simpson's Index of Diversity (1-D) = $1-\sum (n/N)^2$

$$D = \sum_{i-1}^{n} \frac{(n_i (n_i - 1)}{(N(N-1))}$$

Where, n_i = the frequency of the ith clone, frequency being the number of farms in which the clone is found in the district, and N = the total number of farms surveyed in the district.

The Shannon–Weaver diversity index (Shannon and Weaver, 1949) and Evenness measure (E) are commonly used tools that incorporate both richness and the evenness of abundance (Magurran, 1988). The Shannon diversity index (H') is high when the relative abundance of the different species in the sample is even, and is low when few species are more abundant than the others. Shannon–Weaver diversity index takes into account both number and evenness of categories considered and can be increased either by greater evenness or more unique species or clones in this case.

It was calculated using the formula, $H' = -\sum p_i \ln p_i$ (Magurran, 1988).

Where pi, the proportional abundance of the ith clone = $(\frac{ni}{N})$.

Although Shannon's index takes into account evenness of the abundance of clones, evenness can be calculated separately as a measure of the observed diversity to the maximum diversity. It is defined by the function E = H'/lnS, where H' is the Shannon index and S refers to the number of clones described in each zone. A high

Table 1. Distribution of households by number of enset plants.

Number of enset plants/household					Number of households	Percent
≤500					59	21
501-1000					80	29
>1000					141	50
Total					280	100
N	**Minimum**	**Maximum**	**Mean**	**Standard error**		
280	60	15000	2018	147		

Table 2. Variation in the number of enset clones planted per farm in the seven zones.

Number of enset clones per farm	Number of farms								Mean number (%) of farms
	S*	**W**	**GG**	**K**	**H**	**D**	**G**	**Total**	
≤ 5 clones	6	2	9	1	2	3	6	27	4.1(10.3)
6-10 clones	19	22	24	39	31	26	23	184	26.3 (65.8)
11-15 clones	12	14	6	0	2	11	9	54	7.7 (19.3)
≥15 clones	3	2	1	0	5	0	2	10	1.9 (4.8)
Total	40	40	40	40	40	40	40	280	

*S = Sidama; W = Wolaita; GG = Gamo Gofa; K = Kembata; H = Hadiya; D = Dawro; G = Gurage.

evenness, resulting from all clones having equal abundance, is normally equivalent to high diversity (Magurran, 1988).

Measures of similarity/variation are almost as numerous as measures of species diversity. The purpose of these functions is to quantify the similarity between two or more sampling sites. The expected variation in clone composition that exists between sites was analyzed using Sorenson's similarity coefficient (Cs) (Sorenson, 1948):

$$Cs = \frac{2J}{a+b}$$

Where, a is the number of clones at site A, b is the number of clones at site B, and J is the number of clones common to both locations.

Sorenson's similarity coefficient ranges in value from zero (no similarity) to one (complete similarity).

Clone diversities (Simpson's and Shannon-Weaver diversity indices) were measured separately for each zone. Pearson's correlation coefficient was used to compare diversity and distribution values at different sites. A tree diagram was constructed based on Euclidean distances developed by an unweighted pair-group method based on arithmetic averages (Nei, 1987). The SAS computer program (SAS, 2002) was employed for data analysis.

RESULTS AND DISCUSSION

Enset clone richness

The number of enset plants per farm household ranged from 60 to 15,000 and depended on farm size and availability of labor. The mean number of enset plants per household was 2,018 ± 147 (Table 1). Half (50-4%) of farm households have more than 1,000 enset plants on their farm. A farmer with a large number of enset plants and a wide diversity of clones is considered food secure and a model farmer in the locality. This study agrees with the study of Brandt et al. (1997) and Negash (2001) who observed large number of enset plants and clones in wealthy farmers' fields. Majority of the farms surveyed (65.8%) constitute 6-10 enset clones per farm (Table 2).

Based on the total number of different clones recorded (richness of the zone) and the number of enset clones per farm, Hadiya was the richest zone with a total of 59 clones, followed by Kembata (43), Dawro (41) Wolaita (39), Gamo Gofa (34) and Gurage (31) (Table 3). The lowest richness was found in Sidama zone with 30 clones. In previous studies, comparable results were reported by Tsegaye (2002), who described 146 different enset clones from three zones (52 clones from Sidama, 55 clones from Wolaita and 59 clones from Hadiya). Negash (2001) recorded 146 different enset clones from four zones (65 clones from Kefa-Sheka, 30 clones from Sidama, 45 clones from Hadiya and six clones from Wolaita). Moreover, Birmeta (2004) described 111 enset clones from nine growing areas of Ethiopia and Tesfaye (2002) studied 79 clones from the Sidama zone of the southern region. Although two zones (Dawro and Kembata) of our geographical study region were different from previous studies, 23 of the Sidama clones reported in our study were also listed by Tesfaye (2002). Out of the 59 enset clones of the Hadiya zone reported in this study, 36 were also reported by Tsegaye (2002). Of the clones in Wolaita studied by Tsegaye (2002), 18 clones were different from those included in

Table 3. Enset clone diversity in the seven zones, Southern Ethiopia, expressed as richness, Simpson (1-D) and Shannon (H') diversity indices, and Evenness.

Districts	Richness (%)	Mean richness/farm	Minimum richness	Maximum richness	Number of unique landraces	1-D	H'	Evenness
Sidama	30 (10.8*)	9.47	3	18	24	0.97	3.58	0.97
Wolaita	39 (14.02)	10.25	4	19	22	0.98	3.67	0.10
GamoGoffa	34 (12.23)	7.95	3	17	23	0.97	3.59	0.97
Kembata	43 (15.5)	7.53	4	10	24	0.98	3.64	0.99
Hadiya	59 (21.2)	9.3	2	26	33	0.97	3.61	0.98
Dawro	42 (15.1)	8.95	3	15	29	0.97	3.61	0.98
Gurage	31 (11.15)	8.95	2	24	23	0.98	3.63	0.98
Mean±SE	39.7 ± 3.8	8.94 ± 0.94						

*Calculated on the basis of the 278 clones described throughout the study area.

our study.

This indicated that the number of clones in any zone is not fully established and is underestimated by the survey methods used in independent studies. Further study including many enset growing area within the same time is very important. Many studies have been conducted to assess the patterns of genetic diversity in landraces of different crops using different methods and identifying promising accessions for different traits that could be utilized in breeding programmes. Examples include studies on tef (Bekele, 1996); wheat (Negassa, 1985), barley (Demissie and Bjørnstad, 1996), and sorghum (Ayana and Bekele, 1998).

The number of clones cultivated on individual farms ranged from two to 26 (mean of 8.94 ± 0.94) (Table 3). Average number of clones per farm ranged between 10.25 for Woalita to 7.53 for Kembata Sidama (9.47) and Hadiya (9.3) had high farm level diversity, followed by Dawro and Gurage with 8.95 clones per farm. This is because they have many farms with 11-15 clones, while other zones such as Kembata have few such clones, although the total number of clones in the zone was the highest (Table 3).

Diversity indices for the seven zones studied were computed from the numbers of clones present on the 40 farms within the zone (Table 2). Although zones differed in richness, they were similar in diversity. The Simpson's 1-D ranged between 0.971 (Sidama) to 0.977 (Wolaita), H' ranged between 3.58 for Sidama to 3.67 for Wolaita, while evenness also had a very narrow range: 0.97 for Gamo Gofa to 0.99 for Wolaita (Table 3). All these values indicate the high enset diversity in these seven zones.

In the seven zones, a total of 218 clones with distinct names were recorded. During the survey, we were able to confirm that each farmer is determined to maintain as much enset diversity as possible as long as he/she has enough land. It was possible to verify the existence of up to 26 different enset clones maintained by one household. During discussion with the farmers it was also observed that there were more than 100 enset clones grown in each locality a few years back, however, farmers reported that

most of the clones were lost due to disease and wild animals such as mole rat, porcupine and wild pigs. Tesfaye (2002) also found out that in Sidama, farmers reported names of 20 enset clones which were not encountered in any of the farms that were visited. Some enset landraces might have been totally lost from farmers' fields.

Hadiya and Kembata zones shared 17 clones (Table 4), while Wolaita and Gamo Gofa, and Wolaita and Dawro had 11 clones in common. These zones are adjacent to each other and the Kembata and Hadiya, and Wolaita and Dawro zones were until recently one administrative area.

Strong cultural and linguistic similarities exist between Kembata and Hadiya, and between Wolaita, Dawro and Gamo Gofa. This justification was noticeably confirmed by Fleming (1975), who stated that Dawro, Gamo Gofa and Wolaita peoples of the Southern Ethiopia belong to Omotic people who have a dialect of the central Omotic languages.

This may be reflected in the observed high similarity in cultivated clones. Clustering of the seven zones using the Sorenson's similarity index grouped the seven zones into four clusters (Figure 1) as follows: i) Kembata and Hadiya, ii) Wolaita, Dawro and Gamo Gofa, iii) Sidama and iv) Gurage. It is interesting to note that Sidama and Gurage do not share many clones with neighboring zones

These findings, however, noticeably differ from those of Tesfaye (2002), who reported that 52% of enset clones in Sidama zone were shared among farmers of the study area suggesting that informal exchanges among farmers are limited within narrowly defined ethnic groups. The informal exchange of planting material among farmers mainly occurs within the geographical zone occupied by an ethnic group and it is hence difficult, to compare values with results of previous surveys due to differences in the number of locations and ethnic considerations.

In agreement with Tabogie (1997), duplication of clone names was observed. The same enset clone was given different names in different areas and vice versa (different enset clones were given the same name in different localities) (Tabogie, 1997). Tsegaye (2002) also showed that duplication of clone names was related to different

Table 4. Number of shared clones (above diagonal) and Sorenson similarity indices (below diagonal) between pairs of zones.

Zones	Sidama	Wolaita	Gamo Gofa	Kembata	Hadiya	Dawro	Gurage
Sidama	--	3	1	2	2	3	1
Wolaita	*0.06	--	11	1	4	11	1
Gamo Gofa	0.06	0.27	--	0	1	6	0
Kembata	0.03	0.02	0.026	--	17	0	2
Hadiya	0.07	0.08	0.02	0.35	--	2	8
Dawro	0.06	0.3	0.16	0	0.04	--	0
Gurage	0.03	0.03	0	0.05	0.18	0	--

*=Sorenson's similarity index.

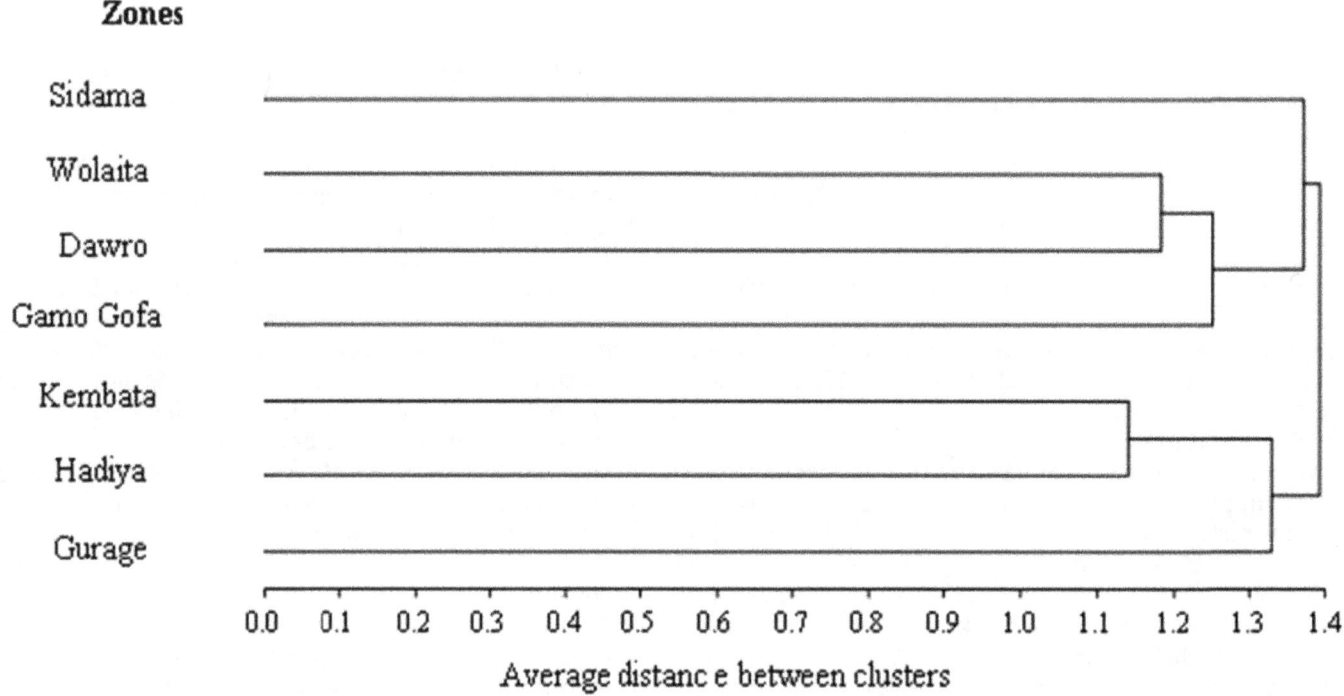

Figure 1. Dendrogram of the seven zones based on Sorenson's similarity index.

utilization purposes of clones and the changing of vernacular names after exchange of clones between different ethnic groups.

Distribution and abundance of clones

Large differences were evident between clones in their abundance and distribution. Some clones had a rather patchy distribution, that is they had a very high local abundance at one or two locations and were absent from the rest. For example, 'Shodedenia' was encountered on all the 40 (100%) of the farms visited in Dawro (Table 5). It was not found in any other zones surveyed. This is an abundant clone with a narrow distribution. The same was true for 'Amerate' in Gurage and Genticha in Sidama

which were recorded on 33 and 27 of the 40 farms, respectively (Table 5). On the other hand a relatively small number of clones played a dominant role in more than one zone. These were 'Mazia', 'Gena', 'Astara' and 'Badedea' (Table 5). Mazia was the most abundant clone as it was recorded on 89 (32%) of all the farms surveyed, and also in a much higher proportion of the 40 farms surveyed in the three zones where it was found: Wolayita, Gamo Gofa and Dawro zones 17 (7%), 35 (12%) and 37 (13%) respectively) (Table 5). However, there was overall a significant correlation between distribution and abundance of the clones (r = 0.66, p ≤ 0.0001). Clones that are used by many famers in any zone tend to be found in other zones and have wider distribution.

There was also a considerable difference among the clones with respect to their distribution across the zones

Table 5. Numbers of farmers growing widely distributed and the most abundant enset clones in each zone.

Clone	H[†]	K	G	W	GG	S	D	TOT	Zones
Astara	13	22	14			10		59	4
Sabara	16	8	16					40	3
Mochea	13	2		5				20	3
Badadea	10		3	8			2	23	4
Gena	1			14	14	20	10	55	5
Katania				11	5		3	19	3
Agena				7		10	9	26	3
Switia				8	4		4	16	3
Kekerwa				7	9		6	22	3
Mazia				17	35		37	89	3
Banga				9	2		7	18	3
Shodedenia							40	40	1
Amerate			33					33	1
Genticha						27		27	1

†= H = Hadiya, K = Kembata, G = Gurage, W = Wolaita, GG = Gamo Gofa, S = Sidama, D = Dawro, TOT = Total number of farmers.

Table 6. Distribution of enset clones across the seven zones

Number of zones	Number of enset clones (%)
One	178 (82)
Two	29 (13)
Three	8 (4)
Four	2 (1)
Five	1 (<1)
Six	0
Seven	0
Total	218

covered by this study. Out of the 218 clones, 178 (82%) were observed in only one zone. Twenty nine (13%) of the clones were present in two zones. Eight clones (4%) were present in three zones. Two clones (1%) were present in four of the seven zones and only one clone (Gena) was present in five of the seven zones (Tables 5 and 6). Household characteristics, distance from one location to another and ethnic preferences in few locations for few number of clones bring high clonal diversity, while for more number of clones that do not fulfill the selection criteria of each ethnic group brings clonal paucity.

To classify the total abundance and distribution of the whole clone into four quadrant plane, a clone having an equal abundance x distribution point was selected. Based on that, if we designate a clone that is present in at least 15 of the 40 farms in a given zone having an average abundance of 0.38, then all the 218 clones can be categorized into four groups on the abundance x distribution plane (Table 7 and Figure 2). The first category (widely distributed and abundant clones) applies only to 'Mazia'. The second category (localized but abundant clones) included 23 enset clones. The highest numbers of enset clones (183) were grouped in the third category which is the localized and rare clones. The fourth group (widely distributed but rare clones) included 11 clones.

The abundance of clones across sites within a zone and the distribution of clones across the seven zones were generally uneven, because of a limited number of widespread and dominant clones. The hierarchical nature of the spatial distribution of enset clones with a small number of highly abundant clones which are also grown throughout the region and a much larger number of moderately common and rare ones has been documented for enset (Tesfaye, 2002), and several other crops including cassava (Boster, 1985) and yam (Tamiru, 2006).

Conclusion

A large number of enset clones was recorded in the southern region. However, the diversity of enset clones is not spread evenly across the region. A small number of highly abundant clones are grown throughout the region, while a much larger number of moderately common and rare clones characterize the distribution-abundance pattern. The widespread distribution of some clones challenges the view that traditional farming systems are isolated and closed, with limited exchange of germplasm. The findings of this study and similar studies depict a system that is rather open and dynamic, where local knowledge exists for exchange of planting materials across wider areas and heterogeneous environments. The unequal distribution and abundance of clones reflect

Table 7. Classification of the 218 clones into four groups based on their abundance and distribution.

Quadrant	Category	Number of clones in the category
I	Widely distributed and abundant clones	1
II	Localized but abundant clones	23
III	Localized and rare clones	183
IV	Widely distributed but rare clones	11

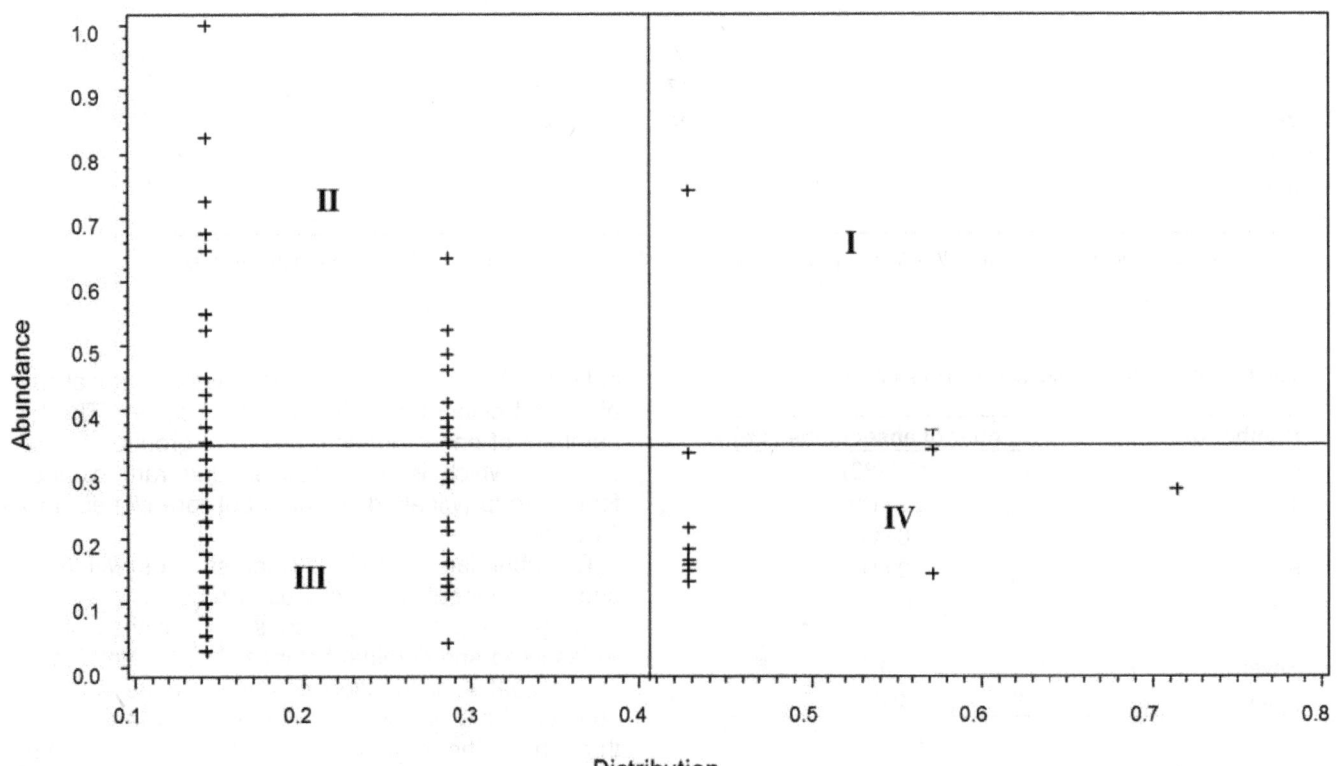

Figure 2. Classification of the 218 enset clones into 4 groups using their scatter in abundance X distribution plane. (a lot of clones are hidden).

their relative importance to farmers and provide strong evidence for selection. Highland regions have a high concentration of diverse and unique enset landraces and should be given priority in efforts aimed at collection and *in situ* germplasm conservation.

ACKNOWLEDGEMENTS

The authors would like to thank the Southern Agricultural Research Institute and the Canadian International Development Agency for their financial support. The authors also thank all staff and farmers who participated at different stages of the field surveys. The zone and district agriculture departments are also acknowledged for their support during the survey.

REFERENCES

Abebe T (2005). Diversity in homegarden agro-forestry systems of southern Ethiopia. PhD dissertation, Wageningen University, The Netherlands. 143 pp.

Alemu K, Sandford S (1991). Enset in North Omo Region. Farmer's Res. Project Technical Pamphlet No. 1. Farm Africa, Addis Abeba. 49 pp.

Ayana A, Bekele E (1998). Geographical patterns of morphological variation in sorghum (*Sorghum bicolor* (L.) Moench) germplasm from Ethiopia and Eritrea: qualitative characters. Hereditas 129:195-205.

Bekele E (1996). Morphological analysis of *Eragrostis teff*: Detection for regional variation. *Sinet*: Ethiopian J. Sci. 19:117-140.

Bezuneh T (1984). Evaluation of some *Ensete ventricosum* clones for food yield with emphasis on the effect of length of fermentation on carbohydrate and calcium content. Trop. Agr. 61:111-116.

Birmeta G (2004). Genetic variability and biotechnological studies for the conservation and improvement of *Ensete ventricosum*. Doctoral thesis, Swedish University of Agricultural Sciences, Alnarp. 91 pp.

Boster J (1985). Selection for perceptual distinctiveness: evidence from Aguarina cultivars of *Manihot esculenta*. Econ Bot. 39:310-325.

Brandt SA, Spring A, Hiebsch C, McCabe ST, Tabogie E, Diro M, Welde-Michael G, Yentiso G, Shigeta M, Tesfaye S (1997). The 'Tree Against Hunger'. Enset-based Agricultural Systems in Ethiopia. American Association for the Advancement of Science. 56 pp.

Demissie A, Bjørnstad A (1996). Phenotypic diversity of Ethiopian barley in relation to geographical region, altitudinal range, and agro-ecological zones: As an aid to germplasm collection and conservation strategy. Hereditas 124:17-29.

Engels J, Hawkes J (1991) .The Ethiopian gene centre and its genetic diversity. In: Engels J, Hawkes J, Worede M (eds). Plant Genetic Resources of Ethiopia. Cambridge University Press, Cambridge

Fleming H (1975). Recent Research in Omotic-Speaking Areas. In: Marcus H.G (ed.). Proceedings of the First United states Conference on Ethiopian Studies, 1973. East Lansing: Michigan State University. pp. 261-278.

Magurran A (1988). Ecological diversity and its measurement. Croom Helm, London. 125 p.

Negash A (2001). Diversity and conservation of enset (*Ensete ventricosum* Welw. Cheesman) and its relation to household food and livelihood security in South-western Ethiopia. PhD dissertation, Wageningen University, The Netherlands. 247pp.

Negash A, Tsegaye A, Van Treuren R, Visser B (2002). AFLP analysis of Ensete clonal diversity in south and southwestern Ethiopia for conservation. Crop Sci. 42:1105-1111.

Negassa M (1985). Patterns of phenotypic diversity in an Ethiopian barley collection and the Arsi-Bale high land as a center of origin of barley. Hereditas 102:139-150.

Nei M (1987). *Molecular evolutionary genetics*. Colombia University Press, New York.

SAS (2002). Statistical Analysis Systems SAS/STAT user's guide Version 9 Cary NC: SAS Institute Inc. USA.

Shannon C, Weaver W (1949). The mathematical theory of communication. University of Illinois, Urbana, IL 117pp.

Shigeta M (1990). Folk *in situ* conservation of enset (*Enset ventricosum* (Welw.) Cheesman): Towards the interpretation of indigenous agricultural science of the Ari, south western Ethiopia. Afr Stud Monogr 10:93-107.

Simpson E (1949). Measurement of diversity. Nature 163:688.

Sorenson T (1948). A method of establishing groups of equal amplitude in plant sociology based on similarity of species content, and its application to analysis of the vegetation on Danish commons. K dan Vidensk Selsk Biol. Skr. 5:1-34.

Tabogie E (1997). Morphological Characterization of enset (Ensete ventricosum (Welw.) Cheesman) clones and the association of yield with different traits. M.Sc. Thesis. Alemaya University of Agriculture, Alemaya, Ethiopia. 89 pp.

Tamiru M (2006). Assessing diversity in yam (*Dioscorea spp.)* from Ethiopia based onmorphology, AFLP marker and tuber quality, and farmers' management of landraces. Ph.D. thesis, George -August University. Germany. 155pp.

Tesfaye B (2002). Studies on landrace diversity, *in vivo* and *in vitro* regeneration of enset (*Ensete ventricosum* Welw.). PhD dissertation, Humboldt University, Berlin, Germany, 129 pp.

Tsegaye A (2002). On indigenous production, genetic diversity and crop ecology of enset (Enset ventricosum (W elw.) Cheesman) . PhD dissertation, W ageningen University, The Netherlands. 197pp.

Yemataw Z, Mohamed H, Diro M, Addis T, Blomme G (2014). Enset (*Ensete ventricosum*) clone selection by farmers and their cultural practices in southern Ethiopia. Genet Resour Crop Ev. 61(3).

21

Vegetation regeneration in formerly degraded hilly areas of Rwampara, South Western Uganda

Juliet Kyayesimira[1] and Julius B. Lejju[2]

[1]Department of Biological Sciences, Kyambogo University, Uganda.
[2]Department of Biology, Mbarara University of Science and Technology, Uganda.

Rwampara hills located in South Western Uganda have long been subjected to intensive degradation due to increased human activities. The hills have been left bare as a result of vegetation clearing for agricultural land, charcoal burning and grazing. In 1998, the National Environmental Management Authority (NEMA) attempted to restore the degraded hilly areas with the aim of establishing the restoration potential. With the cooperation of the local people, NEMA set aside some parts of the hills to allow natural regeneration, while another parts were planted with exotic tree species mainly *Eucalyptus* spp. *and Pinus patula.* This paper presents findings of an assessment on the level of indigenous vegetation regeneration in the three zones namely; restored, planted and areas undergoing degradation due to grazing. The indigenous vegetation was sampled using nested quadrats set along line transects. The results indicate that species richness was different among the three habitat types with the highest number (17 species) recorded in the degraded (grazing) area, followed by the restored area (12 species) and the plantation had the least (10 species). Species density was highest in the restored zone (289.83/ha) and least (80.2/ha) in the plantation zone. The most common indigenous tree species regenerating in all the three study zones were; *Olea europaea* subsp. *africana, Albizia adiathifolia* and *Markhamia lutea.*

Key words: Degraded hills, vegetation regeneration, Western Uganda.

INTRODUCTION

Increased human activities such as agriculture, grazing, firewood collection and charcoal burning aimed at improving livelihoods has caused severe land degradation of marginal lands, especially in hilly areas of western Uganda. Population increase and economic growth are primarily the driving forces behind degradation of these marginal lands (Olson and Berry, 2003).

The Rwampara hills in Western Uganda have a long history of land degradation. Past land use patterns and disturbance regimes have had a profound effect on the abundance, distribution and diversity of vegetation in the area. Due to severe effects of degradation, the area has become prone to agents of erosion. In 1998, the National Environment Management Authority (NEMA) initiated a program to restore the degraded ecosystem of Rwampara hills, with an aim of curbing soil erosion and increase its biological productivity and, local economic benefits and environmental services. Some local farmers volunteered

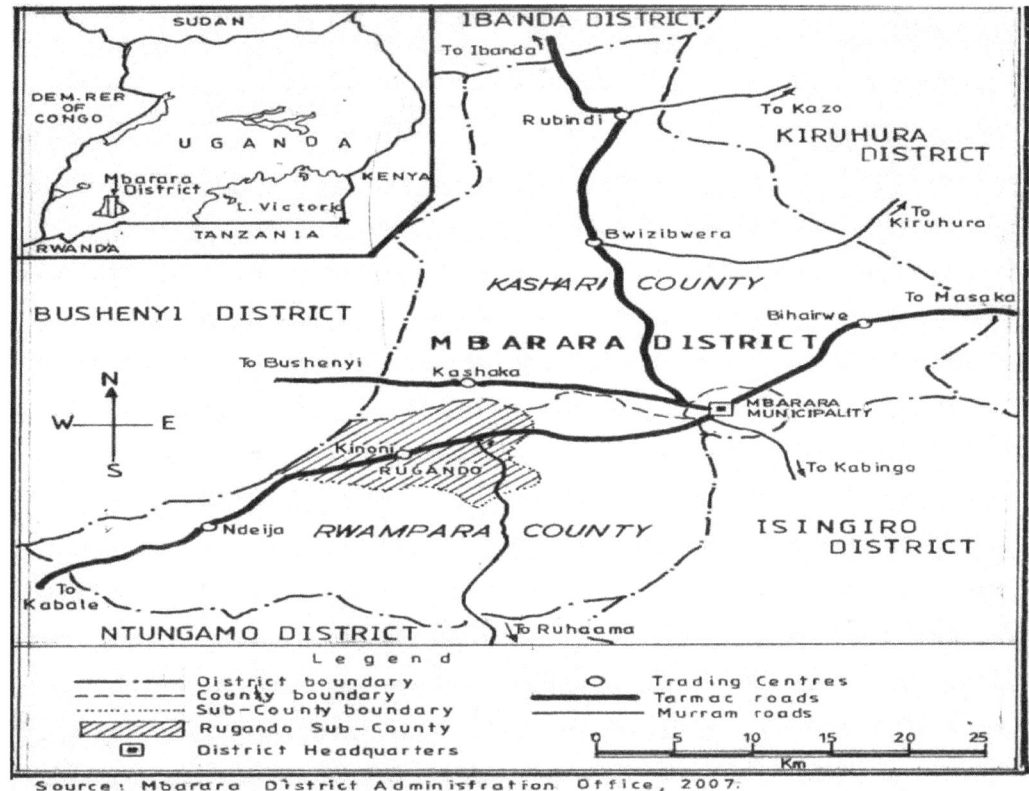

Figure 1. The study area in Rwampara County, South Western Uganda.

portions of their land for the restoration purpose. These portions were set to regenerate naturally, while other parts were planted with eucalyptus and pine. This paper presents the ecological data on a decade of vegetation restoration in the degraded areas of Rwampara hills in western Uganda.

MATERIALS AND METHODS

Study area

Mbarara district in Western Uganda comprises of Rwampara County. The district is located at Latitude:-0.6132; Longitude: 30.6582 (Figure 1). The landscape consists of rolling hills intercepted by long, but shallow valleys with wetlands occurring in the valleys. According to the National Population and Housing Census of 2002, Rwampara County has a population of 132,802 and a land area of 659.8 km². The area receives a moderate rainfall throughout the year with an average rainfall of 1200 m and temperatures ranging from 17 to 30°C. Two rainy seasons occur in the area from March to May and September to December, while the dry spells are experienced from December to February and June to August. The relative humidity ranges from 80-90% in the mornings and 48-60% in the evenings throughout the year.

The vegetation in the area consists of ever green and broad leaves, characteristics of medium attitude tropical rain forests. The current vegetation is dominated by indigenous and exotic tree species. The land has been subjected to intensive agricultural activities, mainly banana cultivation and livestock farming.

Vegetation sampling

Vegetation sampling was conducted in the three zones namely; degraded, restored and plantation zones. A stratified random sampling method was applied within the three zones. The principle of stratification was that the vegetation of the area under investi-gation was divided up before samples are chosen on the basis of major and usually obvious variations within the habitat (Kent and Coker 1996). The reason for this method of stratification was to sample zones of vegetation subject to different gradients and management regimes.

The vegetation was sampled using nested quadrats set along line transects as described in Kent and Coker (1996). The transect lines were set in the three vegetation zones namely; restored zone, degraded zone and plantation zones. A total of 12 line transects, each measuring 100 m long were positioned in such a way that they ran from the bottom of the hill to the top, so as to sample different vegetation starata. Systematic sampling which involved the location of nested sampling points at regular intervals was employed (Kent and Coker, 1996). In each line transect, a series of nested plots were established in an alternating left and right as employed by Kasenene (1987) and Lejju (1999). A series of 15 × 15 m plots were set up at regular intervals of 20 m apart using a measuring tape. Within each plot of 15 × 15m, a series of nested quadrat measuring 10 × 10 m, followed by 5 × 5 m and 2 × 2 m were established to enumerate trees of different size classes that occupy different vegetation strata. Large tree size class measuring ≥ 15 cm dbh (1.3 m) were sampled in 15 × 15 m plot, followed by small trees (dbh 10 -< 15 cm) sampled in 10 m x 10 m plots. Poles (5 – <10 cm, dbh) and saplings (dbh 2 - <5 cm) were sampled in 5 x 5 m plot, whereas seedlings (< 2 cm) were sampled from 1 × 1 m plots.

Table 1. Species richness, R and Shannon Diversity Indices (H) of trees of different size-classes recorded in Restored, Degraded and plantation zones.

Study site	Diversity indices	Diameter size-classes (dbh)				
		Seedlings (<2 cm)	Sapling (2-< 5 cm)	Poles (5-<10 cm)	Small tree (10-<15 cm)	Large trees (≥15 cm)
Restored	R	12	12	6	4	1
	H	1.85	1.89	1.15	1.09	0.00
Degraded	R	12	12	7	8	1
	H	2.26	2.51	1.78	1.99	0.00
Plantation	R	10	7	1	0	0
	H	2.07	1.68	0.00	0.00	0.00

*R= Species richness, H = Diversity index.

Species diversity and richness

Plant species diversity was calculated from Shannon's diversity index, $H = -\sum p_i \log_e p_i$ where P_i = is the proportion of each species in the sample (Bibi and Ali, 2013). Species richness, R was calculated as measure of the number of species found in a sample or zone.

Important value index (IVI)

To obtain importance value index, the frequency, density, and dominance of each species was determined in each zone. Then, IVI was calculated as;

Importance value = Relative frequency + Relative density + Relative dominance

Where;
Relative frequency: Number of occurrences of one species as a percentage of the total number of occurrences of all species.
Relative density: Number of individuals of one species as a percentage of the total number of individuals of all species.
Relative basal area (dominance): Total basal area of one species as a percentage of the total basal area of all species.
Absolute density: This was determined by summing up the number of individuals found in each plot and divide by the number of plots. Average species density = (density in plot 1) + (density in plot 2) + (density in plot X) /total number of plots.

RESULTS

Species richness and diversity

Species richness and diversity indices of different diameter size classes in the three vegetation zones are presented in Table 1. The results show that in the restored area, species richness ranged from 1 species for large tree size class ≥ 15 cm dbh to 12 species for seedlings and poles (Table 1). Similarly in the degraded area, species richness ranged from 1 species for large trees to 12 species for seedlings and poles. However, in

the plantation zone, the highest number of species (10) was recorded for seedlings and none for small trees and large trees.

The Shannon diversity index, H of restored zone was highest (1.89) in saplings (2-< 5 cm dbh) followed by seedlings (1.85) and lowest (0.00) in large trees (≥15 cm, dbh), while in the degraded zone, the H^1 values ranged from 1.78 for poles (5-< 10 cm, dbh) to 2.51 for seedlings (<2 cm dbh). In the plantation zone the diversity index ranged from 1.68 to 2.07.

Among the three vegetation zones, the degraded zone is more diverse (2.51) for saplings than the restored (1.89) and plantation (1.68). Generally, the seedlings (dbh <2 cm) recorded the highest diversity indices, and large trees had the least diversity in all the three study sites.

Tree densities (No/ha) of different size classes

The log density stand curves of different size-classes in plantation, restored and degraded zones are presented in Figure 2. There was a general decrease in log density for all the tree size classes in all the three habitats from seedlings to large trees. In the restored and degraded areas, the log density decrease was gradual, while in the plantation zone, the log density decreased sharply to zero at poles.

Absolute density of tree species in different study zones

The absolute density (No/ha) of tree species recorded in degraded, restored and plantation zones are presented in Table 3. *Markhamia lutea* (had the highest density (1076/ha) in the restored zone followed by *Albizia adiathifolia* (705/ha) and *Olea europaea* subsp. *africana* (600/ha). Within the degraded zone, *Tetrochidium*

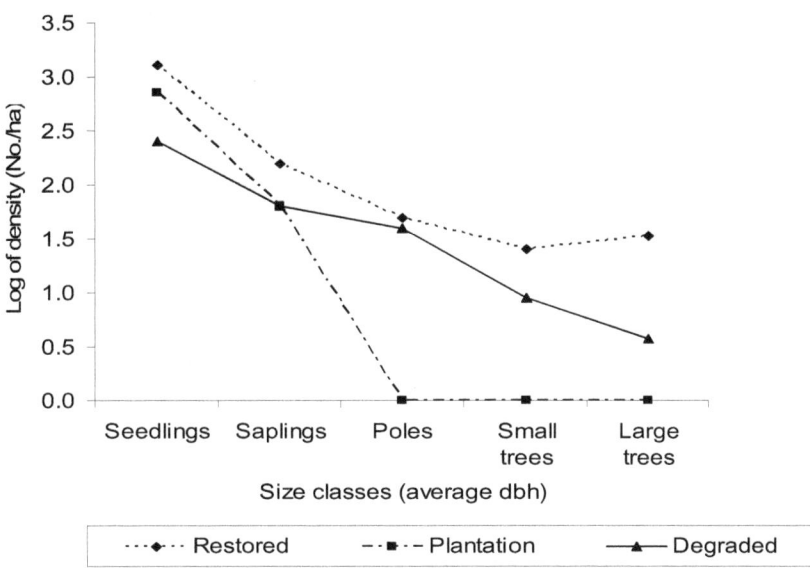

Figure 2. Species area curves showing a cumulative number of plant species recorded in the restored, degraded and plantation zones.

didymostemon had the highest density (64/ha) followed by *Senna didymobotrya* (957/ha) and *Cordia africana* (55/ha). In the plantation zone, *Olea europaea* subsp. *africana* had the highest density (190/ha) followed by *A. adiathifolia* (17/ha) and *Solanecio manii* (117/ha). Three species namely; *O. europaea* subsp. *african*, *A. adiathifolia* and *M. lutea* occurred in the three zones with *M. lutea* contributing the highest density (38/ha) in the restored zone, while *A. adiathifolia* had the least density (7/ha) recorded in the degraded zone. Generally the restored zone had the highest tree density compared to the plantation and degraded zones (H=17.12, df=2, P=0.000) (Table 2).

Principal component scatter diagram in Figure 3 was used to explain the species density in the restored, plantation and degraded zone. As you move towards the right hand side, there is less disturbance and hence high plant densities while as you move towards the left hand side, there is high disturbance and hence lower densities but many varieties of plant species.

Importance Value Indices (IVI) for tree species

Importance value indices (IVI) calculated from relative densities, relative frequency and relative dominance for all the tree species recorded in the natural, degraded and plantation zones are presented in Table 3. In restored zone, *O. europaea* subsp. *Africana* had the highest importance value (271.2) followed by *A. adiathifolia* (147.1) and *M. lutea* (105.4). Whereas in the degraded zone, *A. adiathifolia* had the highest IVI (127.1) followed by *Croton sylvaticus* (96.8) and *Cordia Africana* (50.3), while in the plantation, *O. europaea* subsp. *africana* had

the highest IVI (222.1) followed by *A. adiathifolia* (66.5) and *M. lutea* (47.3).

DISCUSSION

The higher values of species richness recorded in the degraded area in comparison with restored and plantation zones are an indication of high levels of regeneration following disturbance in the degraded hills. The study also indicated that following disturbance, re-growth of new species, which are either from the soil seed store or are dispersed into the site from the outside, occurred efficiently.

Bazzaz (1984) stated that most disturbances create highly heterogeneous habitats that recruit different species and play out different growth scenarios. Some of the agents of disturbances in the study area included fires, overgrazing and cultivation and this caused the creation of gaps. Lamb (1990) found that forests are subject to a number of naturally occurring disturbances that produce a range of different sized gaps (spaces). This led to creation of space for new tree species.

The lower species richness observed in the restored zone compared to the degraded zone could be due to the slow regeneration rates of some species. Studies by Hooper et al. (2004) noted that fire significantly affected species composition and decreased species richness because most species had either their resprouting ability or seed germination inhibited by fire and this could be the case in the restored area. This is in agreement with studies done by Uhl et al. (1988) and Nepstad et al. (1990). Restoration in the area that begun 10 years back after several agents of disturbances, the zone consists of

Table 2. Absolute densities (No/ha) recorded in the restored, degraded and plantation zones.

Species	Restored Zone	Degraded Zone	Plantation zone
Olea europaea subsp. *Africana*	600	12	190
Albizia adiathifolia (Schumach.) W.F. Wight	705	7	162
Markhamia lutea (Benth.) K. Schum	1076	38	110
Syzygium cordatum Hochst.ex Krauss	112	-	76
Acacia hockii De Wild	46	-	-
Peddiea fischeri Engl.	219	-	-
Pittosporum spathicalyx De Wild.	129	-	45
Myriathus holstii Engl.	205	-	-
Bridelia micrantha(Hochst.) Baill	76	-	-
Solanecio manii (Hook.f.) C.Jeffery	217	-	117
Allophylus macrobotrys Gilg.	60	-	-
Olea welwitschii(Knobl.) Gilg& Schellenb	33	-	-
Sapium ellipticum(Hochst. Ex Krauss) Pax	-	31	-
Albizia coriaria Welw.ex Oliv	-	45	-
Senna didymobotrya (Fresen.)Irwin & Barneby	-	57	-
Antiaris toxicoria(Rump h.ex Pers.) Lesch.	-	-	29
Polyscias fulva(Hiern) Harms	-	-	21
Cyphomandra batacea F I. Neotrop.Monogr.	-	-	7
Macaranga kilimandscharica Pax	-	-	45
Senna bicapsularis(L.) Roxb	-	21	-
Maes lanceolata Forsk.	-	2	-
Albizia gumefera Welw.ex Oliv	-	26	-
Cordia Africana Burm.f.	-	55	-
Vernonia amygdalina Del.	-	52	-
Erythrina abyssinica Lam. Ex DC	-	48	-
Ficus natelensis Hochst.	-	17	-
Acanthus pubescens(Thomson ex Oliv.) Engl.	-	19	-
Tetrochidiumdidymostemon (Baill.) Pax & K.Hoffm.	-	64	-
Diospyros abyssinica(Hiern) F. White	-	48	-
Croton sylvaticus Hochst.	-	**43**	-
Ficus natelensis Hochst.		17	-
Acanthus pubescens(Thomson ex Oliv.) Eng	-	19	-
Tetrochidiumdidymostemon (Baill.) Pax & K.Hoffm	-	64	-
Diospyros abyssinica(Hiern) F. White	-	48	-
Croton sylvaticus Hochst.		43	
Total	**289.83**	**34.41**	**80.2**

mostly secondary species. Species regenerated from seeds but primary species owed their presence in regeneration to their ability to reproduce vegetatively. Some seeds require fires to break their dormancy and if the rains come soon after, this enables regeneration. Species richness was low in restored zone because some species like *O. europea* subsp. *africana* have a slow growth rate.

However, the low number of species obtained in the plantation zone could be attributed to suppression of indigenous trees by the eucalyptus trees. This agrees with results obtained by Lejju (1999) that eucalyptus suppresses the growth of native species, hence the reason for low species richness.

In general, the regeneration pattern of tree species varied in each study site and human disturbance could have influenced seed dispersal mechanism, fruiting, germination and regeneration of tree species.

The degraded zone was more diverse and showed higher equitability than other areas. The high diversity of indigenous tree species in the lower size classes (seedlings and saplings) for all the three study sites is an indicator of regeneration. Large size classes showed lower diversity and density indicating low survival rate of seedlings into the large size class. This is in agreement with the results obtained by Grubb (1977) and Lejju

Table 3. Importance value indices for tree species recorded in the restored, degraded and plantation zones.

Tree species	Importance value indices		
	Restored zone	Degraded zone	Plantation zone
Olea europaea subsp. africana	271.2	9.3	222.1
Albizia adiathifolia	147.1	127.1	66.5
Markhamia lutea	105.4	27.1	47.3
Syzygium cordatum	12.5	-	34.5
Acacia hockii	6.64	-	-
Peddiea fischeri	20.5	-	-
Pittosporum spathicalyx	13.5	-	27.1
Myriathus holstii	19.8	-	-
Bridelia micrantha	8.5	-	-
Solanecio manii	30.6	-	52.1
Allophylus macrobotrys	38.6	-	-
Olea welwitschii	25.7	-	-
Sapium ellipticum	-	35.0	-
Albizia coriaria	-	42.8	-
Senna didymobotrya	-	36.3	-
Antiaris toxicoria	-	-	8.2
Polyscias fulva	-	-	3.34
Cyphomandra batacea	-	-	28.7
Macaranga kilimandscharica	-	-	10.0
Senna bicapsularis	-	13.6	-
Maes lanceolata	-	10.9	-
Albizia gumefera	-	29.4	-
Cordia africana	-	50.3	-
Vernonia amygdalina	-	33.4	-
Erythrina abyssinica	-	62.9	-
Ficus natelensis	-	32.3	-
Acanthus pubescens	-	28.1	-
Tetrochidium didymostemon	-	38.4	-
Diospyros abyssinica	-	26.3	-
Croton sylvaticus	-	96.8	-

(1999) who pointed out that the younger size class is usually numerous compared to the older size class. This is due to the fact that mortality rate is high in early stages of life because of predation, desiccation and competition as well as removal by human activities.

Janzen and Vazquez-Yanes (1978) stated that on tropical mainland, more than 90% of all tree species have more than 50% of their seeds killed by animals and fungi between fruit set and seed germination. Some seeds land in places where the seedlings have no chance of survival and so a few may reach maturity. On the other hand, lower number of large trees recorded in the degraded area could probably be due to selective removal of some trees during harvesting, while the low diversity of large size class trees in the restored area could be due to dominance caused by some species.

Bawa (1983) pointed out that tree species in a tropical rainforest display much variation in timing, duration and frequency of flowering and this could be the case in Rwampara. Species vary considerably in duration of flowering which extends from a few days in some species to several months in others. Lower diversity of large size classes in the plantation zone could be attributed to suppression by eucalyptus and Pinus patula in the area. It could be that the gaps are small and so the shade intolerant species begin to die as soon as maturation starts.

Rwampara hilly areashave been influenced by human activities such as agriculture and fire. High abundance of some tree species suggests a form of dispersal and plant utilization. For instance, O. europea subsp. africana, M. lutea and Tetrochidium didymostemon were highly abundant in the restored, plantation and degraded zones respectively. Plants become established either from the

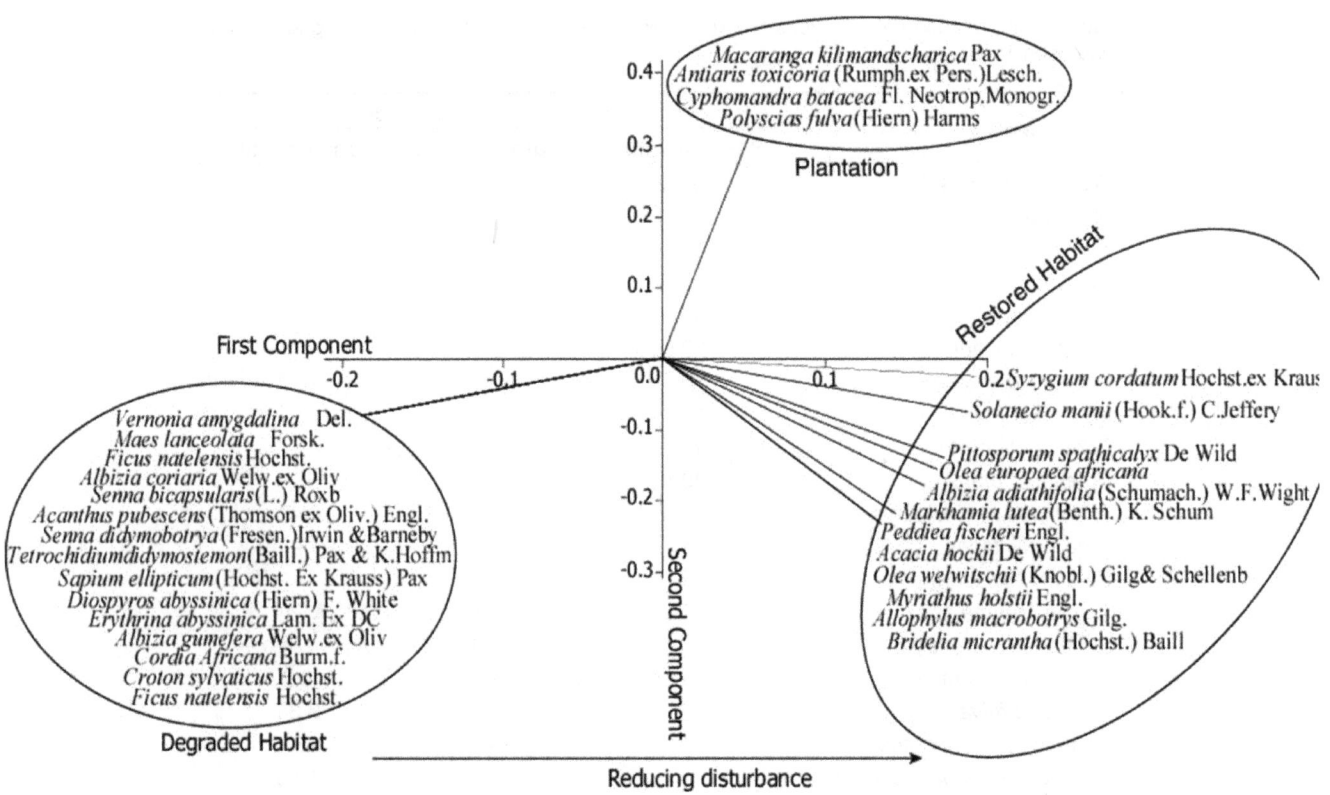

Figure 3. A principal component analysis (PCA) diagrams showing the absolute densities of plant species in the study area.

seedling pool, soil seed store, vegetative re-growth and dispersed seeds. *Olea europea subsp. africana* can regenerate from wildings and seedlings and *M. lutea* from wildings, seedlings and cuttings.

Several tree species like *Sampium ellipticum, M. lutea, Peddiea fischeri* and *Pittosporum spathicalyx*. had several seed or fruit dispersal mechanisms and they may also regenerate from coppice or root suckers apart from wildings, direct sowing and seedlings.

O. europea subsp. africana and *Albizia adiathifolia* had the lowest densities in the degraded area compared to the restored and plantation areas. *Olea europea subsp. africana* is highly used by the local community for firewood, charcoal, medicine, poles, walking sticks, tool handles and environmental purposes such as soil conservation and this could be the reason why it is low in the degraded area. *Olea europea subsp. africana* is also a slow growing tree so once it is harvested; it takes long to re-grow (Katende et al., 1995). Currently the restored and plantation zones are restricted from harvesting.

Maesa lanceolata had the lowest density in the degraded area than other species yet it is a fast growing tree. This could be attributed to the rate it is harvested since the local people use it to treat diseases like ulcers, diarrhoea and febrile convulsions in children. Some species like *Olea europea subsp. africana* and *Peddiea fischeri* are slow growing yet are highly diverse in

restored and plantation zones. This could be because both restored and plantation zones are restricted from harvesting, hence allowing them time to regenerate.

Polyscias fulva had the lowest density in the plantation zone yet it is a fast growing tree species. The reason for this is that it is highly harvested for firewood, timber, and bee hive making. *Polyscias fulva* is a light demanding species (Omeja et al., 2001) but being in the plantation zone it is shedded and so its growth is suppressed.

Vegetation sampling from the study indicated a high number of indigenous tree species in the lower size-class in all the study sites. Large size class trees had low densities and this could be due to the low survival rate of seedlings into large trees. Hartshorn (1978) and Schulz (1960) have shown that size class distribution of tree diameters of tropical forests show a reverse J-shape or negative exponential distribution which is in agreement with the results of this study.

The high densities of seedlings of exotic species are an indication of higher initial recruitment in the lower size class (Richards, 1966). The fewer numbers of large size-class could be due to the high rate of larger tree harvesting by local people. Lejju (1999) also observed a similar trend of size class distribution in Mgahinga Gorilla National Park. The high densities of seedlings in all the three zones indicated the importance of the presence of propagules in determining the composition of early successional communities and their establishment.

Conclusion

The regeneration pattern of the indigenous tree species in Rwampara hills varied in each study site because of human disturbance which could have influenced seed dispersal mechanism, fruiting, germination and regeneration of tree species. The indigenous and exotic trees are very essential to the rural people and this was recognized from the resources harvested.

Conflict of interests

The authors did not declare any conflict of interest.

ACKNOWLEDGEMENTS

I thank my supervisor Dr Bunny Julius Lejju for the tireless guidance and support he gave me throughout my study. I am also very grateful to NEMA- Mbarara branch for the assistance rendered during fieldwork.

A vote of thanks goes to my research assistants in Rwampara who assisted me throughout data collection and necessary information to complete the study. I also thank the staff of Biology Department, Mbarara University of Science and Technology.

My sincere thanks go to Nalwanga Rose Mary who guided me in the review of this document.

Lastly, my thanks go to my husband, my parents, brothers, sisters, relatives and friends who gave financial and moral support towards the completion of this study.

REFERENCES

Bawa KS (1983). Patterns of flowering in tropical plants. In: Jones, C.E & Little, R.J (eds). Handbook of experimental pollination Biology. (Van Nostrand and Reinhold Co. New York). pp. 394-410.

Bazzaz FA (1984). Regeneration of Tropical forests: Physiological responses of pioneer and secondary species. In: Gomez-Pompa, A., Whitmore, T.C. & Hadley, M. (eds.). Rainforest Regeneration and management (Vol 6), New Jersey, USA.

Bibi F, Ali Z (2013). Measurement of diversity indices of avian communities at Taunsa barrage wildlife sanctuary, Pakistan. The J. Animal & Plant Sci. 23(2):469-474.

Grubb PJ (1977). The maintenance of species richness in Plant communities: The Importance of the regeneration niche. Biol. Rev. 52: 107 – 145.

Hartshorn GS (1978). Tree falls and tropical forest dynamics. In: P.B. Tomlinson & M. H. Zimmermannn (eds), Tropical trees as Living systems. Cambridge University Press, Cambridge. pp. 617-638.

Hooper E, Legendre P, Condit R (2004) Factors affecting community composition of forest regeneration in deforested, abandoned land in Panama. Ecology 85(12):3313-3326.

Janzen DH, Vazquez-Yanes C (1978). Aspects of Tropical seed ecology of relevance to management of tropical forested wild lands. In. A Gomez-Pompa, T.C. Whitmore and M. Hadley(eds.), Rain Forest Regeneration and Management, UNESCO, Paris. pp. 137-157.

Kasenene JM (1987). The influence of selective logging, felling Intensity and gap size on the regeneration of a tropical moist forest in Kibale forest reserve, Uganda. PhD dissertation, Michigan State University.

Katende AB, Birnie A, Tengrias BO (1995). Useful trees and shrubs for Uganda. Identification propagation & management for agriculture & Pastoral communities. Regional Soil Conservation Unit. Nairobi, Kenya.

Kent M, Coker P (1996). Vegetation Description and Analysis. A practical approach, John Wiley and Sons Ltd.

Lamb D (1990). Exploiting the Tropical Rainforest: An account of pulpwood logging in Papua New Guinea, University of Queensland, Australia.

Lejju BJ (1999). An assessment of the status of exotic plant species and Natural vegetation types of Mgahinga Gorilla National Park South Western Uganda, Msc Dissertation, Makerere University, Kampala.

Nepstad D, Uhl C, Serrao EA (1990). Surmounting barriers to forest regeneration in abandoned, highy degraded pastures: a case study from Paragominas, Para´, Brazil.

Olson J, Berry L (2003). Land degradation in Uganda: its extent and impact. Kampala, Uganda.

Omeja P, Obua J, Cunningham AB (2001) Regeneration, density and size class distribution of tree species used for drum making in Central Uganda. Makerere University, Kampala, Uganda

Richards PW (1966). The tropical Rain forest. Cambridge University Press, Cambridge.

Schulz JP (1960). Ecological studies on rainforest in northern suriname. Meded. Bot. Herb. Rijks. Univ. Utrecht 163:1-267.

Uhl C, Buschbacher R, Serra¨o EAS (1988). Abandoned pastures in eastern Amazonia. I. Patterns of plant succession. J. Ecol. 76:663-681.

22

A study on ecological distribution and community diversity of spiders in Gulmarg Wildlife Sanctuary of Kashmir Himalaya

Mansoor Ahmad Lone[1], Idrees Yousuf Dar[1] and G. A. Bhat[2]

[1]Department of Environmental Sciences, University of Kashmir, Srinagar -190 006, J&K, India.
[2]Department of Environmental Sciences, Centre of Research for Development, University of Kashmir, Srinagar -190 006, J&K, India.

The present study was an attempt to assess and evaluate the distribution, diversity and occurrence of spider community in Gulmarg Wildlife Sanctuary. India has 59 of the 110 spider families and at least, 1442 formally described species of the 39,000 known worldwide. Documenting spider assemblages assumes greater importance in the context of current rate of loss and degradation of forests which is known to have detrimental effect on many invertebrate groups. In order to assess the diversity and distribution of spiders at four sites during the months of May, June, July, October and December 2012, standard protocol was used to collect the spider community across the study area. The spider community was found to be represented by 18 taxa. Araneidae was dominant family followed by Lycosidae, Linyphiidae, Pholcidae, Salticidae, Sparassidae and Clubionidae. Differences in vegetation cover or human use showed variation in diversity and composition of spiders between different sites. Forest sites showed relatively higher diversity as compared to meadow sites.

Key words: Spider community, diversity, Araneidae, Gulmarg.

INTRODUCTION

Spiders form a diverse group of invertebrates in varied ecosystems and are known to be sensitive indicators of environmental change (Hodge and Vink, 2010). India has 59 of the 110 spider families and at least 1442 formally described species of the 39,000 known worldwide (Siliwal et al., 2009). Spiders also have an added advantage of being conspicuous, amenable to capture by relatively cheap, easily deployable and replicable techniques. These attributes make spiders as a group, suitable for statistical appraisal, comparisons and monitoring of sites or habitats. Arachnids are an important albeit poorly studied group of arthropods that play a significant role in the regulation of other invertebrate populations in most ecosystems (Russell-Smith, 1999). Spiders, which globally include about 42,055 described species (Platnick, 2011), are estimated to be about 60,000-170,000 species (Coddington and Levi, 1991). They include a significant portion of the terrestrial arthropod diversity, being one of the dominant macro invertebrate predator groups in terrestrial environments (35 - 95%) (Specht and Dondale, 1960; Van Hook, 1971; Moulder and Reichle, 1972; Edwards et al., 1976).

Spiders are copious in both natural and cultivated environments, in which their average annual abundance ranges from 50 to 150 individuals per square meter but can periodically reach maximal densities of more than 1000 individuals per square meter (Pearse, 1946; Duffey, 1962). They occupy a wide range of spatial and temporal niches, exhibit taxon and guild responses to environmental change, extreme sensitivity to small changes in habitat structure, primarily vegetation complexity and microclimate characteristics (Uetz, 1991). Furthermore, strong associations exist between plant architecture and species that capture prey without webs (Duffey, 1962; Uetz, 1991). Spiders respond distinctly to altered litter depth, and structural complexity and nutrient content of litter (Uetz, 1991; Bultman and Uetz, 1982). They employ a remarkable variety of predation strategies. As they are generalist predators, they are of immense economic importance to man because of their ability to suppress pest abundance in agro ecosystems. The population densities and species abundance of spider communities in agricultural fields can be as high as that in natural ecosystems (Riechert, 1981). In spite of this, they have not been treated as an important biological control agent since very little is known of the ecological role of spiders in pest control (Riechert and Lockley, 1984). Spiders regulate decomposer populations (Clarke and Grant, 1968) and by doing so, they influence ecosystem functioning (Lawrence and Wise, 2000, 2004). Their high biomass also makes them a critical resource for larger forest predators such as salamanders, small mammals and birds. Spiders can be used as successful biological indicators to assess the 'health' of an ecosystem because they can be easily identified and are differentially responsive to natural and anthropogenic disturbances (Pearce and Venier, 2006). For a species to be identified as an effective ecological indicator, it must meet the primary criteria of being feasible and cost effective to sample, easily and reliably identified, functionally significant, and have ability to respond to disturbance in a consistent manner. Spiders readily meet the first three criteria. Their high relative abundance, ease of collection and diversity in habitat preferences and foraging strategies allow for effective monitoring of site differences (Yen, 1995). Many studies have widely recommended the potential of spiders as bioindicators (Duchesne and McAlpine, 1993; Niemelä et al., 1993; Beaudry et al., 1997; Atlegrim et al., 1997; Churchill, 1997; Duchesne et al., 1999; Bromham et al., 1999; Werner and Raffa, 2000; Heyborne et al., 2003). This paper intends to study the diversity of spiders at different vegetation types.

MATERIALS AND METHODS

Study area

The study was conducted at Gulmarg (Figure 1), Gulmarg literally means 'meadow of flowers'. Gulmarg is a town, a hill station and Kashmir's premier ski resort. It is located 56 km south west of Srinagar. Gulmarg's legendary beauty, prime location and proximity to Srinagar naturally make it one of the premier charming luxury hill resorts in the country. The study sites selected had relatively different vegetation and anthropogenic impacts. Site-1 represented Drang Forest with geographical coordinates of N 34° 02' 04.0" and E 74° 24' 25" and an elevation of about 2328 m. The site was having dominant tree cover of *Pinus wallichiana* and *Picea smithiana*, while *Taxus baccata* was less prominent. The prominent shrubs were *Viburnum grandiflora* and *Geranum wallicianum*. Site-2 represented Drang Meadow (N 34° 03' 35.7" and E 74° 25' 31.7"; Elevation 2328 m). It was dominated by herbaceous vegetation but witnessed grazing and anthropogenic activities. Site-3 represented Gulmarg Forest (N 34° 02' 41.6" and E 74° 23' 09.3"; Elevation 2684 m). This site had a mixed type of vegetation dominated by *Populous migra*, *Rolinia pseudacacia* and dotted with *P. wallichiana* trees also. Site-4 represented Gulmarg meadow (N 34° 02' 51.6" and E 74° 23' 09.3"; Elevation 2687 m).

Spiders have been sampled using many methods, each with its own limitations, such as direct searches, pitfall traps, canopy fogging, vegetation beating, litter shifting or extraction, sweap net and suction sampling (Churchill and Arthru, 1999). Established sampling protocols for spider collection (Sorensen et al., 2002) were adopted in different sampling plots. The study was carried out using belt transects vegetation beating, pitfall traps and leaf litter extraction. Pitfall traps method was used to capture the spiders (Curtis, 1980; Kitching et al., 2000). The belt transects were of 10 m length and 2 m width with sampling restricted to the maximum height of 1 m. At each site, exercise was conducted for 30 min. Vegetation beating method is employed to collect spiders living in the shrub, high herb vegetation, bushes and small trees and branches (Coddington et al., 1996; Coddington and Levi, 1991). Spiders were collected by beating the vegetation with a stick and collecting the samples on a cloth (1 m²). The spiders were preserved in different vials filled with ethyl alcohol (75-80%) and marked using a piece of paper with the sample number.

Statistical analysis

No single index encompasses all characteristics of an ideal index, that is, high discriminate ability, low sensitivity to a sample size, and ease in calculation (Margurran, 1988). Therefore an observation of the different indices reflecting species evenness, dominance and diversity heterogeneity provide some valid viewpoints. Shannon's index of diversity (Price, 1997) reflects both evenness and richness (Colwell and Huston, 1991) and is commonly used in diversity studies (Krebs, 1989). It is calculated as $H = -\sum(ni/N)Ln(ni/N)$; i =1–n; where n is the number of species and Pi is the proportion of the ith species in the total. Index of dominance is calculated as $= \sum(ni/N)^2$ where ni is the number of individuals of a species and N is the total number of individuals of all species. Evenness indicates the degree of homogeneity in abundance between species and is based on the Shannon index of diversity. Shannon evenness [$E = H/H_{max} = H/lnS$; where H is the Shannon diversity index and S the number of species in the community] ranges from 0 to 1.

RESULTS

Taxonomical diversity

The spider community (order Araneae) was found to be represented by 18 taxa. Araneidae was a dominant family followed by Lycosidae, Linyphiidae, Pholcidae, Salticidae,

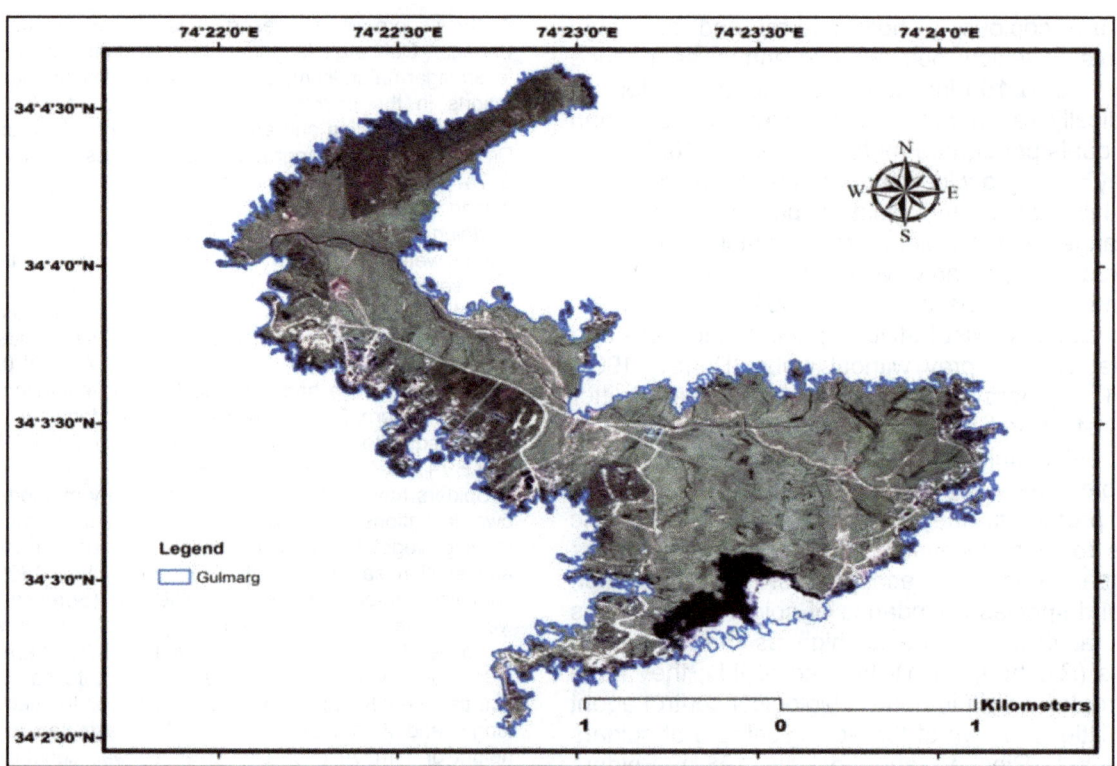

Figure 1. Satellite image of the study *area* (Gulmarg).

Table 1. Monthly variation in spider community density (Ind./m^2) at site I from May 2012-December 2012.

S/N	Taxa	May	June	July	October	December	Mean (n$_i$)
	Site I (Drang Forest)						
1	*Lycosa* sp.	6	4	0	0	1	2.2
2	*Araneus* sp.	2	4	4	3	2	3
3	*Obscuriphantes* sp.	0	2	0	0	0	0.4
4	*Stegodyphus* sp.	0	1	0	3	0	0.8
5	*Sparassus* sp.	0	0	2	4	0	1.2
6	*Lepthyphantes* sp.	0	0	1	0	0	0.2
7	*Pholcus* sp.	2	1	2	0	0	1
8	*Microlinphia* sp.	0	0	0	6	0	1.2
9	*Pardosa* sp.	0	0	0	2	1	0.6
	Total	10	12	9	18	4	10.6

Sparassidae and Clubionidae. Among the four sites selected, site I (Drang forest) showed the maximum number of taxa followed by site III (Gulmarg Forest), II (Drang meadow) and IV (Gulmarg meadow). At site I (Drang Forest) *Araneus* sp. was found to be dominant taxa throughout the study period. *Araneus* sp. recorded its maximum density (4 individual/m^2) in the month of July 2012 and lowered to 2 individual/m^2 in the month of December 2012. While the *Lepthyphantes* sp. was least dominant at site I having a maximum density (1

individual/m^2) in the month of July and was not recorded in the month of December (Table 1). At site II (Drang Meadow) *Lycosa* sp. and *Padosa* sp. were two dominant taxa throughout the sampling. In the month of June, *Lycosa* sp. showed the highest dominance (10 individuals/m^2) and was totally absent in the month of July. While *Salticus* sp. and *Thomisius* sp. were present only in the month of December (Table 2). At site III (Gulmarg Forest), *Lycosa* sp. was found to be dominant taxa throughout the study period. In the month of June;

Table 2. Monthly variation in spider community density (Ind. /m^2) at site II from May 2012-December 2012.

S/N	Taxa	May	June	July	October	December	Mean (n_i)
	Site II (Drang Meadow)						
1	*Lycosa* sp.	4	10	0	2	4	4
2	*Pardosa* sp.	4	6	0	4	0	2.8
3	*Microlinphia* sp.	0	0	3	4	0	1.4
4	*Salticus* sp.	0	0	0	4	0	0.8
5	*Thomisius* sp	0	0	0	6	0	1.2
	Total	8	16	3	20	4	10.2

Table 3. Monthly Variation in Spider Community Density (Ind./m^2) at Site III from May 2012-December 2012.

S/N	Taxa	May	June	July	October	December	Mean (n_i)
	Site III (Gulmarg Forest)						
1	*Lycosa* sp.	3	4	2	1	1	2.2
2	*Araneus* sp.	2	2	4	2	1	2.2
3	*Clubiona* sp.	1	0	2	0	0	0.6
4	*Dictyna* sp.	2	0	2	0	0	0.8
5	*Microlinyphia* sp.	0	0	2	0	0	0.4
6	*Salticus* sp.	0	0	4	0	0	0.8
7	*Loxosceles* sp.	0	0	4	0	0	0.8
8	*Pholcus* sp.	1	2	3	1	0	1.4
	Total	9	8	23	4	2	9.2

Table 4. Monthly variation in spider community density (Ind./m^2) at site IV from May 2012-December 2012.

S/N	Taxa	May	June	July	October	December	Mean (n_i)
	Site IV (Gulmarg Meadow)						
1	*Lycosa* sp.	15	2	2	1	0	4
2	*Pardosa* sp.	4	4	6	2	0	3.2
	Total	19	6	8	3	0	7.2

Lycosa sp. showed the highest dominance (4 individual/m^2) and lowest (1 individual/m^2) in the month of December. While *Clubiona* sp. was least dominant at site III having a maximum density (2 individuals/m^2) in the month of July and lowered to 0 individual/m^2 in the month of December (Table 3). At site IV (Gulmarg Meadow), only *Lycosa* sp. and *Pardosa* sp. were observed, out of which *Lycosa* sp. was found to be more dominant. In the month of May, *Lycosa* sp. showed the highest dominance (15 individual/m^2) but no individuals were recorded during December. *Pardosa* sp. was dominant in the month of July (6 individual/m^2) while no individuals were encountered in the month of December (Table 4). At site I (Drang Forest), *Araneus* sp. was found to be dominant taxa throughout the study period. *Araneus* sp. recorded its

maximum density (4 individual/m^2) in the month of July 2012 and lowered to 2 individual/m^2 in the month of December 2012. While *Lepthyphantes* sp. was least dominant at site I having a maximum density 1 individual/m^2 in the month of July and was not recorded in the month of December.

At site II (Drang Meadow), *Lycosa* sp. and *Padosa* sp. were two dominant taxa throughout the sampling. In the month of June, *Lycosa* sp. showed the highest dominance (10 individual/m^2) and was totally absent in the month of July. While *Salticus* sp. and *Thomisius* sp. were present only in the month of December.

At site III (Gulmarg Forest), *Lycosa* sp. was found to be dominant taxa throughout the study period. In the month of June, *Lycosa* sp. showed the highest dominance

(4 individual/m^2) and lowest (1 individual/m^2) in the month of December. While *Clubiona* sp. was least dominant at site 3 having a maximum density (2 individual/m^2) in the month of July and was absent in the month of June, October and December.

At site IV (Gulmarg Meadow), only *Lycosa* sp. and *Pardosa* sp. were observed, out of which *Lycosa* sp. was found to be more dominant. In the month of May, *Lycosa* sp. showed the highest dominance (15 individual/m^2) and lowered to 0 individual/m^2 in the month of December. While *Pardosa* sp. was dominant in the month of July (6 individual/m^2) and lowered to 0 individual/m^2 in the month of October and December.

Araneus sp. and *Lycosa* sp. were two dominant taxa throughout the study period; they are cosmopolitan in distribution and have high species diversity. However, the families like Lycosidae and Araneidae are more tolerant and overcome harsh climatic conditions and can survive in low temperature.

Also, site I (Drang forest) has high diversity than site III (Gulmarg forest), this may be due to the fact that the site I is away from the dwelling areas and its natural conditions while the site III which is a tourist spot is in a relatively more stress.

Also site II (Drang meadow) showed high diversity than site IV (Gulmarg meadow), the reason may be that in site IV, there is high anthropogenic and more biotic interferences taking place.

DISCUSSION

Spider community of the study area was found to be represented by 18 genera belonging to order Araneae. Araneidae was the dominant family followed by Lycosidae, Linyphiidae, Pholcidae, Salticidae, Sparassidae and Clubionidae. Among arthropods, spiders are the most abundant predators in many terrestrial ecosystems, playing an important role in ecosystem functioning throughout habitats (Van Hook, 1971). While spiders in forest ecosystems contribute to the maintenance of insect community equilibrium, the distribution of species and the composition of assemblages are significantly influenced by environmental conditions (Ziesche and Roth, 2008). Spiders seem well suited to discriminate habitat type and quality, since they play important role as diverse and abundant invertebrate predators in terrestrial ecosystems. Despite their ecological role in many ecosystems, high diversity, documented threats and the known imperilment of some species, spiders have received little attention from the conservation community (Skerl, 1999). While this lack of attention may be related to negative public attitudes towards spiders (Kellert, 1986), a paucity of compiled information on spider conservation status and distribution may be a more important issue. However, it is important that imperiled and vulnerable spiders and other invertebrates are not left out of conservation planning

efforts, as they may have unique ecological requirements or require particular site selection and management activities.

The diversity of spiders in the two forest sites was noted to be higher as compared to the two meadow sites. This may be due to the increased anthropogenic stress in the meadow areas which lead to the decrease in biodiversity and also the less availability of food in the meadow. Meadows are open areas in which there are high chances of predation. There are several other environmental factors that may also affect spider species diversity such as, spatial heterogeneity, competition, predation, habitat type, environmental stability and productivity (Rosenzweig, 1995). On the other hand, forests have large number of microhabitats which help spiders to escape there predators. Availability of food also effects diversity. In forests, food is available in abundance which is another reason why forests show high diversity as compared to meadow.

Also, the results showed that the number of individuals recorded from the sampling sites linearly decreased with the increasing altitude and also found that the family diversity showed a constant negative value with altitude. As spiders are sensitive to even small changes in the environment especially vegetation topography and climatic changes, patterns of linear decline may also be probably related to more severe climatic conditions terrain and landscape of study site. Similar results of spider abundance and declining linearly with elevation were observed in the studies of Otto and Swenson (1982) and McCoy (1990). Diversity is supposed to peak at mid elevation via primary productivity, which is considered to peak at mid elevations. The study provides information on spider community in different ecosystems and the effects of both biotic and a biotic factors, as well as anthropogenic impacts on diversity and distribution of these spiders. Different sites with differences in either vegetation cover or human use showed variation in diversity and composition of spiders. The number of individuals recorded from the sampling sites linearly decreased with the increasing altitude and also found that the family diversity showed a constant negative value with altitude. As was observed from the results of the study, altitude, habitat type and temperature play an important role in distribution and composition of spiders. Forests showed highest diversity as compared to meadow.

Gulmarg Wild Life sanctuary is interestingly diverse in spider fauna. During study, it was found that there have been less attention towards spiders in the state and therefore similar research in other parts of the Kashmir valley will surely provide information in this direction. It is also important to note that spider fauna is ubiquitous in nature and their diversity cannot be explained by quantifying one aspect of the environment. It does depend on many other factors or a combination of factors, apart from altitudinal variation and habitat structure. Looking into these factors would surely bring in more interesting results which can be relevant for maintenance and management of spider diversity of this region.

Conflict of interests

The authors did not declare any conflict of interest.

ACKNOWLEDGEMENT

The authors are highly grateful to Prof. Azra N. Kamili, Head, Department of Environmental Sciences, University of Kashmir, for providing the laboratory and other needed facilities.

REFERENCES

Atlegrim O, Sjoberg K, Ball JP (1997). Forestry effects on a boreal ground beetle community in spring: selective logging and clear-cutting compared. Entomol. Fenn. 8:19-26.

Beaudry S, Duchesne LC, Cote B (1997). Short-term effects of three forestry practices on carabid assemblages in a jack pine forest. Can. J. Forest Res. 27: 2065-2071.

Bromham L, Cardillo M, Bennett AF, Elgar, MA (1999). Effects of stock grazing on the ground invertebrate fauna of woodland remnants. Aust. J. Ecol. 24: 199-207.

Bultman TL, Uetz GW (1982). Abundance and community structure of forest floor spiders following litter manipulation. Oecologia 55:34-41.

Churchill TB (1997). Spiders as ecological indicators in the Australian tropics: family distribution patterns along rainfall and grazing gradients. In P.A. Selden, Ed., Proceedings of the 17th European Colloquium of Arachnology, Edinburgh.

Churchill TB, Arthur JM (1999). Measuring spider richness: effects of different sampling methods and spatial and temporal scales. Journal of Insect Conservation. Kluwer Academic Publishers, 4: 287-295.

Clarke RD, Grant PR (1968). An experimental study of the role of spiders as predators in a forest litter community. Ecology 1152-1154.

Coddington JA, Levi HW (1991). Systematics and evolution of spiders Araneae. Annu. Rev. Ecol. Syst. 22: 565-592.

Coddington JA, Young LH, Coyle FA (1996) Estimating spider species richness in a southern Appalachian cove hardwood forest. J Arachnol 24:111–128

Colwell RK, Huston MA (1991). Conceptual framework and research issues for species diversity at community level. In: Solbrig O.T. (Ed.), From genes to ecosystems: a research agenda for biodiversity. International Union of Biological Sciences, Paris, France, pp. 37-71.

Curtis DJ (1980). Pitfall in spider community. (Arachnida : Aranaea). J. Arachnol. 8: 280-281.

Duchesne LC, Mcalpine RS (1993). Using carabid beetles Coleoptera: Carabidae as a means to investigate the effect of forestry practices on soil diversity. Forestry Canada Petawawa National Forestry Institute, Chalk River, Ontario, Canada Report No. 16.

Duchesne LC, Lautenschlager RA, Bell FW (1999). Effects of clearcutting and plant competition control methods on carabid Coleoptera: Carabidae assemblages in north western Ontario. Environ. Monit. Assess. 56: 87-96.

Duffey E (1962). A population study of spiders in limestone grassland. J. Anim. Ecol. 31: 571–599.

Edwards CA, Butler CG, Lofty JR (1976). The Invertebrate fauna of the park grass plots II. Surface fauna. Rep. Rothamst. Exp. Stn. 1975, Part 2: 63–89.

Heyborne WH, Miller JC, Parsons GL (2003). Ground dwelling beetles and forest vegetation change over a 17-year period, in western Oregon, USA. Forest Ecol. Manage. 179:123-134.

Hodge S, Vink CK (2010). An evaluation of Lycosa hilaris as an

indicator of organophosphate insecticide contamination. New Zealand Plant Prot. 53:226-229

Kellert SR (1986). Social and perceptual factors in the preservation of animal species. In B.G. Norton, Ed., The Preservation of Species: The Value of Biological Diversity. Princeton University Press, Princeton, New Jersey.

Kitching RL, Vickerman G, Laidlaw M, Hurley K (2000). The comparative assessment of Arthropod and tree biodiversity in old world forest. The rainforest CRC/Earthwatch protocol manual. Cooperative Research Centre for Tropical Rainforest Ecology and Management. Technical Report Rainforest. CRC Crains

Krebs CJ (1989). Ecological methodology. Harper and Row, New York, USA.

Lawrence KL, Wise DH (2004). Unexpected indirect effect of spiders on the rate of litter disappearance in a deciduous forest. Pedobiologia 48: 149–157.

Lawrence KL, Wise DH (2000). Spider predation on forest-floor Collembola and evidence for indirect effects on decomposition. Pedobiologia 44: 33–39.

Margurran AE (1988). Ecological diversity and its measurement. Princeton University Press, Princeton, New Jersey, USA.

McCoy ED (1990). The distribution of insects along elevational gradients. Oikos 58: 313-322.

Moulder BC, Reichle DE (1972). Significance of spider predation in the energy dynamics of forest-floor arthropod communities. Ecol. Monogr. 42: 473-498.

Niemela J, Langor D, Spence JR (1993). Effects of clear-cut harvesting on boreal ground-beetle assemblages (Coleoptera: Carabidae) in western Canada. Conserv. Biol. 7: 551-556.

Otto C, Svensson BS (1982). Structure of communities of Ground living spiders along altitudinal gradients. Holarctic Ecol. 5: 35-47

Pearce JL, Venier LA (2006). The use of ground beetles Coleoptera: Carabidae and spiders Araneae as bioindicators of sustainable forest management: A review. Ecol. Indic. 6: 780–793.

Pearse A (1946). Observations on the microfauna of the Duke forest. Ecol. Monogr. 16: 127–150.

Platnick NI (2011).The world spider catalog, version 12.0. American Museum of Natural History.

Price PW (1997). Insect Ecology. 3rd ed. Wiley & Sons, New York.

Riechert SE (1981). The consequences of being territorial: spiders, a case study. Am. Nat. 117: 871-892.

Riechert SE, Lockley TC (1984). Spiders as biological control agents. Ann. Rev. Entomol. 29: 299-320.

Rosenzweig ML (1995). Species diversity in space and time. Cambridge, Cambridge University Press.

Russell-Smith A (1999). The Spiders of Mkomazi Game Reserve. In: M. Coe et al., Eds., Mkomazi: The Ecology, Biodiversity and Conservation of a Tanzanian Savanna. Royal Geographical Society, London.

Siliwal M, Molur S, Biswas BK. (2009). Indian Spiders (Arachnida: Aranaea): Update checklist 2009. Zoos Plant J. 20(10): 1999-2049.

Skerl KL (1999). Spiders in conservation planning: a survey of US natural heritage programs. J. Insect Conserv. 3:341-347.

Specht HB, Dondale CD (1960). Spider populations in New Jersey apple orchards. J. Econ. Entomol. 53: 810–814.

Uetz, GW (1991). Habitat structure and spider foraging. In: S.S. Bell, E.D. McCoy, H.R. Mushinsky, Eds., Habitat Structure: The physical arrangement of objects in space. Chapman and Hall, London, U.K.

Van Hook RI (1971). Energy and nutrient dynamics of spider and orthopteran populations in a grassland ecosystem. Ecol. Monogr. 41: 1–26.

Werner M, Raffa KF (2000). Effects of forest management practices on the diversity of ground occurring beetles in mixed northern hardwood forests of the Great Lakes region. For. Ecol. Manage. 139: 135-155.

Yen AI (1995). Australian spiders: An opportunity for conservation. Records of the Western Australian Museum Supplement 52: 39-47.

Ziesche T, Roth M (2008). Influence of environmental parameters on small-scale distribution of soil-dwelling spiders in forests: What makes the difference, tree species or microhabitat? Forest Ecol. Manage. 255:738-752.

23

Primary conifer succession on a 1915 mudflow in Lassen Volcanic National Park, California

Glenn Clinton Kroh[1], Rebecca Laura Upjohn[1,2] and John Edgar Pinder III[1,3]

[1]Department of Biology, Winton Scott, Room 401, Texas Christian University, Fort Worth, TX 76129, United States of America.
[2]Department of Ecosystem Science and Management, University of Wyoming, Laramie, Wyoming, WY 82070, United States of America.
[3]Department of Environmental and Radiological Health Sciences, Colorado State University, 305 W. Magnolia PMB 231; Fort Collins, CO 80521, United States of America.

Repeated observations of forest development using permanent plots can map pathways and rates of primary succession at the individual plant, the plot and the community level. This study re-measures the trees in 34 100 m^2 plots that were first sampled in 1987 to document recent and to predict continued forest development for a mixed-conifer forest established on a volcanic mudflow formed at a 2000-m elevation in Lassen Volcanic National Park (LVNP) in 1915. In 1987 and 2008, trees ≥ 0.11-m tall were identified to species level, and measured for height (m) and basal area (m^2). The most abundant species in both 1987 and 2008 were *Pinus contorta*, *Abies magnifica* and *Pinus monticola*, and there was no statistically significant difference in species composition despite a 20% increase in tree densities. From 1987 to 2008, the mean (± SE) proportional rates of increase for the number of trees per plot, the mean heights of trees and the total basal area per plot were, respectively, $0.009 \pm 0.002 \ y^{-1}$, $0.023 \pm 0.002 \ y^{-1}$, and $0.055 \pm 0.004 \ y^{-1}$. Despite these increases, canopy closure has not occurred for most of the forest. This lack of closure, in conjunction with the continuing similarity of relative species abundances, suggests that abiotic factors such as snow damage and drought and not biotic interactions such as competition may still be the major limitation to tree growth and forest development. Projecting the current rates of basal area growth for the next 10 to 20 years suggests rapid forest development that includes canopy closure. In expectation of these rapid changes, data on tree positions within plots were collected to allow the fates of individual trees to be monitored and determined.

Key words: *Pinus contorta*, conifers, primary succession, volcanism, mudflow.

INTRODUCTION

Active volcanic mountain ranges from much of the terrain in the American Pacific Northwest (Kiver, 1982; Harris, 1988), and where eruptions of those volcanoes involving pyroclastic blasts, lava flows, lahars, mudflows to landslides that can destroy existing, sometimes very old forests and create newly-formed mineral surfaces

(Dale et al., 2005b). Re-establishment of forests on these newly-formed surfaces through primary succession involves the establishment and growth of tree seedlings on inhospitable, bare surfaces of lava, cobble, gravel, muds, tephra or other rocky debris (Frenzen and Franklin, 1985; Drake, 1993; Larsen and Bliss, 1998; Dale et al., 2005a; Munoz-Jimenez et al., 2005). Although other forms of vegetation, including nitrogen-fixing herbaceous plants (Morris and Wood, 1989; Halpern et al., 1990; Walker et al., 2003; del Moral and Rozzell, 2005; Titus and Bishop, 2014), may either inhibit or facilitate the establishment of tree seedlings as they do for other herbaceous species, but the initial survival and growth of seedling and sapling trees may often be dependent upon their ability to survive harsh abiotic factors such as the physical extremes of temperature, light intensity, drought, snow pack and low nutrient availability (Turner, 1985; Alejandre-Melena et al., 2007; Deligne et al., 2013). Seedling and sapling trees may have to persist and grow slowly for decades before the they become large enough for their crowns to merge into a closed-canopy forest (Heath, 1967b).

As trees increase in size and the canopy of the forest closes, the factors limiting the survival and growth of trees may shift from these abiotic extremes of physical factors to biotic factors including competition for light and nutrients. This shift in the primary limiting factors may cause changes in the relative success of different individuals and different species (Peet and Christiansen, 1980; Walker and Chapin, 1987; Walker and del Moral, 2003). For example, competition for light among shade intolerant species such as pines and other conifers results in self-thinning (Dewar, 1993; Zeide 1995, 2001; Walker and del Moral, 2003) where there are greater mortality rates for shorter individuals whose crowns are in the forest's shaded lower subcanopy than for taller individuals with crowns in the forest's upper canopy. For intraspecific competition for light in some conifers such as *Pinus taeda* (VanderSchaaf, 2010), *Pseudotsuga menziensii* (Weiskittel et al., 2008) and *Pinus contorta* (Dean and Long, 1992), the progressive pattern of declining densities and increasing sizes of surviving individuals may follow predictable patterns where stands of initially different densities converge to similar densities and similar individual tree sizes.

The abilities to predict 1) the path of the plant succession and 2) the rate at which succession will produce forests that are either similar to or different from those that were destroyed is important in understanding: 1) the structure, composition and ecology of the Pacific Northwest forests (Acker et al., 1987) and 2) the processes of primary succession in general. Because of the long time frames required for forest succession, this process of forest development is usually inferred from historic reconstruction based on surviving trees (Johnson et al., 1994; Fastie, 1995) or from chrono sequences where forest development is compared among similar sites of varying ages (McCune and Allen, 1985; Drake and Mueller-Dumbois 1993; Clarkson, 1997; Walker et al., 2010).

It may be that studies involving repeated observations and measurements through time on permanent plots could be preferable to historic reconstructions and comparisons of sites of different ages (Bakker et al., 1996), and as similar studies of permanent plots have proven effective in understanding the processes involved in the establishment of herbaceous communities on volcanic terrains (Dale and Adams, 2003; Halpern et al., 1990; del Moral, 2007; del Moral et al., 2010, 2012; del Moral and Magnusson, 2014). Continued observations on permanent plots may also enable investigators to generate and test hypotheses on potential mechanisms driving succession (Bakker et al., 1996; Herben, 1996; Csecserits et al., 2007). Permanent plots have also been recommended as a possibly preferable method for monitoring changes through time in vegetation (Walker and del Moral, 2003; Chytry et al., 2014). Results from studies of forest development in repeatedly sampled permanent plots could be of fundamental importance in complementing and assessing the accuracies of historic reconstructions and comparisons of different-aged sites, but such long-term, permanent plot studies for tree species (Heath, 1971; Woods, 2000; Harcombe et al., 2002; Kangur et al., 2005; Weber et al., 2006) are less common than those for shorter-lived herbaceous and shrub communities.

This study reports on repeated observations of forest development using permanent plots that were originally established and studied in 1987 (Kroh et al., 2000) on a 1915 volcanic mudflow in Lassen Volcanic National Park (hereafter, LVNP) in northern California, USA (40° 30' N, 121° 26' W). These permanent plot measurements are complemented and supported by a combination of historical photographs, aerial photographs, unpublished National Park Service reports (Swartzlow, 1946; Bailey, 1961), theses and dissertations (Bailey, 1963; Heath, 1967a; Fessenden, 1984; Eppler, 1984; Upjohn, 2009), and published research (Heath, 1967b; Eppler, 1987; Parker, 1993) that have documented tree establishment and growth on the surface of the mudflow since 1940.

The purposes of the present study were: 1) to resample the 1987 permanent plots, 2) determine the rates of change in forest composition and structure, and 3) use these rates of change to predict the development of forest structure and composition that may be expected to occur in the next 10 or more years. This resampling was prompted by visits to the site in 2003 and 2008 which 1) suggested that the forest may be approaching canopy closure and the potential period of transition from primarily abiotic to primarily biotic limiting factors and 2) emphasized the need for documentation of forest structure and composition before the onset of canopy closure and its associated changes in forest structure had begun.

METHODOLOGY

Site description and history

The mudflow site is part of a larger area of volcanic disturbance termed the Devastated Area that was formed within LVNP by eruptions in May 1915. This area is located on the northeast slope of Lassen Peak at elevations > 1900-m amsl where the climate type is dry upland with winter snows and frequent summer droughts (Griffin, 1967). The predominant forest type at these elevations and aspects within LVNP is pine, especially *Pinus contorta* (Douglas; species nomenclature follows Hickman, 1993). Although *P. contorta* and other pines are the predominant forest type, there are local variations due to slopes and aspects (Parker, 1991).

Two separate days of eruptions were involved in the formation of the Devastated Area. During the first eruption on 19 May 1915, magma surfaced at the summit, melted the surrounding snowpack, and caused a fast-moving, water and mud based landslide, termed a lahar (Eppler, 1984, 1987), which flowed northerly along the floodplains of Lost Creek and Hat Creek destroying existing vegetation and leaving ≤ 1.5-m deep sediment deposits (Eppler, 1984). Additional eruptions on 22 May 1915 involved three main components including, in the sequence which they occurred: 1) a lateral blast of hot gasses that uprooted, broke-off or burned trees, seedlings and other vegetation (Swartslow, 1946) in a northeasterly direction for > 5 km; 2) a fast-moving lahar that deposited additional sediments on the previous deposited 19 May Lost Creek lahars; and 3) an apparently more viscous and slower-moving mudflow involving 10^6 m^3 of material (Eppler, 1984, 1987) that came to rest on sloping terrain south of Lost Creek (121° 28' 10" W, 40° 30' 45" N). Whereas, the previous lahars had followed the existing streambed of Lost Creek, most of this mudflow was deposited on top of upland terrains, whose nature is obscured by the depths of the deposits, or the previous lahar deposits. The shallowest deposits occurred on the mudflow's easternmost section of pre-existing soils where forest cover had been blown down by the earlier blast (Figure 1) (Kroh et al., 2000).

This mudflow is characterized by a discrete, lobate terminal margin with leading edges marked by steep-sided deposits ≥ 0.3-m thick (Eppler, 1984, 1987) and being composed of fractured, banded-dacite rocks containing alternating layers of white dacite and black andesite (MacDonald and Katsura, 1965). Although some portions of the site appear to have been deposited as a debris flow rather than a mudflow (Eppler, 1987), the use of the term "mudflow" follows Eppler's (1984) analysis. This nomenclature cannot be resolved with the lahar (including mudflow), debris flow and avalanche terms of del Moral and Grishin (1999) as the initial water content of the flow is unknown.

The mudflow surface is variable with 1) some areas of steep slopes, 2) the presence of up to 1-m diameter dacite blocks at higher elevations, 3) approximately 1-m deep craters formed by outgassing from buried hot boulders (Eppler, 1984), and 4) relatively flat terrains on the lower elevations. The nearby Lost Creek lahar deposits from the 19 and 22 May eruptions and their developing vegetation have been subsequently disturbed by stream erosion and overwash (Eppler, 1984; Heath, 1967b), but the mudflow surface has remained mostly free of these disturbances. In contrast to studies at other sites described as debris flows (Yoshida et al., 1997), mudflows (Frenzen et al., 2005) or lahars (Larsen and Bliss, 1998), this mudflow did not include patches of trees that could provide a ready source of propagules to initiate forest recovery.

Previous studies of forest development on the mudflow

Early studies of plant community development on the mudflow used various plot designs and were mostly conducted on the lower elevations of the mudflow on relatively flat terrains. Swartslow (1946) sampling in a 30.3 x 30.3 m plot in 1936, 1940 and 1946 reported approximately 30,000 forbaceous individuals per ha (3 individuals per m^2) with the species *Ericameria bloomer* (A. Gray) J. F. Macbr. and *Calyptridium umbellatum* ((Torrey) E. Greene each comprising approximately 40% of the individuals. The lupine species, *Lupinus andersonii* (S. Watson) and *Lupinus grayi* (S. Watson), had densities of 839 and 279 individuals per ha, respectively. A single *P. contorta* was observed in the plot. Bailey (1961), sampling during 1960 in a different 30.3 x 30.3 m plot, mapped and measured the heights of 181 conifers (1,970 trees ha[-1]). Relative conifer abundances on this plot decreased in the order *Abies* spp., *P. contorta*, *Tsuja mertensiana* (Bong.) Carriere, *Pinus monticola* (Douglas) and *Pinus jeffreyi* (Grev. & Balf.). Tree heights ranged from 0.15 m to one 9.1-m tall *P. contorta*. Bailey (1961) noted that conifers were approximately four times more numerous than lupines and commented that "shifting wind currents ... could apparently have brought seeds in from all directions...". Bailey (1963) also reported tree densities and maximum tree heights in 1962 for twenty, contiguous 1.83 x 18.3 m sections of a belt transect (CTT4) crossing the lower elevations where tree densities were 1,433 trees per ha with 69% being *P. contorta* < 4-m tall and 18% *Abies* spp. < 2-m tall. Heath (1967a) mapped and measured the heights of 251 conifers in a 3 m by 152 m belt transect in 1963 where the relative abundances of the species were *P. contorta*, *Abies concolor* ((Gordon & Glend.) Lindley), *P. monticola*, *Abies magnifica* (Andr. Murray), *T. mertensiana* and *P. jeffreyi*. Tree heights ranged up to 4.27-m tall for *P. contorta* and *P. monticola*. Although the results of Swartzlow (1946), Bailey (1961, 1963) and Heath (1967a) are based on sampling of single, relatively small square or belt-shaped areas, their results are consistent with the general appearance of the mudflow site in panchromatic aerial photographs from 1941 and 1966, respectively.

To sample the tree, shrub and forb vegetation on the mudflow in 1982, Fessenden (1984) used 5-m radius, circular plots spaced 31-m apart along a series of south to north transects arranged at intervals of 61 m. Fessenden (1984) divided the mudflow into a Dense Forest Margin (hereafter, DFM) and a Sparse Central Forest (hereafter SCF) areas. The SCF corresponds to the area where the mudflow was deposited over previous lahar sediments (Kroh et al., 2000), whereas the DFM corresponds to the area of mudflow deposits over preexisting soils. In the 34 SCF plots, tree densities were 1,946 individuals per ha with *P. contorta*, *A. magnifica* and *P. monticola* accounting for 61, 19 and 13%, respectively, of the trees. Tree heights and stem diameters were not reported.

Initial permanent plot sampling in 1987

In 1987, Kroh et al. (2000) established four parallel transects spaced at intervals of 200 m that 1) crossed the full width of the mudflow and 2) were oriented at right angles across the direction of the flow's path. These transects ranged from the uppermost elevations (H1) down to the lowest elevation (H4) which had been studied by Swartzlow (1946), Bailey (1961, 1963) and Heath (1967a, 1967b), and the area sampled on transects H2, H3 and H4 corresponded to Fessenden's SCF (Kroh et al., 2000). Circular 100-m^2 (radius = 5.64 m) sample plots were established at 20-m intervals along each transect for a total of 70 plots. Global Positioning Systems using post-processing, differential corrections (El-Rabbany, 2002; Dodd, 2011) were used to determine plot locations within the positional accuracies necessary for sequential sampling of permanent plots (Ross et al., 2010; Dodd, 2011). Plot elevations were measured using single-based barometric surveying (Davis et al., 1981). The mean elevations for the plots on transects H1 through H4 were 2033, 2005, 1989 and 1977 m amsl, respectively. More complete descriptions of transects and plots and

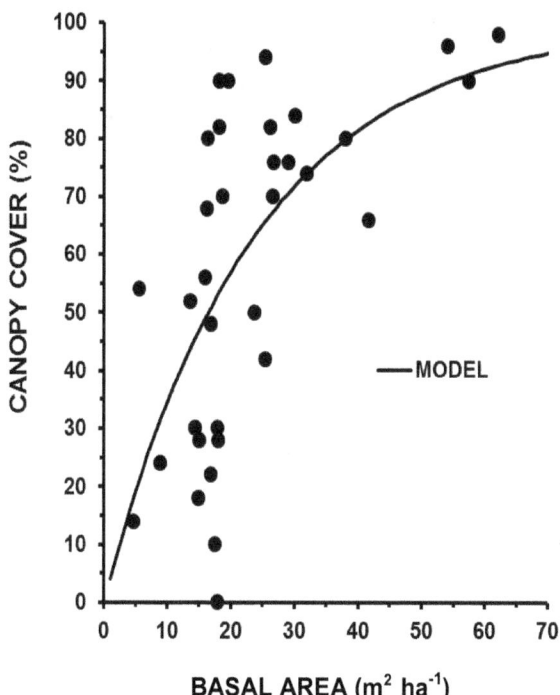

Figure 1. The relationship between plot basal area and canopy cover for 34 plots on the mudflow.

maps of the site are available in Kroh et al. (2000).

Each plot was visited in 1987 with all living trees being identified to species level and measured for height and, for trees ≥ 1.37 m tall, diameter-at-breast height (hereafter, dbh). Tree heights ≤2 m were measured with stadia rods. Heights > 2 m were measured trigonometrically based on line-of-sight angles to the base and the top of the tree from a known distance. Calipers were used to measure dbh. Trees with dbh ≥ 80 mm had a 5.15-mm diameter increment core extracted as near to the ground surface as possible for the determination of age by ring counts (Kroh et al., 2000 for methods of determining ring counts). Trees too small for increment coring were aged by counting whorls that demark annual increments of height growth. Few dead trees were seen, and these were not measured. In all procedures, care was taken to minimize damage to the trees and disturbances to the mudflow surface. A number of trees > 3-m tall were labelled for subsequent identification using numbered, metal tags hung loosely on the tree with rubber-shielded wire.

The 1987 sampling indicated a short-stature (median tree heights < 2 m), dense (3,507 trees ha^{-1}) forest of conifers who were mostly ≤ 50 year-old (Kroh et al., 2000). The species composition was, as listed in order of decreasing abundance, *P. contorta*, *A. magnifica*, *P. monticola*, *A. concolor*, *P. jeffreyi* and *T. mertensiana* (Kroh et al., 2000). This conifer assemblage contains a mix of shade intolerant and shade tolerant species with intolerance to shade increasing in the order *T. mertensiana* , *A. concolor*, *A. magnifica*, *P. monticola* and *P. contorta* (Minore, 1979).

These species differed in their abundances, but their age distributions indicated similar dates of first invasions with each species being present by the mid-1940s (Kroh et al., 2000). The predominance of ≤50-year-old trees was consistent with photographic evidence showing the development of large, presumably seed-producing, trees beginning in 1940 on the previously denuded, blast-effected, upland areas surrounding the mudflow (Kroh et al., 2000). The age distributions measured by

Kroh et al. (2000) were also consistent with the pattern of tree invasion apparent in historical photographs (Stillman and Turnage, 1962) of known dates (Kroh et al., 2000).

For *P. contorta*, there were gradients of increasing densities and increasing basal areas from the upper transects to the lower transects, and these gradients contributed to similar overall gradients for the total forest. There were also other less pronounced gradients in species abundances with elevation that included an increase in densities of *A. concolor* at lower elevations.

Permanent plot sampling in 2008

Forest structure in 2008 was determined for every second plot along these transects because the 20-m interval between plot centers may not prove sufficient to ensure future independence between large trees in neighboring plots. Thus, data are reported for only 34 of the 70 plots. Heights for every tree ≥ 0.11-m tall and dbh's for each tree ≥ 1.37-m tall were measured using methods consistent with those in 1987. The trees were also inspected for damage such as broken, bent or dead upper main stems, and the presence of female cones. In addition, the position of all trees that were ≥ 0.11-m tall were determined by measuring the compass bearing to and the distance from the center of the plot. A few dead trees were observed in the forest, but none of these occurred in the sampled plots. Canopy cover, expressed as the percent of sky obscured by the forest canopy, was determined as the mean of two concave spherical densitometer (Lemmon, 1956) measurements taken from the plot center while alternatively facing east and west.

Statistical procedures

Statistical procedures were performed using version 9.3 of the Statistical Analysis System (hereafter SAS; Der G, 2001) using a probability (P) level of 0.05 for statistical significance. Paired observations of data between plots from 1987 and 2008 were analyzed using t-tests of paired observations. Analyses of non-paired observations were conducted using analysis of variance (hereafter ANOVA; Milliken and Johnson, 1984). Where significant differences among means were indicated by ANOVA F-tests, the differences between means required for statistical significance was computed using the T-method of multiple comparisons for unequal sample sizes (Spjotvoll and Stoline, 1973; Milliken and Johnson, 1984). Correlations were evaluated using Spearman rank correlations which are less sensitive to departures from normality than Pearson product-moment correlations (Conover, 1971). Tests to determine statistically significant differences among proportional abundances were performed using likelihood-ratio Chi-square procedures (Conover, 1971). Nonlinear regressions were performed using PROC NLIN of SAS (SAS Institute Inc., 2003).

RESULTS

To compare species abundances and tree sizes between the 1987 and 2008 plot data, all the < 0.11-m tall trees in the 1987 data were deleted to obtain corresponding ranges in height data from the two sample periods. Only 58 trees < 0.11-m tall, including 39 *P. contorta*, were removed from the 1987 data. There was an increase in the total number of trees ≥ 0.11-m tall from 1,167 in 1987 to 1,398 in 2008 (Table 1), and there were also increases for each of the species (Table 1). There was no statistically significant difference in the relative abundances among the species between years (likelihood ratio χ^2 = 1.55; P > 0.05).

Table 1. The number of trees of each of the six conifer species observed in the 34 100-m2 plots in 1987 and 2008.

Species	Number Observed	
	1987	2008
Abies concolor	25	40
Abies magnifica	172	206
Pinus contorta	797	939
Pinus jeffreyi	24	31
Pinus monticola	130	158
Tsuga mertensiana	19	24

Comparisons of tree densities, heights and basal areas between 1987 and 2008

The mean tree density, mean tree height and mean sum of tree basal areas in the sample plots for the four transects are summarized in Table 2 for 1987 and 2008. The differences among transects for the 1987 data in Table 2 were statistically significant as indicated by differences between means being greater than the T-interval, and there were gradients of increasing density and increasing basal area from H1 to H4. For the 2008 samples, there were no statistically significant differences among transects for any of these variables, and the ratios of largest to smallest means were all < 2, whereas similar ranges among transect means for these 34 plots in the 1987 data were all > 2 with the ratio for basal area being > 4.

Tree densities in 2008 were greater than those in 1987 for all but 5 plots. The mean (± SE) tree densities per plot for 1987 and 2008 were, respectively, 34.3 ± 2.9 trees/100 m^2 (3,430 ± 290 trees ha^{-1}) and 41.2 ± 3.2 trees/100 m^2 (4,120 ± 320 trees ha^{-1}) and mean (± SE) densities per plot increased significantly (t-test of paired observations t = 4.29; df = 33; P < 0.01) by 6.8 ± 1.6 trees (680 ± 160 trees ha^{-1}).

The changes in densities may reflect 1) the mortality of trees that were ≥ 0.11-m tall in 1987, 2) the growth of 1987 trees that were < 0.11-m tall to ≥ 0.11-m tall in 2008 or 3) the germination and growth of trees that were not present in 1987. The relative importance of these processes cannot be completely evaluated from the currently available data because quantifying these alternative processes requires being able to identify the fates of individual trees. However, because 1) the number of trees increased by 231 (1398 − 1167) while 2) only 58 trees < 0.11-m tall were removed from the 1987 data, most (≤ 231 − 58) of the increase in densities must represent new individuals who germinated and grew to ≥ 0.11-m tall during the intervening 21 years.

Although all conifer species had increased numbers of individuals, female cones were only observed on a single *P. monticola* and 22% of the *P. contorta*. The female cones on the *P. contorta* included those produced in previous years, but it is not clear what proportion, if any, of these cones had released seeds.

The mean (±SE) proportional rate of increase in density among plots, estimated as the mean of the logarithms of density in 2008 minus that in 1987 divided by 21 years, for each of the plots, was 0.009 ± 0.002 y^{-1}, and there were no statistically significant differences in mean proportional rates of increasing density among transects (Table 3; ANOVA F = 0.80; df = 3, 30; P > 0.05). The proportional rates of increase for the more abundant species were also similar to this mean value.

The mean height of the trees per plot in 2008 was greater than in 1987 for all of the plots. Mean (± SE) heights per plot in 2008 and 1987 were, respectively 3.23 ± 0.35 m and 1.98 ± 0.23 m, and the mean (± SE) height increase per plot of 1.25 ± 0.03 m was statistically significant as indicated by 1) a t-test of paired observations (t = 7.87, df = 33; P < 0.01) and 2) the increase in all of the 34 plots. The mean (± SE) proportional rate of height increases (y^{-1}), estimated as the mean of the logarithm of mean height in 2008 minus that in 1987 divided by 21 years for each of the plots, was 0.023 (± 0.002) y^{-1}, and there were no statistically significant differences in mean proportional height growth rates among the four transects (Table 3; ANOVA F = 0.53; df = 3, 30; P > 0.05).

There were increases in mean tree heights between 1987 and 2008 for all species except *T. mertensiana*. The median, mean and maximum tree heights for each species in each year of sampling are summarized in Table 4, and the 2008 mean heights are generally > 1.3 times those in 1987. The largest increase in mean heights of 1.07 m occurred for *P. contorta*, but this increase was similar to those for the other two most abundant species *P. monticola* (0.90 m) and *A. magnifica* (0.87 m). A more involved interpretation of these changes in mean heights is complicated by the addition of trees that were established after 1987, but the data do suggest similar increases in mean heights among the most abundant species over the 21-year period.

This similarity in height growth among species was also observed for 96 tagged trees located within the 34 plots. Mean proportional rates of height growth (Table 5) ranged from 0.019 y^{-1} for *P. jeffreyi* to 0.031 y^{-1} for a small number of *A. concolor*, and there was no significant difference in rates among species in either height growth (F = 1.09; df = 4, 91; P > 0.05) or basal area (F = 2.41; df = 4, 91; P > 0.05). The mean (± SE) proportional height growth rate for all 96 trees was 0.022 (± 0.001) y^{-1} which is similar to the mean proportional growth rates per plot (Table 3).

Mean tree heights increased in all plots, but decreases in mean heights could have occurred due to 1) additions of numerous newer and smaller trees or 2) damage to existing trees. There was a statistically significant Spearman Rank Correlation of -0.38 (n = 34; P < 0.05) between the increases in tree numbers and the increases in mean tree heights among plots, and a plot where the

Table 2. Means ± standard errors of tree density (number per m²), mean tree height (m), and total basal area per plot (m² ha⁻¹) for transects H1 though H4.

Transect	Number of plots	Tree densities Mean ± SE	Tree heights Mean ± SE	Plot basal areas Mean ± SE
			1987	
H1	7	22.7 ± 1.9	1.29 ± 0.18	3.14 ± 0.63
H2	10	319 ± 5.0	1.86 ± 0.26	6.73 ± 0.92
H3	8	31.6 ± 5.6	2.96 ± 0.73	11.6 ± 2.6
H4	9	48.4 ± 5.7	1.87 ± 0.40	12.9 ± 3.8
T-interval		21.5	1.62	22.6

Transect	Number of plots	Tree densities Mean ± SE	Tree heights Mean ± SE	Plot basal areas Mean ± SE
			2008	
H1	7	30.0 ± 3.1	2.29 ± 0.41	14.6 ± 1.7
H2	10	39.2 ± 5.9	3.31 ± 0.59	21.6 ± 2.2
H3	8	38.8 ± 5.5	4.37 ± 0.98	30.8 ± 5.1
H4	9	54.1 ± 6.9	2.89 ± 0.66	26.5 ± 6.3
T-interval		29.9	3.58	22.6

Table 3. Mean ± SE of proportional rates of increase (y^{-1}) in tree densities, mean tree heights per plot, and total basal areas per plot for transects H1 though H4.

Transect	Number of plots	Tree Densities Mean ± SE	Tree Heights Mean ± SE	Plot Basal Areas Mean ± SE
H1	7	0.012 ± 0.005	0.026 ± 0.003	0.079 ± 0.008
H2	10	0.009 ± 0.005	0.025 ± 0.004	0.057 ± 0.005
H3	8	0.011 ± 0.003	0.022 ± 0.003	0.051 ± 0.007
H4	9	0.005 ± 0.002	0.021 ± 0.002	0.039 ± 0.006
T-interval		0.042	0.009	0.017

density increased from 29 to 72 had an increase in mean height of only 0.14 m. Also, approximately 32% of all trees showed some indication of damage that may have affected height growth. Damage, which was primarily in the form of bent or broken upper stems, suggested physical damage or stress from wind, deep snow accumulation, or down slope movement of the upper portions of the winter snow pack. Fessenden (1984) noted wind and snow damage and reported two episodes of avalanches on the mudflow in 1972 and 1982. Damage varied significantly among species (likelihood ratio χ^2 = 246.9; df = 12; P < 0.01) with the most abundant species *P. contorta*, *A. magnifica* and *P. monticola* having damage percentages of 36, 22 and 32%, respectively. Although the additions of new small trees and the damage to existing trees were not sufficient to result in reductions in the mean heights for plots, they probably affected mean height growth rates.

The total basal area per plot, computed as the sum of the basal areas for all the sufficiently tall trees, was greater in 2008 than in 1987 for all the plots (Table 4). The mean (± SE) total basal area per plot in 2008 and 1987 were, respectively, 23.6 ± 2.3 m² ha⁻¹ and 8.8 ± 1.3 m² ha⁻¹, and the mean (± SE) increase in basal area of 14.9 ± 1.2 m² ha⁻¹ was statistically significant as indicated by 1) a t-test of paired observations (t = 12.3; df = 33; P < 0.01) and 2) an increase in all of the 34 plots. The increase in basal areas represents 1) the continual growth of trees with measurable dbh in 1987 and 2) the development of measurable basal area for trees that were < 1.37-m tall in 1987 but were > 1.37-m tall in 2008. The number of trees with measurable basal areas increased from 315 in 1987 to 733 in 2008 with all species contributing to this increase. The mean (± SE) proportional rate of increase in total basal area for the plots was 0.055 ± 0.004 y^{-1}, and varied significantly among transects (Table 3; ANOVA F = 6.77; df = 3, 30; P < 0.01).

The mean (± SE) proportional rate of increase in basal areas for the 96 tagged trees (Table 5) was

Table 4. Tree heights (m) for the species in the 34 plots on the mudflow in 1987 and 2008. SD = standard deviation.

Species	Number of trees	Minimum	Median	Mean ± SD	Maximum
1987					
A. concolor	25	0.19	0.99	1.48 ± 2.57	6.20
A. magnifica	172	0.12	1.12	1.95 ± 2.76	14.7
P. contorta	797	0.11	0.82	1.51 ± 1.98	13.2
P. jefferyi	24	0.12	1.98	2.76 ± 2.51	8.90
P. monticola	130	0.11	1.80	2.83 ± 2.01	15.1
T. mertensiana	19	0.12	1.40	1.27 ± 1.54	5.10
2008					
A. concolor	40	0.11	0.92	1.98 ± 2.76	10.7
A. magnifica	206	0.11	1.54	2.82 ± 3.44	19.2
P. contorta	940	0.11	1.49	2.56 ± 2.92	21.8
P. jefferyi	31	0.14	2.43	3.66 ± 3.67	13.6
P. monticola	158	0.11	1.52	3.73 ± 4.00	19.0
T. mertensiana	24	0.29	1.53	1.23 ± 1.54	5.74

Table 5. Mean (± SE) proportional height and basal area growth rates from 1987 to 2008 for 96 tagged trees from the 34 plots. Proportional height and basal area growth rates have units of y^{-1}. There were no statistically significant differences among species for either height growth rates (ANOVA F = 1.09; df = 4, 91; P > 0.05) or basal area growth rates (ANOVA F = 2.41; df = 4, 91; P > 0.05), and T-intervals are not reported because of the large variation in sample sizes.

Species	No. of Trees	Mean (± SE) Height	Proportional growth rates Basal Area
A. concolor	2	0.031 ± 0.006	0.041 ± 0.009
A. magnifica	16	0.021 ± 0.002	0.043 ± 0.005
P. contorta	57	0.022 ± 0.001	0.031 ± 0.002
P. jefferyi	6	0.019 ± 0.004	0.032 ± 0.006
P. monticola	15	0.021 + 0.002	0.029 ± 0.005

Table 6. Mean ± SE of canopy cover measured as percent of the sky obscured by conifer foliage for the four transects

Transect	Number of plots	Canopy cover Mean ± SE
H1	7	25.1 ± 7.8
H2	10	55.6 ± 7.4
H3	8	75.5 ± 7.4
H4	9	70.7 ± 8.4
T-interval		40.3

Canopy cover

Canopy cover ranged from 0 to 98% with a mean (± SE) cover per plot of 58 ± 5% and a median of 67% (Table 6). Canopy cover differed significantly among transects (Table 5; ANOVA on arcsin transformations of percentages F = 6.89, df = 3, 30; P < 0.01) with canopy closure increasing from H1 through H4. There was also a statistically significant (F = 251.4; df = 1, 33; P < 0.001; r^2 = 0.884) asymptotic, nonlinear relationship between basal area per plot and plot canopy cover (Figure 1) of the form,

$$Y = 100\% * [1 - e^{(-b * x)}] \tag{1}$$

where Y = percent canopy closure, x = basal area in m^2 ha^{-1} and b (± SE) = 0.0421 ± 0.005. Most (29 of 34) predicted percent covers were within a factor of 2 of the observed percent covers.

There were changes in plot basal area growth rates

0.033 (± 0.002) y^{-1} with no significant differences in rates among species (F = 2.41; df = 4, 19; P > 0.05). The smaller proportional rate of basal area increase for the tagged trees relative to the increases for the plot basal areas (Table 3) reflects the increase in the number of trees that contributed to the total basal area in the plots.

associated with increasing canopy cover. The proportional rate of increase in basal area per plot declined with increasing canopy cover (r_s = -0.640; n = 34; P < 0.001). For every 1% greater canopy cover, there was a 0.00047 (SE = ± 0.0001) y^{-1} decline in the proportional rate of basal area increase. To what extent this decline in rates may be due to 1) shading between larger trees or 2) shading effects of larger trees on smaller trees cannot be evaluated from the present data and would require growth data for individual trees. There was no significant correlation between canopy cover and the proportional rates of increases for densities (r_s = 0.176; n = 34; P > 0.05) or mean tree heights (r_s = -0.145; n = 34; P > 0.05).

Spherical densitometer measures of canopy cover include the attenuation of light passing through the canopy at oblique angles. Thus, they can substantially overestimate cover when compared to devices which measure the proportion of open canopy directly above a number of separate positions (Brunnel and Vales, 1990; Cook et al., 1995). However, Nuttle (1997) has argued that spherical densitometers may provide more appropriate measures of the various influences of the canopy on affecting the light that does not penetrate to near the ground surface. It is the light that does not reach the ground surface that will limit the success of seedlings and smaller trees. Therefore, the use of spherical densitometer readings, which account for the attenuation of light approaching at oblique angles, is more appropriate for this study. Furthermore, the correspondences of the densitometer readings with plot basal areas and proportional rates of increase in plot basal areas suggests that these readings are informative measures of the impact of the larger trees on the attenuation of light resources for the smaller trees.

DISCUSSION

The changes between the 1987 forest and the 2008 forest involved mostly greater increases in mean heights (by a factor of 1.6 = 3.23 m / 1.98 m) and basal areas per plot (by a factor of 2.7 = 23.6 m^2 ha^{-1} / 8.8 m^2 ha^{-1}) and less pronounced, smaller increases in tree densities (by a factor of 1.2 = 4,120 trees ha^{-1} / 3,430 trees ha^{-1}). Thus, the 2008 forest is primarily a larger, rather than a denser, version of the 1987 forest with a relatively unchanged species composition. The forest also appears to be becoming more uniform as suggested by the reductions in differences among transects in mean density, height and basal area from 1987 to 2008 (Table 2).

There was little indication that the forest becoming larger involved the competitive displacement or inhibition of one species by another. There was little evidence that increases in densities or heights by one of the more abundant species resulted in declining densities or (with the possible exception of T. mertensiana) notably slower rates of height growth for the less abundant species. Perhaps an appropriate description of the forest develop-

ment over these 21 years is that it approximates a transient equilibrium (or steady state) where ratios of abundances among species and ratios of sizes among species are remaining relatively constant, but the magnitudes of abundances and sizes, are not remaining constant.

There was little indication in 2008 that approaching canopy closure had begun to modify forest composition. Even where canopy cover was nearly complete, there was no indication of declining densities due to self-thinning. There was little evidence of tree mortality from any cause as dead trees were not observed in the sampled plots.

Kroh et al. (2000) described the 1987 forest composition as being typical of the Upper Montane Coniferous forests which commonly occur at elevations from 1800 to 2400 m in the Northern Sierra Mountains and the Southern Cascade Range (Rundel et al., 1990), and the relatively small changes in species compositions between 1987 and 2008 suggest a continuing similarity to this forest type. These forests typically have basal areas ranging from 50 to 70 m^2 ha^{-1} (Barbour, 1988; Parker, 1991), and the 2008 mean basal area of 23.6 m^2 ha^{-1} illustrates the remaining forest development that may be expected to occur.

The species compositions of the 1987 and 2008 forests are similar to 1) the 1963 forests described by Heath (1967a) and 2) the 1982 data for Fessenden's (1984) SCF. They are similar in that P. contorta is clearly more abundant than the other conifer species. The major difference in composition between Heath's (1967a) 1963 transect data and the 1987 and 2008 samplings is that A. concolor is more abundant than A. magnifica in Heath's data, whereas A. magnifica is the more abundant species in 1987 and 2008. However, this does not necessarily imply a changing forest composition because Heath's 1963 transect was near the lower elevation transects H3 and H4 where A. concolor is relatively more abundant than in the upper transects.

Whatever the factors were that established the initial species composition of the forest; this initial composition has largely persisted with mostly minor changes involving the relative abundances of species. In the various measures of species composition from on the mudflow from 1960 through 2008, the percent composition of P. contorta 1) ranged from 43 to 69%, and 2) there was indication of any consistent increasing or decreasing trend. The 1) similar proportional increases in the densities of the conifer species (Table 1) despite 2) a lack of obvious evidence of seed production by trees on the mudflow suggests that the increases in the tree numbers may continue to be determined more by external seed deposition from neighboring tree stands. There are patches of large, seed-producing trees which occur at higher elevations within several hundred meters south of the mudflow forest (Kroh, pers. obs.). These trees are observable in aerial photography beginning in 1940.

Thus, it is possible that 1) the combination of external seed input and 2) the absence of pronounced tree mortality may be responsible for the continuing similarity in conifer species composition across years.

Expectations of possible changing abundances for *P. contorta* and the *Abies* species

While the canopy cover by 2008 had not produced the anticipated mortality and declining densities indicative of increasing competition, these types of declines have been documented for nearby sites on more fertile lahar deposits. Parker (1993), using plots of 50.6 x 50.6 m, has demonstrated this expected shift from abiotic to biotic factors by contrasting *P. contorta* populations on the Hat Creek lahar deposits formed during the 19 May 1915 eruption (site 5 in Parker, 1993) with those on a part of the mudflow located near transect H4 (site 8 in Parker, 1993). These two sites are about 2-km apart with the Hat Creek site being approximately 100-m lower in elevation. The lahar site is predominantly sand whereas the mudflow site is primarily gravel-sized dacite, and the lahar site had at least 25% more soil nitrogen and phosphorus than the mudflow site (Parker, 1993).

Although both these sites were approximately 70 years-old, at the time of sampling, there were pronounced differences in their forest structures. The forest on the lahar site had 1) a density of 2,200 live *P. contorta* trees per ha, 2) a basal area of 15.3 m^2 ha^{-1}, 3) a visually estimated canopy closure of 91%, and 4) 97% of the *P. contorta* trees were > 1.4-m tall. In addition to the live trees, there were 1,100 standing-dead *P. contorta* trees per ha. In contrast, the mudflow site had 1) a density of 840 live *P. contorta* per ha, 2) a canopy closure of 16%, and 3) less than 10% of the *P. contorta* were ≥ 1.4-m tall, and 4) there were no standing dead trees. Thus, the lahar site represents a forest with canopy closure where competition for light is likely causing mortality among the smaller, shade intolerant *P. contorta*.

Factors contributing to shade-induced mortality in *P. contorta* may include 1) the lower ability of shaded crowns to draw water to the leaves (Reid et al., 2003), 2) wind related physical damage to trees with shade-induced spindly growth forms (Long and Smith, 1992; Rudnicki et al., 2001), and 3) snow damage to smaller and more spindly trees (Teste and Lieffors, 2011). In shaded conditions, *P. contorta* show a proportionately greater reduction in lateral branch growth relative to vertical growth than other similar species (Chen et al., 1996), and these changes result in a more spindly growth form than that for more shade tolerant species which have proportionately smaller reductions in lateral branch growth relative to vertical growth in shaded conditions (Chen et al., 1996). This greater proportional reduction in lateral branch growth for *P. contorta* results in a more spindly growth form and greater susceptibility to damage from wind and snow and contributes to the tree's common name of lodgepole pine

which reflects its use by native American cultures.

Bailey (1961) also noted crowding-induced mortality for *P. contorta* in a 30.3 x 30.3 m plot on the Lost Creek lahar. The plot contained 2,630 4.5 to 9 m tall, ≤ 30-y old *P. contorta* per ha that were ≤ 30-years old and > 3,000 small *Abies* spp. per ha. From this plot structure, Bailey (1961) postulated an on-going shift in forest composition from *P. contorta* to the numerous, relatively shade-tolerant *A. concolor* and *A. magnifica* in the plot.

Parker (1993) also reports sites where older *P. contorta* stands have *A. magnifica* understories and suggests that *P. contorta* is an initial invader of disturbed locations that will only remain in the forest canopy at sites of chronic disturbance or low fertility. The mudflow is clearly a low fertility site as indicated by the composition of the mudflow materials (Eppler, 1984), the comparison of soil properties with other sites of more robust *P. contorta* (Parker, 1993), and the predominance of short-stature, slow growing trees (Kroh et al., 2000).

Thus, two alternative paths of forest development could be suggested for this mudflow site. Either continuing tree growth will result in a closing canopy with declining abundances of *P. contorta* and increasing abundances of the *Abies* spp., or the low fertility of the mudflow will result in a continuing forest composed of mostly small *P. contorta*.

Projected near-term forest development on the mudflow

Although the changes in forest structure between 1987 and 2008 have been relatively minor, they suggest that the mudflow forest is not likely to continue as an open stand of mostly small *P. contorta* but rather may be on the verge of imminent changes that could result in 1) a closing canopy, 2) a self-thinning decline in the abundances of the shade intolerant *P. contorta*, and 3) a shift to greater relative abundances of the more shade tolerant *A. concolor* and *A. magnifica*. The potential imminent occurrence of these changes is based on a per plot projection of forest growth using 1) the measured plot basal areas in 1987 and 2008, 2) the basal area proportional growth rates measured for the plot between 1987 and 2008, 3) equation 1 relating canopy cover to basal area, and 4) the 0.00047 y^{-1} reduction in proportional basal area growth rates for each 1% increase in canopy cover. The details of the procedures used in making these projections are presented in Appendix I with an example and discussions of assumptions and possible inaccuracies in the projected basal areas and canopy covers.

The distributions of plot basal areas and canopy covers for 2008 and the projected distributions of basal areas and canopy covers among individual plots for 2018 and 2028 are shown in Figures 2 and 3, respectively. Mean basal areas for 2018 and 2028 are projected to be 37.1 and 56.6 m^2 ha^{-1}, respectively, and median canopy covers for 2018 and 2028 are projected to be 74 and

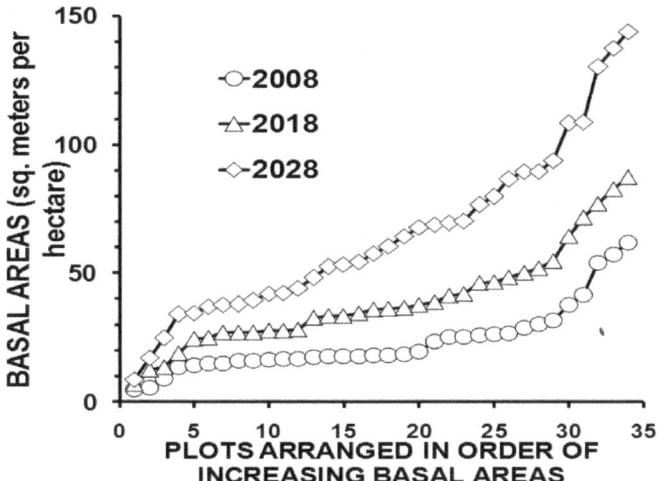

Figure 2. The frequency distribution of basal area ($m^2\ ha^{-1}$) in 2008 and the projected basal areas expected for 2018 and 2028 as estimated from the plot's 2008 basal area and the proportional rate of increase for the plot for the interval 1987 to 2008. Because of variations among plots for 1) 1987 basal areas per plot and 2) proportional rates of basal area increase between 1987 and 2008, the sequence of plots may vary among years.

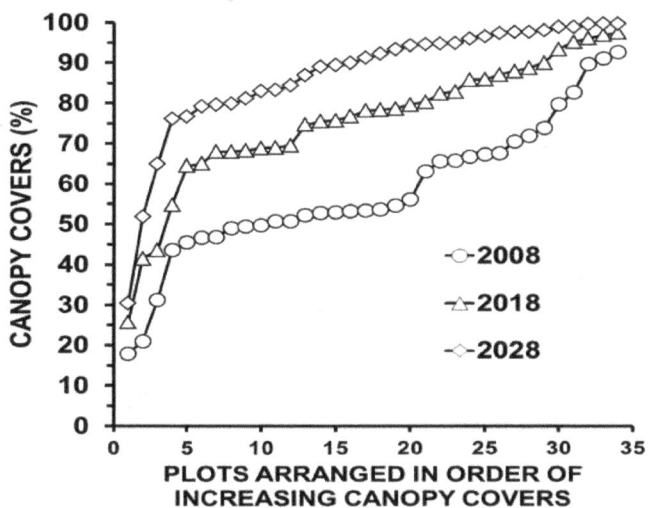

Figure 3. The frequency distribution of the canopy covers (%) in 2008, 2018, and 2028 for the 34 plots. For consistency among the years, all canopy covers are predicted from either measured (2008) or projected (2018 and 2028) plot basal areas (Figure 2) using Equation 1.

88%, respectively. These mean basal areas are consistent with the range of 50 to 70 $m^2\ ha^{-1}$ reported for mature Upper Montane Coniferous forests by Barbour (1988) and Parker (1991). By 2018 almost 40% of the canopy covers will be > 80%. By 2028, more than 70% of canopy covers will be >80% and almost 50% of the canopy covers will be >90%.

It may be argued that these projections are merely extensions of trends measured between only the two points in

time of 1987 and 2008, but these are the projections that can be made from the available data and the processes that they suggest. Without projections whose accuracy can be quantitatively tested by future measurements, there can be no rigorous test of how well current knowledge forecasts future forest development. Moreover, two aspects of these projections should be emphasized. First, there is not just one projection but one projection for each of the 34 plots based on the unique, specific data for that plot, and relating the accuracy of each plot's prediction to its initial conditions may identify the factors contributing to inaccurate predictions. Second, that these projections are based on a 21-year interval that encompasses 30% of the forest's existence (21/(2008 – 1940)).

These projected increases in canopy covers imply a progressive and rapid (rapid at least for a forest) transition from a mostly open to a mostly closed canopy. Whether the rate of this transition will be 1) faster or 2) slower than these projections remains to be determined from continuing studies of these plots. Moreover, the variations among plots, and to some extent variations among transects, may lead to varying rates of change and varying results of change across the mudflow. When the future development of plots meets these projections and expectations, it may confirm our assumptions and interpretations, but add little to our current knowledge.

However, the questions that may arise if these projections are not met, or are met earlier or later than expected, have the potential to increase our understanding of the processes active in the structuring of these forests.

Projected long-term forest development on the mudflow

If the canopy on the mudflow continues to close as projected in Figure 3 and the expected self-thinning begins, tree densities may then decline, but the declines in density may not be uniformly distributed among the species. This may be especially the case if shading is the primary cause of mortality. Reductions in sub-canopy light intensities may have less impact on the relatively shade-tolerant and relatively long-lived *A. magnifica*, which is the second most abundant species on the mudflow. This shift from a primarily *P. contorta* forest to a primarily *A. magnifica* forest may be a prolonged process for Parker (1993) reports on LVNP forests of > 100 year-old forests that have continuing greater abundances of *P. contorta* than *A. magnifica*. However, the tree age distributions in the forests reported by Parker (1993) suggest that the *A. magnifica* were 1) later colonizers of the area than the *P. contorta*, and 2) are consequently mostly smaller than the *P. contorta*. In contrast, the transition from *P. contorta* to *A. magnifica* on the mudflow may be more rapid because 1) the age distributions of these two species are similar (Kroh et al., 2000) and 2) their current mean heights are also similar (Table 4).

The continuing importance of the permanent plots

The transition from a *P. contorta* to *an A. magnifica* forest may be known or inferred from large-scale plot studies (Parker, 1993), but the scale of these large plots may not reveal 1) which individuals have the greatest potential for surviving and thriving in this transition or 2) to what extent this potential for surviving and thriving is a function of the properties of the individual or of the individual's immediate surroundings. Because the locations, as well as the species, heights and basal areas, have been measured for every individual ≤ 0.11-m tall in 2008, continuing studies of the mudflow plots can move from a plot-based analysis to an individual-based analysis of forest development that can 1) evaluate the success of each individual measured as either its survival and/or growth relative to other individuals and 2) to relate this success to the properties of the individual (species, size) or to the properties of the individual's surroundings (nearness, sizes and species of neighboring individuals). This individual-based analysis can address questions such as, is it merely the largest individuals in the plots that will be successful, or is the process more complex?

Or, are more isolated individuals more likely to persist than merely larger individuals (Fangliang and Duncan, 2000)? There may also be an interaction between size and neighborhood effects as has been demonstrated by Wyckoff and Clark (2000) who found that growth rate for a number of tree species could be predicted from their initial size and the area of the tree's crown that was exposed in the forest canopy.

Moreover, questions also remain concerning the possible inhibitory and facultative interactions among these species. For example, are the effects of *P. contorta* shading only inhibitory or may there be subtle facultative effects for shade tolerant species? Shade may limit growth but may also mediate temperature and moisture regimes with the result that *Abies* sp., which are more drought sensitive (Minore, 1979), are more likely to become established in the future forest from shaded rather than from exposed plots or sites.

If there are both inhibitory and facilitative effects among these species, their relative importance may be affected by stress levels on the mudflow. Callaway and Walker (1997) reviewed the relative effects of competition and facilitation in plant communities and discuss: 1) model studies by Holmgren et al. (1997) that suggest the alleviation of moisture stress in xeric habitats is more important than light limitation; and 2) the results of Callaway (1998) that suggest that facilitation between *Pinus albicaulis* and the shade-tolerant (Minore, 1979), but relatively drought intolerant (Minore, 1979) *Abies lasiocarpa*, is more important than inhibition in high stress environments.

As Herben (1996) discusses, monitoring the fates of individuals of different species and noting their differential survivorship in permanent plots may suggest 1) interspecific interactions, 2) more complex interrelationships among species, and 3) causative mechanisms for subsequent experimentation. The fates of individual trees would not identify mechanisms but would suggest which mechanisms and which intraspecific and interspecific interactions are more probable than others and, consequently, are more appropriate for experimental studies. Moreover, the long time frame for forest succession in these plots may also allow hypotheses, projections and models derived from one time frame to be tested in subsequent time frames.

Connell et al. (1987) postulated that the effects of one species upon another in succession could involve both inhibition and facilitation whose net effects could be positive, negative or zero. Nearly zero net observable effects could be indistinguishable from the no effect tolerance model and could make the unravelling of the contrasting effects of one species on another a difficult proposition. A balance of shading effects which 1) reduces photosynthetic rate but 2) moderates moisture stress could result in no effects or in difficult to discern non-zero effects. In this regard, the development of the mudflow forest from 1987 to 2008, which demonstrates an apparently transient equilibrium, raises the issue of whether this represents 1) a general level of interspecific tolerance among most of the species which may not persist as the trees grow larger or 2) complex interactions of inhibition and facilitation.

The investigator's anxiety of very long-term research

Clearly the delayed forest establishment and the slow tree growth rates on the mudflow substrate have slowed forest development. This has allowed 1) the scientific discipline of ecology to develop the techniques, tools and questions to utilize this resource and 2) the Mt. St. Helens catastrophic eruption to demonstrate the need for a better understanding of forest development in the Pacific Northwest.

However, the slow development is also problematic as the current aging generation of investigators may not witness the accuracy of their projections, the answers to current questions, or the more detailed analyses of the fates of individuals that may lead to testable hypotheses. They may even wonder if there will be those who do.

Conflict of Interests

The author(s) have not declared any conflict of interests.

ACKNOWLEDGEMENTS

This study was conducted in Lassen Volcanic National Park with the permission and assistance of Gilbert Blinn, Alan Dennison, Louise Johnson, Sara Koenig and Steven Zachary. Financial and logistical support was provided by the U. S. Park Service. Funding for the 2008 sampling was provided by the Texas Christian University Research Foundation. Data analysis resources were provided by the Rocky Mountain Center for Nuclear Computations of the Department of Environmental and Radiological Health Sciences, Colorado State University. We thank Ms. Deborah Flynn and Mr. Cameron Pool for their assistance with the 2008 field work.

REFERENCES

Acker SS, McKee WA, Harmon ME, Franklin J (1987) Long-term research on forest dynamics in the Pacific Northwest: a network of permanent forest plots, pp. 93-106. In F. Dallmeier & J. A. Comiskey (eds), Forest biodiversity in North, Central and South America and the Caribbean. The Parthenon Publishing Group Inc. Pearl River, NY. p. 792.

Alejandre-Melena N, Lindig-Cisneros R, Saenz-Romero C (2007). Response of P*inus pseuodstrobus* (Lindl.) to fertile growing medium and tephra-layer depth under greenhouse conditions. New Forests34:25-30.

Bailey WH (1961). A repopulation study of the Devastated Area Lassen Volcanic National Park. An unpublished Special Problem Study for Stanford University. Lassen Volcanic National Park Archives, Mineral California. p. 50.

Bailey WH (1963). Revegetation in the 1914–15 Devastated Area of Lassen Volcanic National Park. Ph.D. Thesis. Oregon State University, Corvallis, Oregon. p. 150.

Bakker JP, Olff H, Willems JH, Zobel M (1996). Why do we need permanent plots inthe study of long-term vegetation dynamics? J. Veg. Sci. 7:147-156.

Barbour MG (1988). California upland forests and woodlands, p. 131-164. In: Barbour MG and Billings WD (eds.). North American terrestrial vegetation. Cambridge University Press. Cambridge, United Kingdom. p. 724.

Brunnel FL, Vales DJ (1990). Comparisons of methods for estimating forest overstory cover: Differences among techniques. Can. J. Forest Res. 20:101-107.

Callaway RM (1998). Competition and facilitation on elevational gradients in subalpine forests of the northern Rocky Mountains, USA. Oikos 82:561-573.

Callaway RM, Walker LR (1997). Competition and facilitation: A synthetic approach to interactions in plant communities. Ecology 78:1958-1965.

Chen HYH, Klinka K, Kayahara GJ (1996). Effects of light on growth, crown architecture and specific leaf area for naturally established *Pinus contorta* var. *latifolia* and *Pseudotsuga menziessi* var. *glauca* saplings. Can. J. For. Res. 26:1149-1157.

Chytry M, Tichy T, Hennekens S M, Schaminee JHJ (2014). Assessing vegetation change using vegetation-plot databases: a risky business. Appl. Veg. Sci. 17:32-41.

Clarkson DB (1997). Vegetation succession (1967-1989) on five recent montane lava flows, Mauna Loa, Hawaii. N. Z. J. Ecol. 22:1-9.

Connell JH, Noble IR, Slayter RO (1987). On the mechanisms producing successional change. Oikos 50: 136-137.

Conover WJ (1971). Practical nonparametric statistics. John Wiley and Sons, New York. p. 462.

Cook JG, Stutzman TW, Bowes CW, Brenner KA, Irwin LL (1995). Spherical densitometers produce bias estimates of forest canopy cover. Wildlife S. B. 23:711-717.

Csecserits A, Szabo R, Halassy M. Redei T (2007). Testing the validity of successional predictions on an old-field chronosequence in Hungary. Community Ecol. 8:195-207.

Dale VH, Adams WM (2003). Plant establishment 15 years after the debris avalanche at Mount St. Helens, Washington. Sci. Total Environ. 313:101-113.

Dale VH, Cambell DR, Adams WM, Crisafuli CM, Dains VI., Frenzen PM, Holland RF(2005a). Plant succession on the Mount St. Helens debris-avalanche deposit, p. 59-73. In: Dale VH, Swanson FJ, Crisafulli, CM (eds.) Ecological responses to the 1980 eruption of Mount St. Helens. Springer Science + Business Media, Inc. New York. p. 342.

Dale VH, Swanson FJ, Crisafulli CM (eds.) (2005b). Ecological responses to the 1980 eruption of Mount St. Helens. Springer Science + Business Media, Inc. New York. p. 342.

Davis RR, Foote FS, Anderson JM, Mikhail EM (1981). Surveying theory and practice. Sixth Edition. McGraw Hill Book Co. New York. p. 992.

Dean TJ, Long JN (1992). Influence of leaf area and canopy structure on size-density relationships in even-ages lodgepole pine stands. Forest Ecol. Manag. 49:109-117.

del Moral R (2007). Limits to convergence of vegetation during early primary succession. J. Veg. Sci. 18:479-488.

del Moral R, Grishin SY (1999). Volcanic disturbances and ecosystem recovery, p. 137-160. In: L. R. Walker (ed.). Ecosystems of disturbed ground. Elsevier, Amsterdam. p. 868.

del Moral R, Magnusson B (2014). Surtsey and Mount St. Helens: a comparison of early successional rates. Biogeosciences 11:2099-2111.

del Moral R, Rozzell LR (2005). Long-term effects of *Lupinus lepidus* on vegetation dynamics at Mount St. Helens. Plant Ecol. 181:203-215.

del Moral R, Thomason LA, Wenke AC, Lozanoff N, Abata MD (2012). Primary succession trajectories on pumice at Mt. St. Helens, Washington. J. Veg. Sci. 23:73-85.

del Moral R., Saura JM, Emenegger JN (2010). Primary succession trajectories on a barren plain, Mount. St.Helens, Washington. J. Veg. Sci. 21:857-867.

Deligne NI, Cashman KV, Roering JJ (2013). After the lava flow: The importance of external soil sources for plant colonization of recent lava flows in the central Oregon Cascades, USA. Geomorphology 202:15-32.

Der G EB (2001). Handbook of statistical analyses using SAS. CRC Press, Bocca Raton, USA. p. 376.

Dewar RD (1993). A mechanistic analysis of self-thinning in terms of the carbon balance of trees. Ann. Bot. London 71:147-159.

Dodd M (2011). Where are my quadrats? Positional accuracy in fieldwork. Methods Ecol. Evol. 2:576-584.

Drake DR (1993). Germination requirements of *Metrosideros polymorpha*, the dominant tree of Hawaiian lava flows and rain forests. Biotropica 25:461-467.

Drake DR, Mueller-Dombois D (1993). Population development of rain forest trees on a chrono sequence of Hawaiian lava flows. Ecology 74:1012-1019.

El-Rabbany A (2002). Introduction to GPS: the global positioning system. Arctech House, Norwood. p. 196.

Eppler DB (1984). Characteristics of volcanic blasts, mudflows and rockfall avalanches in Lassen Volcanic National Park, California. M. S. Thesis. Arizona State University, Tempe, Arizona. p. 261.

Eppler DB (1987). The May 1915 eruptions of Lassen Peak II: May 22 volcanic blast effects, sedimentology and stratigraphy of deposits, and characteristics of the blast cloud. J. Volcanology Geothermal Res. 31:65-85.

Fangliang HE, Duncan RP (2000). Density-dependent effects on tree survival in an old-growth Douglas fir forest. J. Ecol. 88:676-688.

Fastie CL (1995). Causes and ecosystem consequences of multiple pathways of primary succession at Glacier bay, Alaska. Ecology 76: 1899-1916.

Fessenden JE (1984). Forest biomass and production estimates for the devastated area, Lassen Volcanic National Park, California. M. S. Thesis. Humbolt StateUniversity, Arcata, California. p. 49.

Frenzen PM, Franklin JP (1985). Establishment of conifers from seed on tephra depositied by the 1980 eruptions of Mt. Saint Helens, Washington. Am. Midl. Nat. 114:84-97. In: Dale VH, Swanson FJ, Crisafulli, CM (eds.) Ecological responses to the 1980 eruption of Mount St. Helens. Springer Science + Business Media, Inc. New York. p. 342.

Frenzen, PM, Hadley KS, Major JJ, Weber MH, Franklin JF, Hardison JH III, Stanton SM (2005). Geomorphic change and vegetation development on the Muddy River mudflow deposit, p. 75-91. In: Dale VH, Swanson FJ, Crisafulli, CM (eds.) Ecological responses to the

1980 eruption of Mount St. Helens. Springer Science + Business Media, Inc. New York. p. 342.

Griffin JR (1967). Soil moisture and vegetation patterns in northern California forests.Pacific Southwest Forest and Range Experiment Station, Berekey, California. U. S. Forest Service Research paper PSW-46. p. 22.

Halpern CB, Frenzen PM, Means JE, Franklin JF (1990). Plant succession in areas of scorched and blown-down forest after the 1980 eruption of Mount St. Helens, Washington. J. Veg. Sci. 1: 181-194.

Harcombe PA, Bill CJ, Fulton M, Glitzenstein JS, Marks PL, Elisk IS (2002). Stand dynamics over 18 years in a southern mixed-hardwood forest, Texas, USA. J. Ecol. 90:947-957.

Harris SL (1988). Fire mountains of the west: The Cascade and Mono Lake volcanoes. Mountain Press Publishing, Missoula, Montana. p. 379.

Heath JP (1967b). Primary conifer succession, Lassen Volcanic National Park. Ecology 48:270-275.

Heath JP (1971). Changes in thirty-one years in a Sierra Nevada ecotone. Ecology 52:1090-1092.

Heath JP (1967a). Field data to accompany conifer succession, Lassen Volcanic National Park. San Jose State University Special Collections& Archives. Dr. Martin Luther King Library, San Jose State University, San Jose, California, p. 107.

Herben T (1996). Permanent plots as tools for community ecology. J. Veg. Sci. 7:195-202.

Hickman JC (ed.) (1993). The Jepson manual: Higher plants of California. University of California Press, Berkeley, California. p. 1400.

Holmgren M, Scheffer M, Huston MA (1997). The interplay of facilitation and competition in plant communities. Ecology 78:1966-1975.

Johnson EA, Miyanishi K, Kleb H (1994). The hazards of interpretation of static age structure as shown by stand reconstructions in a *Pinus contorta-Picea englemanni*. J. Ecol. 82:923-931.

Kangur A, Korjus H, Jogiste K, Kiviste A (2005). A conceptual model of forest stand developmemt based on permanent sample-plot data in Estonia. Scand. J. Forest Res. 20 (Suppl 6):94-101.

Kiver EP (1982). The Cascade volcanoes: Comparison of geological and historical records, p. 3-12. In: Keller SAC (ed.). Mount St. Helens: one year later. Eastern Washington University Press, Cheney, Washington. p. 243 .

Kroh GC, White JD, Heath SK, Pinder JE III (2000). Colonization of a volcanic mudflow by an upper montane coniferous forest at Lassen Volcanic National Park, California. Am Midl. Nat. 143:126-140.

Larsen DR, Bliss LC (1998). An analysis of structure of tree seedling populations on a lahar. Landsc. Ecol. 13:307-323.

Lemmon PE (1956). A spherical densitometer for estimating forest overstory density. For. Sci. 2:314-320.

Long, JN, Smith FW (1992). Volume increment in *Pinus contorta* var. *latifolia*: the influence of stand development and crown dynamics. For. Ecol. Manage. 53:53-64.

MacDonald GA, Katsura T (1965). Eruptions of Lassen Peak, Cascade Range, California, in 1915: Example of mixed magmas. Geol. Soc. Am. Bull. 76:475-482.

McCune B, Allen TFH (1985). Will similar forests develop on similar sites. Can. J. Bot. 63:367-376.

Milliken GA, Johnson DE (1984). Analysis of messy data. Volume I. Designed experiments. Van Nostran Reinhold Company, New York. p. 473.

Minore D (1979). Comparative autecological characteristics of northwestern tree species – a literature review. U. S. Dept. Agr. For. Serv. Gen. Tech. Rep. PNW-87. p. 72.

Morris WF, Wood DM (1989). The role of lupine in succession on Mount St. Helens, facilitation or inhibition. Ecology 70:679-703.

Munoz-Jimenez J, Rangel-Rios K, Garcia-Romero A (2005). Plant colonization of recent lahar deposition on Popocatepetl Volcano, Mexico. Phys. Geogr. 26:192-215.

Nuttle T (1997). Densiometer bias? Are we measuring the forest or the trees. Wildl. Soc. B. 25: 610-611.

Parker AJ (1991). Forest/environmental relationships in Lassen Volcanic National Park, California, U.S.A. J. Biogeogr. 18:543-552.

Parker AJ (1993). Structural variation and dynamics of lodgepole pine

forests in Lassen Volcanic National Park, California. Ann. Assoc. Am. Geogr. 83:613-629.

Peet RL, Christiansen NL (1980). Succession: a population process. Vegetatio 43:131-140.

Reid DEB, Silinus U, Lieffers VT (2003). Stem sapwood permeability in relation to crown dominance and site quality in self-thinning fire-origin lodgepole pine stands. Tree Physiol. 23:833-840.

Ross LC, Woodin SJ, Hester A, Thompson DBA, Birks HJB (2010). How important is plot relocation accuracy when interpreting revisitation studies of vegetation change. Plant Ecol. Divers. 3:1-8.

Rudnicki M, Silnus U, Lieffers VJ, Josi G (2001). Measure of simultaneous tree sways and estimation of crown interactions among a group of tree. Trees 15:83-90.

Rundel PW, Parsons DJ, Gordon DT (1990). Montane and subalpine vegetation of the Sierra Nevada and Cascade ranges, p. 559-599. In: Barbour MG and Major J (ed.) Terrestrial vegetation of California. California Native Plant Society. p. 1002.

Spjotvoll E, Stoline MR (1973). An extension of the *T*-method of multiple comparisons to include the cases with unequal sample sizes. J. Am. Stat. Assoc. 68:975-978.

Stillman AG, Turnage WA (eds.) (1962). Ansel Adams, our national parks. Little Brown and Co., Boston. p.127.

Swartzlow CR (1946). The Devastated Area: A preliminary study in natural reforestation. An unpublished National Park Service Report. Lassen Volcanic National Park Archives, Mineral California. p. 7.

Teste FP, Lieffers VJ (2011). Snow damage in lodgepole pine standings brought into thinning and fertilization regimes. For. Ecol. Manag. 261:2096-2104.

Titus JH, Bishop JG (2014). Propagule limitation and competition with nitrogen fixers during primary succession. J. Veg. Sci. 24: 990-1003.

Turner DP (1985). Successional relationships and a comparison of biological characteristics among six northwestern conifers. B. Torrey Bot. Club 112:421-428.

Upjohn RL (2009). Primary conifer succession in the Devastated Area in Lassen Volcanic National Park, California. M.S. Thesis. Texas Christian University, Fort Worth, Texas. p. 29.

Valverde T, Silvertown J (1997). Canopy closure rate and forest structure. Ecology 78:1555-1562.

Vanderschaaf CL (2010). Estimating individual stand size-density trajectories and a maximum size-density relationship species boundary line slope. For. Sci. 56: 327-335.

Walker LR, Chapin FS III (1987). Interactions among processes controlling successional change. Oikos 50:131-135.

Walker LR, Clarkson BD, Silvester WB, Clarkson BR (2003). Colonization dynamics and facultative impacts of a nitrogen-fixing shrub in primary succession. J. Veg. Sci. 14:277-290.

Walker LR, del Moral R (2003). Primary succession and ecosystem rehabilitation. Cambridge University Press, Cambridge, United Kingdom. p. 442.

Walker LR, Wardle DA, Bardgett RD, Clarkson BD (2010). The use of chronosequences in studies of ecological succession and soil development. J. Ecol. 98:725-736.

Weber, MH, Hadley KS, Frenzen PM, Franklin JF (2006). Forest development following mudflow deposition, Mount St. Helens, Washington. Can. J. For. Res. 36: 437-449.

Weiskittel A, Gould P, Temesgen H (2009). Sources of variation in the self-thinning boundary line for three species with varying levels of shade tolerance. For. Sci. 55: 84-93.

Woods KD (2000). Dynamics in late-successional hemlock-hardwood forests over three decades. Ecology 81:110-126.

Wyckoff PH, Clark JS (2005). Tree growth prediction using size and exposed crown area. Can. J. For. Res. 35:13-20.

Yoshida K, Kikuchis S, Nakamura F, Noda M (1997). Dendrochronological analysis of debris flow disturbance on Rishiri Island. Geomorphology 20:135-145.

Zeide B (1995). A relationship between size of trees and their number. Forest Ecol. Manag. 72:265-272.

Zeide B (2001). Natural thinning and environmental change: an ecological process model. Forest Ecol. Manag. 154:165-177.

APPENDIX I

Computational procedures for projecting a plot's forest development from 2008 to 2018

The computations to project a plot's forest development from 2008 to 2028 involve two phases. The first phase involves estimating the proportional rate of basal area increase (hereafter, PRBAI) appropriate for the plot from 2008 to 2009 and computing the increases in basal area and canopy cover for 2009. The second phase involves calculating the yearly increases in basal areas and canopy covers from 2009 to 2028.

Phase I

The first step in projecting current basal areas and canopy covers across the interval from 2008 to 2028 for each plot was to convert the PRBAI measured across the interval 1987 to 2008, which is an average of the likely declining rates across these years, to a PRBAI appropriate for the year 2008. Declining values of PRBAI can to be expected to have occurred during the interval from 1987 to 2008 due to the negative correlation between canopy cover and basal area increase. Thus, a plot's PRBAI for the 1987 to 2008 interval is a likely overestimate of the PRBAI for the 2008 to 2009 interval.

To estimate a rate for 2008 to 2009, 1) Equation 1 was used to predict the canopy covers for 1987 and 2008 from the measured basal areas for those dates; 2) an average rate of increasing canopy covers (% per year) for this interval was computed as the predicted cover for 2008 minus that for 1987 divided by 21 years; and 3) the predicted PRBAI for 2008 was then computed as the average PRBAI for the interval minus 10.5 times the average yearly rate of increasing canopy covers times the $0.00047\ y^{-1}$ reduction in PRBAI per percent increase in canopy cover. These adjustments for the 34 plots resulted in an average reduction in the PRBAI from the 1987 to 2008 interval of 13% with a range from 3 to 17%.

Example of phase I computations

As an example of the Phase I procedures, the following computations are for a plot whose 1987 basal area was near the median for that year.

First, for the 1987 and 2008 basal areas of 4.42 and 18.1 $m^2\ ha^{-1}$, respectively, the projected canopy covers for 1987 and 2008, as estimated by Equation 1, are 17.0 and 53.3%.

Second, the average yearly rate of canopy cover increase for this 21 year period was: $1.729\%\ y^{-1} = (53.3\% - 17.0\%) / 21\ y$.

Third, the measured PRBAI for the plot in the interval was $0.07042\ y^{-1}$, and the adjusted PRBAI for 2008 becomes: $0.0619\ y^{-1} = 0.07042\ y^{-1} - 10.5\ y\ (1.729\%\ y^{-1} \times 0.00047\%\ y^{-1})$

There are numerous sources for possible errors in estimating these adjusted PRBAI for 2008; however, the alternative of using the PRBAI for the 1987 to 2008 interval, which is likely an overestimate, could introduce a positive bias which is the basal area and canopy cover growth rates. In the above computations, it would have been preferable to have densitometer measurements of percent cover for each plot in 1987, but these data were not collected.

Phase II

The projected yearly increases in basal areas and canopy covers were computed for the years 2009 through 2028 by using the plot's measured 2008 basal area and estimated PRBAI for the 2008 to 2009 interval as a base, 2) using that estimated PRBAI to predict a basal area for the following year; 2) using that predicted basal area to estimate canopy cover for that year; 3) computing the increase in canopy cover% between the two years; 4) using this increase in canopy cover to estimate a reduction in the PRBAI as $0.00047\ y^{-1}$ times the increase in canopy cover between years; and 5) repeating these steps successively for each year from 2008 through 2028 for each of the 34 plots.

Example phase II computations

As an example of the Phase II procedures, the following computations are for the plot used in Phase I.

First, using the $0.0619\ y^{-1}$ PRBAI for the interval 2008 to 2009, the projected basal area for 2009 becomes: 18.1 $m^2\ ha^{-1} = 17.0\ m^2\ ha^{-1} \times e^{(0.0619\ y^{-1} \times 1\ y)}$

Second, the projected canopy cover for 2009 from equation 1 is 55.5%.

Third, the increase in canopy cover from 2008 to 2009 becomes: 2% = 55.5 - 53.3%

Fourth, the projected PRBAI for 2010 becomes: $0.0608\ y^{-1} = 0.0619\ y^{-1} - (2\% \times 0.00047\ y^{-1}\%^{-1})$

Continuing these yearly computations through 2018 and 2028 result in projected basal areas of 31.2 and 52.2 $m^2\ ha^{-1}$, respectively, with canopy covers of 74 and 89% in these same years. The pattern of declining exponential rates of canopy cover increase is consistent with the analyses of canopy closure using Markov-chain models by Valverde and Silvertown (1997).

These computations assume 1) that the proportional basal area growth rates from 1987 to 2008 are applicable to the following 20 years and 2) that Equation 1 is applicable to estimating plot canopy covers in the years preceding 2008 and will continue to be applicable in estimating covers in the succeeding 20 years. The current basal area proportional growth rates have been computed over a 21-y span. The length of this span suggests that barring any major disturbance, severe climate change, or pronounced slowing in basal area growth rates due to crowding, these rates should be applicable the next 10 to 20 years following the 2008 sampling. The Equation 1 relationship between basal area and canopy cover may be complicated by continuing increases in mean tree heights which could contribute to greater canopy closure as measured by spherical densitometers. If this occurs, this complication would mean that projected canopy covers using Equation 1 may likely be underestimates of actual covers.

Permissions

List of Contributors

George Efthimiou
Department of Forestry and Natural Environment Management, Technological Educational Institute of Larissa, 34100, Karditsa, Greece

Longonje N. Simon
Department of Environment, University of Buea, P. O. Box 63 Buea, Cameroon

Dave Raffaelli
Environment Department, University of York, YO 105 DD United Kingdom

Nidhi Lohani
Department of Botany, D.S.B. Campus, Kumaun University, Nainital Uttarakhand, India

Lalit. M. Tewari
Department of Botany, D.S.B. Campus, Kumaun University, Nainital Uttarakhand, India

Ravi Kumar
RRIHF CCRAS, Ranikhet, Uttarakhand, India

G. C. Joshi
RRIHF CCRAS, Ranikhet, Uttarakhand, India

Jagdish Chandra
RRIHF CCRAS, Ranikhet, Uttarakhand, India

Kamal kshore
Department of Botany, D.S.B. Campus, Kumaun University, Nainital Uttarakhand, India

Sanjay Kumar
Department of Botany, D.S.B. Campus, Kumaun University, Nainital Uttarakhand, India

Brij Mohan Upreti
Department of Botany, D.S.B. Campus, Kumaun University, Nainital Uttarakhand, India

Reza Hamidi
Crop Production and Plant Breeding Department, College of Agriculture, Shiraz University, Shiraz, Iran

Tchobsala
Department of Biological Sciences, Faculty of Science, University of Ngaoundéré, P. O. Box 454, Cameroon

M. Mbolo
Department of Biology and Plant Physiology, Faculty of Science, University of Yaoundé I, P. O. Box 812, Yaoundé, Cameroon

Minerva Singh
School of Geography and Environment, University of Oxford, UK and Department of Plant Sciences, University of Cambridge, CB23EA, UK

Tomohiro Fujita
Graduate School of Asian and African Area Studies, Kyoto University, Japan

B. G. Oguntuase
Department of Ecotourism and Wildlife Management, Federal University of Technology Akure, Ondo State, Nigeria

E. A. Agbelusi
Department of Ecotourism and Wildlife Management, Federal University of Technology Akure, Ondo State, Nigeria

Muhammad Farrukh Nisar
College of Bioengineering, Chongqing University, Chongqing 400044, China

Farrukh Jaleel
College of Chemistry and Chemical Engineering, Chongqing University, Chongqing 400044, China

Muhammad Waseem
Department of Biology, Allama Iqbal Open University (AIOU), Islambad (44000), Pakistan

Sajil Ismail
Department of Botany, Govt. Sadiq Egerton (SE) College, Bahawalpur (63100), Pakistan

Muhammad Arfan
Department of Biology, Lund University, Sweden

Mathewos Hailu
Ziway Fisheries Resources Research Center, Ethiopia

Bing-Hua Liao
The Key Laboratory of Ecological Restoration in Hilly Areas, Forestry Department of Henan Province, Ping-ding-shan University, Ping-ding-shan, Henan Province,China,467000
Department of Environment and Geography, Ping-ding-shan, Henan Province, China, 467000
Institute of Ecological Science and Technology, College of Life Sciences, Henan University, Kaifeng, China, 475001

Pei-Song Liu
The Key Laboratory of Ecological Restoration in Hilly Areas, Forestry Department of Henan Province, Ping-ding-shan University, Ping-ding-shan, Henan Province,China,467000

Zhen-Zhong Wen
The Key Laboratory of Ecological Restoration in Hilly Areas, Forestry Department of Henan Province, Ping-ding-shan University, Ping-ding-shan, Henan Province,China,467000

Sheng-Yan Ding
Institute of Ecological Science and Technology, College of Life Sciences, Henan University, Kaifeng, China, 475001

Hai-Long Yu
Department of Environment and Geography, Ping-ding-shan, Henan Province, China, 467000

Zhi-Chao Wang
Department of Environment and Geography, Ping-ding-shan, Henan Province, China, 467000

Zhong-Kai Li
Department of Environment and Geography, Ping-ding-shan, Henan Province, China, 467000

Huan-Xin Chu
Department of Environment and Geography, Ping-ding-shan, Henan Province, China, 467000

Wen-Liang Li
Department of Environment and Geography, Ping-ding-shan, Henan Province, China, 467000

Yi Shen
Department of Environment and Geography, Ping-ding-shan, Henan Province, China, 467000

Zahid Hussain Malik
Department of Botany, University of Azad Jammu and Kashmir Muzaffarabad, Pakistan

Muhammad Shoaib Amjad
Department of Botany, University of Azad Jammu and Kashmir Muzaffarabad, Pakistan

Sidra Rafique
Department of Botany, University of Azad Jammu and Kashmir Muzaffarabad, Pakistan

Nafeesa Zahid Malik
Department of Botany, University of Azad Jammu and Kashmir Muzaffarabad, Pakistan

A. Rahimi
Department of Biology, Bojnourd Branch, Islamic Azad University, Bojnourd, Iran

M. Atri
Department of Biology, Bu Ali-Sina University, Hamedan, Iran

Isaac Mapaure
Department of Biological Sciences, University of Namibia, P. Bag 13301, Windhoek, Namibia

Nishita Giri
Forest Ecology and Environment Division, FRI, Dehradun, India

Laxmi Rawat
Forest Ecology and Environment Division, FRI, Dehradun, India

A. MISRA
Central Institute of Medicinal and Aromatic Plants, P. O. CIMAP, Kukrail Picnic Spot Road Lucknow – 226015, India

N. K. SRIVASTAVA
Central Institute of Medicinal and Aromatic Plants, P. O. CIMAP, Kukrail Picnic Spot Road Lucknow – 226015, India

A. K. SRIVASTAVA
Central Institute of Medicinal and Aromatic Plants, P. O. CIMAP, Kukrail Picnic Spot Road Lucknow – 226015, India

Aidin Parsakhoo
Department of Forestry, Faculty of Forest Sciences, Gorgan University of Agricultural Sciences and Natural Resources Gorgan, Iran

Mohammad Hadi Moayeri
Department of Forestry, Faculty of Forest Sciences, Gorgan University of Agricultural Sciences and Natural Resources Gorgan, Iran

Majid Poursadeghi
Department of Forestry, Faculty of Forest Sciences, Gorgan University of Agricultural Sciences and Natural Resources Gorgan, Iran

Kwaku Brako Dakwa
Department of Entomology and Wildlife, School of Biological Sciences.University of Cape Coast, Cape Coast, Ghana

Kweku Ansah Monney
Department of Entomology and Wildlife, School of Biological Sciences.University of Cape Coast, Cape Coast, Ghana

Daniel Attuquayefio
Department of Animal Biology and Conservation Science, University of Ghana, Legon, Accra, Ghana

Mikolo Yobo Christian
Department of Bioengineering Science, Graduate School of Bioagricultural Sciences, Division of International Cooperation in Agricultural Science Laboratory of Project Development, Nagoya University, Furo-cho, Chikusa-ku, Nagoya, 464-8601, Japan

I. T. O. Kasumi
International Cooperation Center for Agricultural Education, Nagoya University, Furo-cho, Chikusa-ku, Nagoya, 464- 8601, Japan

Z. Yemataw
Southern Agricultural Research Institute, Areka Agricultural Research Center, P. O. Box 79, Areka, Ethiopia

H. Mohamed
Hawassa University, Awassa College of Agriculture, P. O. Box 05, Hawassa, Ethiopia

M. Diro
Capacity building for scaling up of evidence-based best practices in agricultural production in Ethiopia (CASCAPE), Addis Ababa, Ethiopia

T. Addis
Southern Agricultural Research Institute, Awassa Agricultural Research Center, P. O. Box 06, Hawassa, Ethiopia

G. Blomme
Bioversity International Uganda Office, P. O. Box 24384, Kampala, Uganda

Juliet Kyayesimira
Department of Biological Sciences, Kyambogo University, Uganda

Julius B. Lejju
Department of Biology, Mbarara University of Science and Technology, Uganda

Mansoor Ahmad Lone
Department of Environmental Sciences, University of Kashmir, Srinagar -190 006, J&K, India

Idrees Yousuf Dar
Department of Environmental Sciences, University of Kashmir, Srinagar -190 006, J&K, India

G. A. Bhat
Department of Environmental Sciences, Centre of Research for Development, University of Kashmir, Srinagar -190 006, J&K, India

Glenn Clinton Kroh
Department of Biology, Winton Scott, Room 401, Texas Christian University, Fort Worth, TX 76129, United States of America

Rebecca Laura Upjohn
Department of Biology, Winton Scott, Room 401, Texas Christian University, Fort Worth, TX 76129, United States of America
Department of Ecosystem Science and Management, University of Wyoming, Laramie, Wyoming, WY 82070, United States of America

John Edgar Pinder III
Department of Biology, Winton Scott, Room 401, Texas Christian University, Fort Worth, TX 76129, United States of America

Department of Environmental and Radiological Health Sciences, Colorado State University, 305 W. Magnolia PMB 231; Fort Collins, CO 80521, United States of America